Lecture Notes in Computer Science 3241

Commenced Publication in 1973
Founding and Former Series Editors:
Gerhard Goos, Juris Hartmanis, and Jan van Leeuwen

Dieter Kranzlmüller   Peter Kacsuk
Jack Dongarra (Eds.)

Recent Advances in Parallel Virtual Machine and Message Passing Interface

11th European PVM/MPI Users' Group Meeting
Budapest, Hungary, September 19-22, 2004
Proceedings

 Springer

Volume Editors

Dieter Kranzlmüller
Joh. Kepler Universität Linz
Institut für Graphische und Parallele Datenverarbeitung (GUP)
Altenbergerstr. 69, 4040 Linz, Austria
E-mail: kranzlmueller@gup.jku.at

Peter Kacsuk
Hungarian Academy of Sciences
Computer and Automation Research Institute, MTA SZTAKI
Victor Hugo u. 18-22, 1132 Budapest, Hungary
E-mail: kacsuk@sztaki.hu

Jack Dongarra
University of Tenneessee, Computer Science Departement
1122 Volunteer Blvd., Knoxville, TN-37996-3450, USA
E-mail: dongarra@cs.utk.edu

Library of Congress Control Number: 2004111520

CR Subject Classification (1998): D.1.3, D.3.2, F.1.2, G.1.0, B.2.1, C.1.2

ISSN 0302-9743
ISBN 3-540-23163-3 Springer Berlin Heidelberg New York

Springer is a part of Springer Science+Business Media

springeronline.com

© Springer-Verlag Berlin Heidelberg 2004
Printed in Germany

Typesetting: Camera-ready by author, data conversion by Olgun Computergrafik
Printed on acid-free paper   SPIN: 11324188   06/3142   5 4 3 2 1 0

Preface

The message passing paradigm is the most frequently used approach to develop high-performance computing applications on parallel and distributed computing architectures. Parallel Virtual Machine (PVM) and Message Passing Interface (MPI) are the two main representatives in this domain.

This volume comprises 50 selected contributions presented at the 11th European PVM/MPI Users' Group Meeting, which was held in Budapest, Hungary, September 19–22, 2004. The conference was organized by the Laboratory of Parallel and Distributed Systems (LPDS) at the Computer and Automation Research Institute of the Hungarian Academy of Sciences (MTA SZTAKI).

The conference was previously held in Venice, Italy (2003), Linz, Austria (2002), Santorini, Greece (2001), Balatonfüred, Hungary (2000), Barcelona, Spain (1999), Liverpool, UK (1998), and Krakow, Poland (1997). The first three conferences were devoted to PVM and were held in Munich, Germany (1996), Lyon, France (1995), and Rome, Italy (1994).

In its eleventh year, this conference is well established as the forum for users and developers of PVM, MPI, and other message passing environments. Interactions between these groups have proved to be very useful for developing new ideas in parallel computing, and for applying some of those already existent to new practical fields. The main topics of the meeting were evaluation and performance of PVM and MPI, extensions, implementations and improvements of PVM and MPI, parallel algorithms using the message passing paradigm, and parallel applications in science and engineering. In addition, the topics of the conference were extended to include cluster and grid computing, in order to reflect the importance of this area for the high-performance computing community.

Besides the main track of contributed papers, the conference featured the third edition of the special session "ParSim 04 – Current Trends in Numerical Simulation for Parallel Engineering Environments". The conference also included three tutorials, one on "Using MPI-2: A Problem-Based Approach", one on "Interactive Applications on the Grid – the CrossGrid Tutorial", and another one on "Production Grid Systems and Their Programming", and invited talks on MPI and high-productivity programming, fault tolerance in message passing and in action, high-performance application execution scenarios in P-GRADE, an open cluster system software stack, from PVM grids to self-assembling virtual machines, the grid middleware of the NorduGrid, next-generation grids, and the Austrian Grid initiative – high-level extensions to grid middleware. These proceedings contain papers on the 50 contributed presentations together with abstracts of the invited and tutorial speakers' presentations.

The 11th Euro PVM/MPI conference was held together with DAPSYS 2004, the 5th Austrian-Hungarian Workshop on Distributed and Parallel Systems. Participants of the two events shared invited talks, tutorials, the vendors' session, and social events, while contributed paper presentations proceeded in separate tracks in parallel. While Euro PVM/MPI is dedicated to the latest developments of PVM and MPI, DAPSYS was a major event to discuss general aspects of distributed and parallel systems. In this way the two events were complementary to each other and participants of Euro PVM/MPI could benefit from the joint organization of the two events.

The invited speakers of the joint Euro PVM/MPI and DAPSYS conference were Jack Dongarra, Gabor Dozsa, Al Geist, William Gropp, Balazs Konya, Domenico Laforenza, Ewing Lusk, and Jens Volkert. The tutorials were presented by William Gropp and Ewing Lusk, Tomasz Szepieniec, Marcin Radecki and Katarzyna Rycerz, and Peter Kacsuk, Balazs Konya, and Peter Stefan.

We express our gratitude for the kind support of our sponsors (see below) and we thank the members of the Program Committee and the additional reviewers for their work in refereeing the submitted papers and ensuring the high quality of Euro PVM/MPI. Finally, we would like to express our gratitude to our colleagues at MTA SZTAKI and GUP, JKU Linz for their help and support during the conference organization.

September 2004   Dieter Kranzlmüller
                                                              Peter Kacsuk
                                                              Jack Dongarra

Program Committee

General Chair

Jack Dongarra   University of Tennessee, Knoxville, USA

Program Chairs

Peter Kacsuk   MTA SZTAKI, Budapest, Hungary
Dieter Kranzlmüller   GUP, Joh. Kepler University, Linz, Austria

Program Committee Members

David Abramson	Monash University, Australia
Vassil Alexandrov	University of Reading, UK
Ranieri Baraglia	CNUCE Institute, Italy
Arndt Bode	Technical Univ. of Munich, Germany
Marian Bubak	AGH, Cracow, Poland
Jacques Chassin	
de Kergommeaux	LSR-IMAG, France
Yiannis Cotronis	Univ. of Athens, Greece
Jose C. Cunha	New University of Lisbon, Portugal
Marco Danelutto	Univ. of Pisa, Italy
Frederic Desprez	INRIA, France
Erik D'Hollander	University of Ghent, Belgium
Beniamino Di Martino	UNINA, Italy
Jack Dongarra	University of Tennessee, Knoxville, USA
Graham Fagg	University of Tennessee, Knoxville, USA
Thomas Fahringer	University of Innsbruck, Austria
Al Geist	Oak Ridge National Laboratory, USA
Michael Gerndt	Technical Univ. of Munich, Germany
Andrzej Goscinski	Deakin University, Australia
William Gropp	Argonne National Laboratory, USA
Rolf Hempel	DLR, Simulation Aerospace Center, Germany
Ladislav Hluchy	Slovak Academy of Sciences, Slovakia
Peter Kacsuk	MTA SZTAKI, Hungary
Dieter Kranzlmüller	Joh. Kepler University, Linz, Austria
Jan Kwiatkowski	Wroclaw University of Technology, Poland
Domenico Laforenza	CNUCE, Italy
Laurent Lefevre	INRIA, France
Thomas Ludwig	University of Heidelberg, Germany
Emilio Luque	Universitat Autonoma de Barcelona, Spain

Ewing Lusk	Argonne National Laboratory, USA
Tomas Margalef	Universitat Autonoma de Barcelona, Spain
Barton Miller	University of Wisconsin, USA
Shirley Moore	University of Tennessee, USA
Wolfgang Nagel	Dresden University of Technology, Germany
Salvatore Orlando	University of Venice, Italy
Benno J. Overeinder	University of Amsterdam, The Netherlands
Raffaele Perego	ISTI, Italy
Neil D. Pundit	Sandia National Labs, USA
Rolf Rabenseifner	University of Stuttgart, Germany
Andrew Rau-Chaplin	Dalhousie University, Canada
Jeff Reeve	University of Southampton, UK
Ralf Reussner	University of Oldenburg, Germany
Yves Robert	ENS Lyon, France
Casiano Rodriguez-Leon	Universidad de La Laguna, Spain
Michiel Ronsse	University of Ghent, Belgium
Wolfgang Schreiner	Joh. Kepler University, Linz, Austria
Miquel Senar	Universitat Autonoma de Barcelona, Spain
Joao Gabriel Silva	University of Coimbra, Portugal
Vaidy Sunderam	Emroy University, USA
Francisco Tirado	Universidad Complutense, Spain
Bernard Tourancheau	SUN Labs Europe, France
Jesper Larsson Träff	NEC Europe, Germany
Pavel Tvrdik	Czech Technical University, Czech Republic
Umberto Villano	University of Sannio, Italy
Jens Volkert	Joh. Kepler University, Linz, Austria
Jerzy Wasniewski	Technical University Denmark
Roland Wismüller	Technical Univ. of Munich, Germany

Sponsoring Institutions

(as of A ugust 3, 2004)

IBM
Intel
NEC
EGEE project
Hungarian Ministry of Education, OM
Hungarian Academy of Sciences, MTA
Foundation for the Technological Progress of the Industry, IMFA (Hungary)

Table of Contents

Algorithms

Applications

Tools and Environments

Cluster and Grid

PVM Grids to Self-assembling Virtual Machines

Al Geist

Oak Ridge National Laboratory
PO Box 2008, Oak Ridge, TN 37831-6016
gst@ornl.gov
http://www.csm.ornl.gov/~geist

Abstract. Oak Ridge National Laboratory (ORNL) leads two of the five big Genomes-to-Life projects funded in the USA. As a part of these projects researchers at ORNL have been using PVM to build a computational biology grid that spans the USA. This talk will describe this effort, how it is built, and the unique features in PVM that led the researchers to choose PVM as their framework. The computations such as parallel BLAST are run on individual supercomputers or clusters within this P2P grid and are themselves written in PVM to exploit PVM's fault tolerant capabilities.

We will then describe our recent progress in building an even more adaptable distributed virtual machine package called Harness. The Harness project includes research on a scalable, self-adapting core called H2O, and research on fault tolerant MPI. Harness software framework provides parallel software "plug-ins" that adapt the run-time system to changing application needs in real time. This past year we have demonstrated Harness' ability to self-assemble into a virtual machine specifically tailored for particular applications.

Finally we will describe DOE's plan to create a National Leadership Computing Facility, which will house a 100 TF Cray X2 system, and a Cray Red Storm at ORNL, and an IBM Blue Gene system at Argonne National Lab. We will describe the scientific missions of this facility and the new concept of "computational end stations" being pioneered by the Facility.

1   Genomics Grid Built on PVM

The United States Department of Energy (DOE) has embarked on an ambitious computational biology program called Genomes to Life[1]. The program is using DNA sequences from microbes and higher organisms, for systematically tackling questions about the molecular machines and regulatory pathways of living systems. Advanced technological and computational resources are being employed to identify and understand the underlying mechanisms that enable living organisms to develop and survive under a wide variety of environmental conditions.

ORNL is a leader in two of the five Genomes to Life centers. As part of this effort ORNL is building a Genomics Computational Grid across the U.S. connecting ORNL, Argonne National Lab, Pacific Northwest National Lab, and

D. Kranzlmüller et al. (Eds.): EuroPVM/MPI 2004, LNCS 3241, pp. 1–4, 2004.

Lawrence Berkley National Lab. The software being deployed is called The Genome Channel[2]. The Genome Channel is a computational biology workbench that allows biologists to run a wide range of genome analysis and comparison studies transparently on resources at the grid sites. Genome Channel is built on top of the PVM software. When a request comes in to the Genome Channel, PVM is used to track the request, create a parallel virtual machine combining database servers, Linux clusters, and supercomputer nodes tailored to the nature of the request, spawning the appropriate analysis code on the virtual machine, and then returning the results.

The creators of this Genomics Grid require that their system be available 24/7 and that analyses that are running when a failure occurs are reconfigured around the problem and automatically restarted. PVM's dynamic programming model and fail tolerance features are ideal for this use. The Genome Channel has been cited in "Science" and used by thousands of researchers from around the world.

2   Harness: Self-assembling Virtual Machine

Harness[3] is the code name for the next generation heterogeneous distributed computing package being developed by the PVM team at Oak Ridge National Laboratory, the University of Tennessee, and Emory University. The basic idea behind Harness is to allow users to dynamically customize, adapt, and extend a virtual machine's features to more closely match the needs of their application and to optimize the virtual machine for the underlying computer resources, for example, taking advantage of a high-speed I/O. As part of the Harness project, the University of Tennessee is developing a fault tolerant MPI called FT-MPI. Emory has taken the lead in the architectural design of Harness and development of the H2O core[4].

Harness was envisioned as a research platform for investigating the concepts of parallel plug-ins, distributed peer-to-peer control, and merging and splitting of multiple virtual machines. The parallel plug-in concept provides a way for a heterogeneous distributed machine to take on a new capability, or replace an existing capability with a new method across the entire virtual machine. Parallel plug-ins are also the means for a Harness virtual machine to self-assemble. The peer-to-peer control eliminates all single points of failure and even multiple points of failure. The merging and splitting of Harness virtual machines provides a means of self healing.

The project has made good progress this past year and we have demonstrated the capability for a self-assembling virtual machine with capabilities tuned to the needs of a particular chemistry application. The second part of this talk will describe the latest Harness progress and results.

3   DOE National Leadership Computing Facility

In May of 2004 it was announced that Oak Ridge National Laboratory (ORNL) had been chosen to provide the USA's most powerful open resource for capability computing, and we propose a sustainable path that will maintain and extend national leadership for the DOE Office of Science in this critical area.

The effort is called the National Leadership Computing Facility (NLCF) and engages a world-class team of partners from national laboratories, research institutions, computing centers, universities, and vendors to take a dramatic step forward to field a new capability for high-end science. Our team offers the Office of Science an aggressive deployment plan, using technology designed to maximize the performance of scientific applications, and a means of engaging the scientific and engineering community.

The NLCF will immediately double the capability of the existing Cray X1 at ORNL and further upgrade it to a 20TF Cray X1e in 2004. The NLCF will maintain national leadership in scientific computing by installing a 100TF Cray X2 in 2006. We will simultaneously conduct an in-depth exploration of alternative technologies for next-generation leadership-class computers by deploying a 20TF Cray Red Storm at ORNL and a 50TF IBM BlueGene/L at Argonne National Laboratory. These efforts will set the stage for deployment of a machine capable of 100TF sustained performance (300TF peak) by 2007.

NLCF has a comparably ambitious approach to achieving a high level of scientific productivity. The NLCF computing system will be a unique world-class research resource, similar to other large-scale experimental facilities constructed and operated around the world. At these facilities, scientists and engineers make use of "end stations"-best-in-class instruments supported by instrument specialists-that enable the most effective use of the unique capabilities of the facilities. In similar fashion the NLCF will have Computational End Stations (CESs) that offer access to best-in-class scientific application codes and world-class computational specialists. The CESs will engage multi-institutional, multidisciplinary teams undertaking scientific and engineering problems that can only be solved on the NLCF computers and who are willing to enhance the capabilities of the NLCF and contribute to its effective operation. All CESs will be selected through a competitive peer-review process. It is envisioned that there will be computational end stations in climate, fusion, astrophysics, nanoscience, chemistry, and biology as these offer great potential for breakthrough science in the near term.

The last part of this talk describes how the NLCF will bring together world-class researchers; a proven, aggressive, and sustainable hardware path; an experienced operational team; a strategy for delivering true capability computing; and modern computing facilities connected to the national infrastructure through state-of-the-art networking to deliver breakthrough science. Combining these resources and building on expertise and resources of the partnership, the NLCF will enable scientific computation at an unprecedented scale.

References

1. G. Heffelfinger, et al, "Carbon Sequestration in Synechococcus Sp.: From Molecular Machines to Hierarchical Modeling", OMICS Journal of Integrative Biology. Nov. 2002 (www.genomes-to-life.org)
2. M. Land, et al, "Genome Channel", (http://compbio.ornl.gov/channel)
3. G. A. Geist, et al, "Harness", (www.csm.ornl.gov/harness)
4. D. Kurzyniec, et al, "Towards Self-Organizing Distributed Computing Frameworks: The H2O Approach", International Journal of High Performance Computing (2003).

The Austrian Grid Initiative – High Level Extensions to Grid Middleware

Jens Volkert

GUP, Joh. Kepler University Linz
Altenbergerstr. 69, A-4040 Linz, Austria/Europe
volkert@gup.uni-linz.ac.at
http://www.gup.uni-linz.ac.at/

Abstract. The Austrian Grid initiative is the national effort of Austria to establish a nation-wide grid environment for computational science and research. The goal of the Austrian Grid is to pursue a variety of scientific users in utilizing the Grid for their applications, e.g. medical sciencs, high-energy physics, applied numerical simulations, astrophyscial simulations and solar observations, as well as meteorology and geophysics. All these applications rely on a wide range of diverse computer science technologies, composed from standard grid middleware and sophisticated high-level extensions, which enable the implementation and operation of the Austrian Grid testbed and its applications.

One of these high-level middleware extensions is the Grid Visualization Kernel (GVK), which offers a means to process and transport large amounts of visualization data on the grid. In particular, GVK addresses the connection of grid applications and visualization clients on the grid. The visualization capabilities of GVK are provided as flexible grid services via dedicated interfaces and protocols, while GVK itself relies on the grid to implement and improve the functionality and the performance of the visualization pipeline. As a result, users are able to exploit visualization within their grid applications similar to how they would utilize visualization on their desktop workstations.

D. Kranzlmüller et al. (Eds.): EuroPVM/MPI 2004, LNCS 3241, p. 5, 2004.

Fault Tolerance in Message Passing and in Action

Jack J. Dongarra[1,2]

[1] Innovative Computing Laboratory
University of Tennessee
Knoxville TN, 37996-3450, USA
[2] Oak Ridge National Laboratory
Oak Ridge, TN
dongarra@cs.utk.edu
http://www.cs.utk.edu/~dongarra

Abstract. This talk will describe an implementation of MPI which extends the message passing model to allow for recovery in the presence of a faulty process. Our implementation allows a user to catch the fault and then provide for a recovery.

We will also touch on the issues related to using diskless checkpointing to allow for effective recovery of an application in the presence of a process fault.

D. Kranzlmüller et al. (Eds.): EuroPVM/MPI 2004, LNCS 3241, p. 6, 2004.

MPI and High Productivity Programming

William D. Gropp*

Mathematics and Computer Science Division
Argonne National Laboratory
Argonne, IL
gropp@mcs.anl.gov
http://www.mcs.anl.gov/~gropp

Abstract. MPI has often been called the "assembly language" of parallel programming. In fact, MPI has succeeded because, like other successful but low-level programming models, it provides support for both performance programming and for "programming in the large" – building support tools, such as software libraries, for large-scale applications. Nevertheless, MPI programming can be challenging, particularly if approached as a replacement for shared-memory style load/store programming. By looking at some representative programming tasks, this talk looks at ways to improve the productivity of parallel programmers by identifying the key communities and their needs, the strengths and weaknesses of the MPI programming model and the implementations of MPI, and opportunities for improving productivity both through the use of tools that leverage MPI and through extensions of MPI.

* This work was supported by the Mathematical, Information, and Computational Sciences Division subprogram of the Office of Advanced Scientific Computing Research, Office of Science, U.S. Department of Energy, under Contract W-31-109-ENG-38.

D. Kranzlmüller et al. (Eds.): EuroPVM/MPI 2004, LNCS 3241, p. 7, 2004.

High Performance Application Execution Scenarios in P-GRADE*

Gábor Dózsa

MTA SZTAKI
Computer and Automation Research Institute of the Hungarian Academy of Sciences
Kende u. 13-17, H-1111 Budapest, Hungary/Europe
dozsa@sztaki.hu
http://www.lpds.sztaki.hu/

Abstract. The P-GRADE system provides high level graphical support for development and execution of high performance message-passing applications. Originally, execution of such applications was supported on heterogeneous workstation clusters in interactive mode only. Recently, the system has been substantially extended towards supporting job execution mode in Grid like execution environments. As a result, P-GRADE now makes possible to run the same HPC application using different message-passing middlewares (PVM,MPI or MPICH-G2) in either interactive or job execution mode, on various execution resources controlled by either Condor or Globus-2 Grid middlewares.

In order to support such heterogeneous execution environments, the notion of *execution host* has been replaced with the more abstract *execution resource* in P-GRADE context. We distinguish between two basic types of execution resources: interactive and Grid resources. An interactive resource can be, for instance, a cluster on which the user has an account to login by the help of ssh. Contrarily, a Grid resource can be used to submit jobs (e.g. by the help of Globus-2 GRAM) but it does not allow interactive user's sessions. Each resource may exhibit a number of facilities that determine the type and level of support P-GRADE can provide for executing an application on that particular resource. Most notably, such facilities include the available message-passing infrastructures (e.g. PVM, MPI) and Grid middlewares (e.g. Condor, GT-2). For instance, P-GRADE can provide automatic dynamic load-balancing if the parallel application is executed on a LINUX cluster with PVM installed and interactive access to the cluster is allowed.

A resource configuration tool has also been developed for P-GRADE in order to facilitate the creation of all the necessary configuration files during the installation process. By the help of this configuration tool, even novice users (i.e biologists, chemists) can easily set up a customized execution resource pool containing various kinds of resources (like desktops, clusters, supercomputers, Condor pools, Globus-2 sites). Having finished the configuration, applications can be launched and controlled from P-GRADE in a uniform way by means of a high-level GUI regardless of the actual executing resource.

---

* This work was partially supported by the Hungarian Scientific Research Fund (OTKA T-042459), Hungarian Ministry of Informatics and Communications (MTA IHM 4671/1/2003) and National Research and Technology Office (OMFB-00495/2004).

D. Kranzlmüller et al. (Eds.): EuroPVM/MPI 2004, LNCS 3241, p. 8, 2004.

An Open Cluster System Software Stack

Ewing Lusk*

Mathematics and Computer Science Division
Argonne National Laboratory
Argonne, IL
lusk@mcs.anl.gov

Abstract. By "cluster system software," we mean the software that turns a collection of individual machines into a powerful resource for a wide variety of applications. In this talk we will examine one loosely integrated collection of open-source cluster system software that includes an infrastructure for building component-based systems management tools, a collection of components based on this infrastructure that has been used for the last year to manage a medium-sized cluster, a scalable process-management component in this collection that provides for both batch and interactive use, and an MPI-2 implementation together with debugging and performance analysis tools that help in developing advanced applications.

The component infrastructure has been developed in the context of the Scalable Systems Software SciDAC project, where a number of system management tools, developed by various groups, have been tied together by a common communication library. The flexible architecture of this library allows systems managers to design and implement new systems components and even new communication protocols and integrates them into a collection of existing components. One of the components that has been integrated into this system is the MPD process manager; we will describe its capabilities. It, in turn, supports the process management interface used by MPICH-2, a full-featured MPI-2 implementation, for scalable startup, dynamic process functionality in MPI-2, and interactive debugging. This combination allows significant components of the systems software stack to be written in MPI for increased performance and scalability.

* This work was supported by the Mathematical, Information, and Computational Sciences Division subprogram of the Office of Advanced Scientific Computing Research, Office of Science, U.S. Department of Energy, under Contract W-31-109-ENG-38.

D. Kranzlmüller et al. (Eds.): EuroPVM/MPI 2004, LNCS 3241, p. 9, 2004.

Advanced Resource Connector (ARC) – The Grid Middleware of the NorduGrid

Balázs Kónya

NorduGrid Collaboration
Dept. of High Energy Physics, Lund University
Box 118, 22100 Lund, Sweden
balazs.konya@hep.lu.se
http://www.nordugrid.org/

Abstract. The Advanced Resource Connector (ARC), or the NorduGrid middleware, is an open source software solution enabling production quality computational and data Grids. Since the first release (May 2002) the middleware is deployed and being used in production environments. Emphasis is put on scalability, stability, reliability and performance of the middleware. A growing number of grid deployments chose ARC as their middleware, thus building one of the largest production Grids of the world.

The NorduGrid middleware integrates computing resources (commodity computing clusters managed by a batch system or standalone workstations) and Storage Elements, making them available via a secure common grid layer. ARC provides a reliable implementation of the fundamental grid services, such as information services, resource discovery and monitoring, job submission and management, brokering and data management.

The middleware builds upon standard open source solutions like OpenLDAP, OpenSSL, SASL and Globus Toolkit 2 (GT2) libraries. NorduGrid provides innovative solutions essential for a production quality middleware: Grid Manager, ARC GridFTP server, information model and providers (NorduGrid schema), User Interface and broker (a "personal" broker integrated into the user interface), extended Resource Specification Language (xRSL), and the monitoring system.

ARC solutions are replacements and extensions of the original GT2 services, the middleware does not use most of the core GT2 services, such as the GRAM, the GT2 job submission commands, the WUftp-based gridftp server, the gatekeeper, the job-manager, the GT2 information providers and schemas. Moreover, ARC extended the RSL and made the Globus MDS functional. ARC is thus much more than GT2 – it offers its own set of Grid services built upon the GT2 libraries.

D. Kranzlmüller et al. (Eds.): EuroPVM/MPI 2004, LNCS 3241, p. 10, 2004.
© Springer-Verlag Berlin Heidelberg 2004

Next Generation Grid: Learn from the Past, Look to the Future

Domenico Laforenza

High Performance Computing Laboratory
ISTI, CNR
Via G. Moruzzi 1, 56126 Pisa, Italy
Domenico.Laforenza@isti.cnr.it
http://miles.isti.cnr.it/~lafo/domenico/domenico.html

Abstract. The first part of this talk will be focused on the Grid Evolution. In fact, in order to discuss "what is a Next Generation Grid", it is important to determine "with respect to what". Distinct phases in the evolution of Grids are observable. At the beginning of the 90's, in order to tackle huge scientific problems, in several important research centers tests were conducted on the cooperative use of geographically distributed resources, conceived as a single powerful computer. In 1992, Charlie Catlett and Larry Smarr coined the term "Metacomputing" to describe this innovative computational approach [1].
The term Grid Computing was introduced by Foster and Kesselman a few years later, and in the meanwhile several other words were used to describe this new computational approach, such as Heterogeneous Computing, Networked Virtual Supercomputing, Heterogeneous Supercomputing, Seamless Computing, etc., Metacomputing could be considered as the 1st generation of Grid Computing, some kind of "proto-Grid".
The Second Grid Computing generation starts around 2001, when Foster at al. proposed Grid Computing as an important new field, distinguished from conventional distributed computing by its focus on large-scale resource sharing, innovative applications, and, in some cases, high-performance orientation [2].
With the advent of multiple different Grid technologies the creativity of the research community was further stimulated, and several Grid projects were proposed worldwide. But soon a new question about how to guarantee interoperability among Grids was raised. In fact, the Grid Community, mainly created around the Global Grid Forum (GGF) [3], perceived the real risk that the far-reaching vision offered by Grid Computing could be obscured by the lack of interoperability standards among the current Grid technologies.
The marriage of the Web Services technologies [4] with the Second Generation Grid technology led to the valuable GGF Open Grid Services Architec-ture (OGSA) [5], and to the creation of the Grid Service concept and specifica-tion (Open Grid Service Infrastructure - OGSI). OGSA can be considered the mile-stone architecture to build Third Generation Grids.
The second part of this talk aims to present the outcome of a group of independent experts convened by the European Commission with the objective to identify potential European Research priorities for Next Generation Grid(s) in 2005 – 2010 [6]. The Next Generation Grid Properties ("The NGG Wish List") will be presented. The current Grid implementations do not individually possess all of the properties reported in the NGG document. However, future Grids not possessing

D. Kranzlmüller et al. (Eds.): EuroPVM/MPI 2004, LNCS 3241, pp. 11–12, 2004.

them are unlikely to be of significant use and, therefore, inadequate from both research and commercial perspectives. In order to real-ise the NGG vision much research is needed.

During the last few years, several new terms such as Global Computing, Ubiquitous Computing, Utility Computing, Pervasive Computing, On-demand Computing, Autonomic Computing, Ambient Intelligence [7], etc., have been coined. In some cases, these terms describe very similar computational approaches. Consequently, some people are raising the following questions: Are these computational approaches facets of the same medal? The last part of this talk will explore the relationship of these approaches with Grid.

References

1. Smarr, L., Catlett, C.: Metacomputing, Communications of the ACM, **35**(6) (1992) 45–52
2. Foster, I., Kesselman, C., Tuecke, S.: The Anatomy of the Grid: Enabling Scalable Virtual Organizations, Int. Journal of High Performance Computing Applications, **15**(3) (2001) 200–222
3. Global Grid Forum, www.ggf.org
4. W3C Consortium, www.w3c.org
5. Foster, I., Kesselman, C., Nick, J., Tuecke, S.: The Physiology of the Grid: An Open Grid Services Architecture for Distributed Systems Integration (2002)
6. EU Expert Group Report, Next Generation Grid(s) 2005 – 2010, Brussels, June 2003 ftp://ftp.cordis.lu/pub/ist/docs/ngg_eg_final.pdf
7. Aarts, E., Bourlard, H., Burgelman, J-C., Filip, F., Ferrate, G., Hermenegildo, M., Hvannberg, E., McAra-McWilliam, I., Langer, J., Lagasse, P., Mehring, P., Nicolai, A., Spirakis, P., Svendsen, B., Uusitalo, M., Van Rijsbergen, K., Ayre, J.: Ambient Intelligence: From Vision to Reality. IST Advisory Group (ISTAG) in FP6: WG116/9/2003 (2003) Final Report.

Production Grid Systems and Their Programming

Péter Kacsuk[1], Balázs Kónya[2], and Péter Stefán[3]

[1] MTA-SZTAKI
Kende str. 13-17, H-1111 Budapest, Hungary
peter.kacsuk@sztaki.hu
[2] Lund University
Department of Experimental High Energy Physics
P.O.Box 118, SE-22100, Lund, Sweden
balazs.konya@hep.lu.se
[3] NIIF/HUNGARNET
Victor Hugo str. 18-22., H-1132 Budapest, Hungary
peter.stefan@niif.hu

Abstract. There are a large variety of Grid test-beds that can be used for experimental purposes by a small community. However, the number of production Grid systems that can be used as a service for a large community is very limited. The current tutorial provides introduction to three of these very few production Grid systems. They represent different models and policies of using Grid resources and hence understanding and comparing them is an extremely useful exercise to everyone interested in Grid technology.

The Hungarian ClusterGrid infrastructure connects clusters during the nights and weekends. These clusters are used during the day for educational purposes at the Hungarian universities and polytechnics. Therefore a unique feature of this Grid the switching mechanism by which the day time and night time working modes are loaded to the computers. In order to manage the system as a production one the system is homogeneous, all the machines should install the same Grid software package.

The second even larger production Grid system is the LHC-Grid that was developed by CERN to support the Large Hydron Collider experiments. This Grid is also homogeneous but it works as a 24-hour service. All the computers in the Grid are completely devoted to offer Grid services. The LHC-Grid is mainly used by physists but in the EGEE project other applications like bio-medical applications will be ported and supported on this Grid.

The third production Grid is the NorduGrid which is completely heterogeneous and the resources can join and leave the Grid at any time as they need. The NorduGrid was developed to serve the Nordic countries of Europe but now more and more institutions from other countries join this Grid due to its large flexibility.

Concerning the user view an important question is how to handle this large variety of production Grids and other Grid test-beds. How to develop applications for such different Grid systems and how to port applications among them? A possible answer for these important questions is the use of a Grid portal technology. The EU GridLab project developed the GridSphere portal framework that was the basis of developing the P-GRADE Grid portal. By the P-GRADE portal users can develop workflow-like applications including HPC components and can run such workflows on any of the Grid systems in a transparent way.

D. Kranzlmüller et al. (Eds.): EuroPVM/MPI 2004, LNCS 3241, p. 13, 2004.

Tools and Services for Interactive Applications on the Grid – The CrossGrid Tutorial

Tomasz Szepieniec[2], Marcin Radecki[2], Katarzyna Rycerz[1],
Marian Bubak[1,2], and Maciej Malawski[1]

[1] Institute of Computer Science, AGH, al. Mickiewicza 30, 30-059 Kraków, Poland
{kzajac,bubak,malawski}@uci.agh.edu.pl
[2] Academic Computer Centre – CYFRONET, Nawojki 11, 30-950 Kraków, Poland
Phone: (+48 12) 617 39 64, Fax: (+48 12) 633 80 54
{T.Szepieniec,M.Radecki}@cyfronet.krakow.pl

Abstract. The CrossGrid project aims to develop new Grid services and tools for interactive compute- and data-intensive applications. This Tutorial comprises presentations and training exercises prepared to familiarize the user with the area of Grid computing being researched by the CrossGrid. We present tools aimed at both users and Grid application developers. The exercises cover many subjects, from user-friendly utilities for handling Grid jobs, through interactive monitoring of applications and infrastructure, to data access optimization mechanisms.

1 Applications and Architecture

The main objective of this Tutorial is to present the CrossGrid's achievements in development of tools and grid services for interactive compute- and data-intensive applications. The Tutorial presentations and exercises are done from three different perspectives:

- user's perspective,
- application developer's perspective,
- system administrator's perspective.

The Tutorial starts with a presentation of demos of the CrossGrid applications which are: a simulation and visualization for vascular surgical procedures, a flood crisis team decision support system, distributed data analysis in high energy physics and air pollution combined with weather forecasting [7].

Next, we give an overview of the architecture of the CrossGrid software. The architecture was defined as the result of detailed analysis of requirements from applications [4], and in its first form it was recently presented in the overview [2]. During the progress of the Project the architecture was refined [5]. The usage of Globus Toolkit 2.x was decided at the beginning of the Project for stability reasons and because of close collaboration with the DataGrid project. Nevertheless, the software is developed in such a way that it will be easy to use in future Grid systems based on OGSA standards. The tools and services of the CrossGrid are complementary to those of DataGrid, GridLab (CG has a close collaboration) and US GrADS [8].

D. Kranzlmüller et al. (Eds.): EuroPVM/MPI 2004, LNCS 3241, pp. 14–17, 2004.

2   Tool Environment

The tools developed in CrossGrid aim to provide new solutions for Grid application users and developers.

The Migrating Desktop together with its backend – Roaming Access Server [11] – provides an integrated user frontend to the Grid environment. It is extensible by means of application and tool plugins. The various CrossGrid applications and tools are integrated with Migrating Desktop by means of plugins that are designed to act similar to those used in popular browsers. The Portal which provides a graphical user interface, in comparison with the Migrating Desktop, is the lightweight client to CrossGrid resources. It requires only the Web browser and offers more simple functionality of job submission and monitoring.

The MARMOT [12] MPI verification tool enables not only the strict correctness of program code compliance with the MPI standard, but also helps locate deadlocks and other anomalous situations in the running program. Both the C and Fortran language binding of MPI standard 1.2 are supported.

The GridBench tool [16] is an implementation of a set of Grid benchmarks. Such benchmarks are important both for site administrators and for the application developers and users, who require an overview of performance of their applications on the Grid without actually running them. Benchmarks may provide reference data for the high-level analysis of applications as well as parameters for performance prediction.

The PPC performance prediction tool [3] is tackling the difficult problem of predicting performance in the Grid environment. Through the analysis of application kernels it is possible to derive certain analytical models and use them later for predictions. This task is automated by PPC.

The G-PM, Performance Measurement tool using the OCM-G application monitoring system [1], can display performance data about the execution of Grid applications. The displayed measurements can be defined during the runtime. The G-PM consists of three components: a performance measurement component, a component for high level analysis and a user interface and visualization component.

3   Grid Services

Roaming Access Server (RAS) [11] serves the requests from the Migrating Desktop or Portal and forwards them to appropriate Grid services (scheduling, data access, etc.). RAS exposes its interface as a Web Service. It uses Scheduler API for job submission and GridFTP API for data transfer.

Grid Visualisation Kernel [15] provides a visualization engine running in the Grid environment. It distributes the visualization pipeline and uses various optimization techniques to improve the performance of the system.

The CrossGrid scheduling system [10] extends the EDG resource broker through support for MPICH-G2 applications running on multiple sites. Scheduler exposes its interface what allows for job submission and retrieval of its status and output. The Scheduler provides interface to RAS by Java EDG JSS API and extends DataGrid code for providing additional functionality and uses postprocessing interface. The main objectives of the postprocessing is to gather monitoring data from the Grid, such as cluster load and data transfers between clusters, to build a central monitoring service to analyze the above data and provide it in the format suitable for the scheduler to forecast the collected Grid parameters.

The application monitoring system, OCM-G [1], is a unique online monitoring system for Grid applications with requests and response events generated dynamically and toggled at runtime. This imposes much less overhead on the application and therefore can provide more accurate measurements for the performance analysis tools such as G-PM using OMIS protocol.

The infrastructure monitoring system, JIMS (JMX-based Infrastructure Monitoring System) [13] provides information that cannot be acquired from standard Grid monitoring systems. JIMS can yield information on all cluster nodes and also routers or any SNMP devices.

SANTA-G can provide detailed data on network packets and publishes its data into DataGrid R-GMA [6]. Santa-G does not use external components of CrossGrid, because it is a low layer of architecture.

The data access optimization is achieved with a component-expert subsystem. It is used to select optimal components to handle various file types and storage systems (e.g. tape libraries). The access time estimator extends the functionality of the Reptor replica manager developed in DataGrid by providing access time for various storage systems [14].

4   Testbed

All Tutorial exercises are run on the CrossGrid testbed which is composed of 18 sites in 9 countries. The current basic middleware is LCG-2. Most sites offer storage capacities around 60 GB. The hardware type ranges mostly from Pentium III to Pentium Xeon based systems, with RAM memories between 256MB and 2GB. Many sites offer dual CPU systems. The operating system in most sites is RedHat 7.3 [9].

Acknowledgments

The authors wish to thank the CrossGrid Collaboration for their contribution as well as Dieter Kranzmueller and Peter Kacsuk for their help with arranging the Tutorial at EuroPVM/MPI'2004. This research is partly funded by the European Commission IST-2001-32243 Project "CrossGrid" and the Polish Committee for Scientific Research SPUBM 112/E-356/SPB/5.PR UE/DZ224/2002-2004.

References

1. Baliś, B., Bubak, M., Funika, W., Szepieniec, T., Wismüller, R.: Monitoring and Performance Analysis of Grid Applications, in: Sloot, P. M. A., et al. (Eds.), Proceedings of Computational Science - ICCS 2003, International Conference Melbourne, Australia and St. Petersburg, Russia, June 2003, vol. I, no 2657, LNCS, Springer, 2003, pp. 214-224
2. Berman, F., Fox, G., and Hey, T. (Eds): Grid Computing. Making the Global Infrastructure a Reality. Wiley 2003; pp. 873-874

3. Blanco, V., Gonzalez, P., Cabaleiro, J.C., Heras, D.B., Pena, T.F., Pombo, J.J., Rivera, F.F.: Avispa: Visualizing the Performance Prediction of Parallel Iterative Solvers. Future Generation Computer Systems, 19(5):721-733, 2003
4. Bubak, M., Malawski, M., Zając, K., Towards the CrossGrid Architecture, in: Kranzlmüller, D., Kascuk, P., Dongarra, J., Volkert, J. (Eds.), Recent Advances in Parallel Virtual Machine and Message Passing Interface - 9th European PVM/MPI Users' Group Meeting Linz, Austria, September 29 – October 2, 2002, LNCS 2474, Springer, 2002, pp. 16-24
5. Bubak, M., Malawski, M., Zając, K., Architecture of the Grid for Interactive Applications, in: Sloot, P. M. A., et al. (Eds.), Proceedings of Computational Science - ICCS 2003, International Conference Melbourne, Australia and St. Petersburg, Russia, June 2003, LNCS Vol. 2657, Part I, Springer, 2003, pp. 207-213
6. Cooke, A., Gray, A., Ma, L., Nutt, W., Magowan, J., Taylor, P., Byrom, R., Field, L., Hicks, S., Leake, J. et al.: R-GMA: An Information Integration System for Grid Monitoring, in Proceedings of the 11th International Conference on Cooperative Information Systems, 2003
7. The CrossGrid Project Homepage: www.eu-crossgrid.org
8. Gerndt, M. (Ed.): Performance Tools for the Grid: State of the Art and Future. APART White Paper, Shaker Verlag 2004, ISBN 3-8322-2413-0
9. Gomes, J., David, M., Martins, J., Bernardo, L., Marco, J., Marco, R., Rodriguez R., Salt, J., Gonzalez, S., Sanchez J., Fuentes, A., Hardt, M., Garcia A., Nyczyk, P., Ozieblo, A., Wolniewicz, P., Bluj, M., Nawrocki, K., Padee, A., Wislicki, W., Fernandez, C., Fontan, J., Gomez, A., Lopez, I., Cotro, Y., Floros, E., Tsouloupas, G., Xing, W., Dikaiakos, M., Astalos, J., Coghlan, B., Heymann, E., Senar, M., Merino, G., Kanellopoulos, C. and van Albada G.D.: First Prototype of the CrossGrid Testbed. In: Fernandez Rivera, F., Bubak, M., Gomez Tato, A., Doallo, R. (Eds.) 1st European Across Grids Conference. Santiago de Compostela, Spain, February 2003, LNCS 2970, 67-77, Springer, 2004
10. Heymann, E., Senar, M.A., Fernández A., Salt, J.: Managing MPI Applications in Grid Environments in Proceedings of 2nd European Across Grids Conference, Nicosia, Cyprus, January 28-30 (to appear)
11. Kosiedowski, M., Kupczyk, M., Lichwala, R., Meyer, N., Palak, B., Plociennik, M., Wolniewicz, P., Beco S.: Mobile Work Environment for Grid Users. Grid Applications Framework. in: Sloot, P. M. A., et al. (Eds.), Proceedings of Computational Science - ICCS 2003, International Conference Melbourne, Australia and St. Petersburg, Russia, June 2003, vol. I, LNCS 2657 , Springer, 2003
12. Krammer, B., Bidmon, K., Muller, M.S., Resch, M.M.: MARMOT: An MPI Analysis and Checking Tool in: Procedings of Parallel Computing 2003, Dresden, Germany, September 2-5, 2003
13. Ławniczek, B., Majka, G., Słowikowski, P., Zieliński, K., Zieliński S.: Grid Infrastructure Monitoring Service Framework Jiro/JMX Based Implementation. Electronic Notes in Theoretical Computer Science 82(6): (2003)
14. Stockinger, K., Stockinger, H., Dutka, Ł., Słota, R., Nikolow, D., Kitowski, J., Access Cost Estimation for Unified Grid Storage Systems, 4th International Workshop on Grid Computing (Grid2003), Phoenix, Arizona, Nov 17, 2003, IEEE Computer Society Press, 2003
15. Tirado-Ramos, A., Ragas, H., Shamonin, D., Rosmanith, H., Kranzmueller D.: Integration of Blood Flow Visualization on the Grid: the FlowFish/GVK Approach, In: Proceedings of 2nd European Across Grids Conference, Nicosia, Cyprus, January 28-30, 2004
16. Tsouloupas, G., Dikaiakos, M.: GridBench: A Tool for Benchmarking Grhmarking Grids. In 4th International Workshop on Grid Computing (Grid2003), pages 60-67, Phoenix, Arizona, 17 November 2003, IEEE Computer Society

Verifying Collective MPI Calls

Jesper Larsson Träff and Joachim Worringen

C&C Research Laboratories, NEC Europe Ltd.
Rathausallee 10, D-53757 Sankt Augustin, Germany
{traff,worringen}@ccrl-nece.de

Abstract. The collective communication operations of MPI, and in general MPI operations with non-local semantics, require the processes participating in the calls to provide consistent parameters, eg. a unique root process, matching type signatures and amounts for data to be exchanged, or same operator. Under normal use of MPI such exhaustive consistency checks are typically too expensive to perform and would compromise optimizations for high performance in the collective routines. However, confusing and hard-to-find errors (deadlocks, wrong results, or program crash) can happen by inconsistent calls to collective operations. We suggest to use the MPI profiling interface to provide for more extensive semantic checking of calls to MPI routines with collective (non-local) semantics. With this, exhaustive semantic checks can be enabled during application development, and disabled for production runs. We discuss what can reasonably be checked by such an interface, and mention some inherent limitations of MPI to making a fully portable interface for semantic checking. The proposed *collective semantics verification interface* for the full MPI-2 standard has been implemented for the NEC proprietary MPI/SX as well as other NEC MPI implementations.

1 Introduction

To be useful in high-performance parallel computing, the Message Passing Interface (MPI) standard is carefully designed to allow highly efficient implementations on "a wide variety of parallel computers" [3, 6]. For this reason, the error behavior of MPI is largely unspecified, and the standard prescribes little in terms of mandatory error checking and reporting. MPI is also a complex standard with many possibilities for making mistakes during application development. Thus there is a trade-off for the MPI implementer between high performance and exhaustive error checking. In this note we mention several useful checks for MPI functions with collective (non-local) semantics that are typically *not* performed by an MPI implementation, and suggest using the *profiling interface* of MPI [6, Chapter 8] to implement such more exhaustive checks.

One motivation for our *collective semantics verification interface* comes from the possibly confusing way the collective communication operations are described in the MPI standard [6, Chapter 4]. On the one hand, it is required that processes participating in a collective communication operation specify the same amount of data for processes sending data and the corresponding receivers: "in contrast

D. Kranzlmüller et al. (Eds.): EuroPVM/MPI 2004, LNCS 3241, pp. 18–27, 2004.

to point-to-point communication, the amount of data sent must exactly match the amount of data specified by the receive" [6, page 192]. On the other hand, the semantics of many of the collective communication operations are explained in terms of point-to-point send and receive operations, which do not impose this restriction, and many MPI implementations indeed implement the collective operations on top of point-to-point communication. Highly optimized collective communications will, as do the point-to-point primitives, use several different protocols, and for performance reasons the decision on which protocol to use in a given case is taken locally: the MPI standard is naturally defined to make this possible. For such MPI implementations non-matching data amounts can lead to crashes, deadlocks or wrong results. The increased complexity of the "irregular" MPI collectives like MPI_Gatherv, MPI_Alltoallw and so on, where different amounts of data can be exchanged between different process pairs aggravates these problems, and in our experience even expert MPI programmers sometimes make mistakes in specifying matching amounts of data among communicating pairs of processes.

Analogous problems can occur with many other MPI functions with collective (non-local) completion semantics. For instance the creation of a new communicator requires all participating processes to supply the same process group in the call. If not, the resulting communicator will be ill-formed, and likely to give rise to completely unspecified behavior at some later point in the program execution.

It would be helpful for the user to be able to detect such mistakes, but detection obviously requires extra communication among the participating processes. Thus, extensive checking of MPI calls with non-local semantics in general imposes overhead that is too high for usage in critical high-performance applications, and unnecessary once the code is correct.

A solution is to provide *optional* extended checks for consistent arguments to routines with collective semantics. This can either be built into the MPI library (and controlled via external environment variables), or implemented stand-alone and in principle in a portable way by using the MPI profiling interface. For the latter solution, which we have adopted for MPI/SX and other NEC MPI implementations [1], the user only has to link with the verification interface to enable the extensive checks. For large parts of MPI, a verification interface can be implemented in terms of MPI without access to internals of a specific implementation. However, certain MPI objects are not "first-class citizens" (the gravest omission is MPI_Aint, but this is a different story) and cannot be used in communication operations, eg. process groups (MPI_Group) and reduction operators (MPI_Op), or are not immediately accessible, like the communicator associated with a window or a file. Thus, a truly portable verification interface independent of a specific MPI implementation would have to duplicate/mimic much of the work already performed by the underlying MPI implementation, eg. tracking creation and destruction of MPI objects [4, 9]. Implementing our interface for a specific MPI implementation with access to the internals of the implementation alleviates these problems, and is in fact straightforward. We recommend other MPI implementations to provide checking interfaces, too.

1.1   Related Work

The fact that MPI is both a widely accepted and complex standard has motivated other work on compile- and run-time tools to assist in finding semantic errors during application development. We are aware of three such projects, namely Umpire [9], MARMOT [4], and MPI-CHECK [5]. All three only (partly) cover MPI-1, whereas our collective semantics verification interface addresses the full MPI-2 standard. The first two are in certain respects more ambitious in that they cover also deadlock detection, which in MPI/SX is provided independently by a suspend/resume mechanism, Umpire, however, being limited to shared-memory systems only. Many of the constant time checks discussed in this note, and that in our opinion should be done by any good MPI implementation, are performed by these tools, whereas the extent to which checks of consistent parameters to irregular collective communication operations are performed is not clear.

The observations in this note are mostly obvious, and other MPI libraries may already include checks along the lines suggested here. One example we are aware of is ScaMPI by Scali (see www.scali.com) which provides optional checking of matching message lengths.

The checks performed for MPI operations with non-local semantics by our verification interface are summarized in Appendix A.

2   Verifying Communication Buffers

For argument checking of MPI calls we distinguish between conditions that can be checked locally by each process in *constant time*, conditions that can be checked locally but require more than constant time (say, proportional to some input argument, eg. size of communicator), and conditions that cannot be checked locally but require communication between processes.

A communication buffer in MPI is specified by a triple consisting of a `buffer` address, a repetition `count`, and an MPI `datatype`.

2.1   Conditions That Can Be Checked Locally in Constant Time

General conditions on communication buffers include:

- `count` must be non-negative.
- `datatype` must be committed.
- the target communication buffer of a one-sided communication operation must lie properly within the target window
- for receive buffers, `datatype` must define non-overlapping memory segments
- for buffers used in MPI-IO, the displacements of the type-map of `datatype` must be monotonically non-decreasing
- for buffers used in `MPI_Accumulate` the primitive datatypes of `datatype` must be the same basic datatype.

The count condition is trivial, and probably checked already by most MPI implementations. For individual communication operations like MPI_Send and MPI_Put which take a rank and communicator or window parameter, it can readily be checked that rank is in range. For the one-sided communication operations the MPI_Win_get_group call allows to get access to the group of processes over which the window is defined, and from this the maximum rank can easily be determined.

The datatype conditions are much more challenging (listed in roughly increasing order of complexity), and checking these independently of the underlying MPI implementation requires the verification interface to mimic all MPI datatype operations. Internally datatypes are probably flagged when committed, thus the first check is again trivial. In MPI/SX datatype constructors record information on overlaps, type map, and basic datatypes (but does no explicit flattening), and derived datatypes are flagged accordingly. Thus also these more demanding checks are done in constant time.

For the one-sided communication calls which must also supply the location of data at the target, the MPI standard [3, Page 101] recommends it be checked that target location is within range of the target window. In MPI/SX information about all target windows is available locally [8], and thus the condition can be checked efficiently in constant time. If this information is not present, the profiling interface for MPI_Win_create can perform an all-gather operation to collect the sizes of all windows.

Some collective operations allow the special MPI_IN_PLACE argument, instead of a communication buffer. When allowed only at the root, like in MPI_Reduce, correct usage can be checked locally, otherwise this check requires communication.

2.2   Conditions That Can Be Checked Locally but Require More Time

Actual memory addresses specified by a communication buffer must all be proper, eg. non-NULL. Since MPI permits absolute addresses in derived datatypes, and allows the buffer argument to be MPI_BOTTOM, traversal of the datatype is required to check validity of addresses. This kind of check thus requires time proportional to the amount of data in the communication buffer. In MPI/SX certain address violations (eg. NULL addresses) are caught on the fly.

For irregular collectives like MPI_Gatherv that receive data, no overlapping segments of memory must be specified by the receiving (root) process. Since the receive locations are determined both by the datatype and the displacement vector of the call, checking this requires at least traversal of the displacement vector.

MPI operations involving process groups pose conditions like that one group argument must be a subgroup of another (MPI_Comm_create), that a list of ranks must contain no duplicates (MPI_Group_incl), and so on. With a proper implementation of the group functionality, all these checks can be performed in time proportional to the size of the groups involved, in time no longer than the time needed to perform the operation itself, such that an MPI implementation is not hurt by performing these checks. MPI/SX performs such checks extensively.

Other examples of useful local, but non-constant time checks are found in the topology functionality of MPI. For instance checking that the input to the MPI_Graph_create call is indeed a symmetric graph requires $O(n + m)$ time, n and m being the number of nodes and edges in the graph.

2.3   Conditions Requiring Communication

The semantics of the collective communication operations of MPI require that data to be sent strictly matches data to be received (type matching is explained in [6, Page 149]). In particular the *size* of data sent which can be computed as count times the size of datatype must match the size of data sent. This is a non-local condition, which requires at least one extra (collective) communication operation to verify. To perform the stricter test for matching *datatype signatures* of sent and received data, the data type signature must be sent together with the data. The size of the signature is, however, proportional to the size of the data it describes, so an overhead of at most a factor of 2 would be incurred by this kind of check. Gropp [2] addresses the problem of faster signature checking, and proposes an elegant, cheap, and relatively safe method based on hash functions. At present, our verification interface does not include signature checking.

An easy to verify condition imposed by some of the collectives is the consistent use of MPI_IN_PLACE. For instance, MPI_Allreduce and MPI_Allgather require that either all or no processes give MPI_IN_PLACE as a buffer argument.

3   Verifying Other MPI Arguments

Many MPI operations with collective semantics are *rooted* in the sense that a given root process play a particular role. In such cases all processes must give the same root as argument. If this is violated by mistake deadlock (in the case of collective communication operations like MPI_Bcast) or wrong results or worse (could be the case with MPI_Comm_spawn) will be the result. Verifying that all processes in a collective call have given the same root obviously require communication.

The collective reduction operations require that if one process uses a built-in operation, all processes use the same operation. In this case it is also stipulated that the same predefined datatype be used. Verifying this again requires communication. Note that formally it is not possible to send MPI_Op and MPI_Datatype handles to another process, but these handles are (in most MPI implementations) just an integer. A portable verification interface would have to translate these handles into a representation that can be used in MPI communication.

The MPI_Comm_create call requires that all processes passes identical groups to the call. Groups can also not be used as objects for communication, making it necessary for a portable interface to translate MPI_Group objects into a representation suitable for communication. With MPI_Group_translate_ranks each process can extract the order of the processes in the argument group relative to the group of the communicator passed in the call (extracted with MPI_Comm_group), and this list (of integers) can be used for verification (namely that the processes all compute the same list).

MPI-2 functionality like one-sided communication and I/O makes it possible to control or modify subsequent execution by providing extra information to window and file creation calls. Like above, it is often strictly required that all processes provide the same MPI_Info argument, and again verifying this require communication. Our verification interface internally translate the info strings into an integer representation suitable for communication.

The use of run-time *assertions* that certain conditions hold, also applicable to one-sided communication and MPI-IO, gives the MPI implementation a handle for certain worthwhile optimizations, eg. by saving communication. The consistency requirements for using assertions must therefore be strictly observed, and violations will for MPI implementations which do exploit assertions most likely lead to undesired behavior. On the other hand, checking for consistent assertion usage requires communication, and is thus contradictory to the very spirit of assertions. However, a collective semantics verification interface should at least provide the possibility for assertion checking.

Creation of virtual topologies are collective operations, and must (although not explicitly said in the standard) provide identical values for the virtual topology on all processes. For graph topologies it is required that all processes specify the whole communication graph. This is in itself not a scalable construct [7], and even checking that the input is correct requires $O(n + m)$ time steps for graphs with n nodes and m edges (this check is performed by MPI/SX).

A final issue is the consistent call of the collective operations themselves. When an MPI primitive with collective semantics is called, all processes in the communicator must eventually perform the call, or in other words the sequence of collective "events" must be the same for all processes in the communicator. Deadlock or crash is most likely to occur if this is not observed, eg. if some processes call MPI_Barrier while other call MPI_Bcast. In the MPI/SX verification interface each collective call on a given communicator first verifies that all calling processes in the communicator perform the same call. Mismatched communicators, eg. calling the same collective with different communicators, is a deadlock situation that cannot readily be caught with a profiling interface. In MPI/SX deadlocks are caught separately by a suspend/resume mechanism.

4   Performing Checks Requiring Communication

By transitivity of equality all checks of conditions requiring the processes to provide the same value for some argument, eg. root, MPI_IN_PLACE, but also that all processes indeed invoke the same collective in a given "event", can be performed by a broadcast from process 0, followed by a local comparison to the value sent by the root. To verify matching send and receive sizes one extra collective communication operation suffices. For instance, MPI_Gatherv and MPI_Scatterv can be verified by the root (after having passed verification for consistent root) scattering its vector of receive- or send-sizes to the other processes. For MPI_Allgatherv a broadcast of the vector from process 0 suffices, whereas MPI_Alltoallv and MPI_Alltoallw require an all-to-all communication.

The overhead for collective semantics verification is 2 to 4 extra collective operations, but with small data.

5 Controlling the Verification Interface

The NEC proprietary MPI implementations by default perform almost all the local, constant-time checks discussed in Section 2. In Appendix A we summarize the verification of all MPI calls with non-local semantics. Errors detected in MPI calls both with local and non-local semantics lead to abortion via the error handler associated with the communicator (usually MPI_ERRORS_ARE_FATAL). To give the possibility to use the interface with other error handlers, the checks for MPI calls with non-local semantics (in most cases) ensure that all calling processes are informed of the error condition and that all call the error handler.

Although not implemented on top of MPI, the NEC *collective semantics verification interface* appears to the user as just another MPI profiling interface. The MPI_Pcontrol(level,...) function is used to control the level of verification. We define the following levels of checking:

- level = 0: Checking of collective semantics disabled.
- level = 1: Collective semantic verification enabled, error-handler abortion on violation with concise error message.
- level = 2: Extended diagnostic explanation before abort.
- level = 3: MPI_Info argument checking enabled.
- level = 4: MPI assertion checking enabled.

The other extreme to collective checking discussed here, is to perform no checks at all. It would be possible to equip our interface with a "danger" mode (level = -1) with no checks whatsoever performed. This might lead to a (very small) latency improvement for some MPI calls, but since the overhead of the possible constant-time checks is indeed very small, we have decided against this possibility.

6 Examples

We give some examples of common mistakes in the usage of MPI routines with non-local semantics, and show the output produced by the checking interface (level > 1).

Example 1. In a gather-scatter application the root process mistakenly calls MPI_Gather whereas the other processes call MPI_Scatter:

```
> mpirun -np 3 example1
VERIFY MPI_SCATTER(2): call inconsistent with call to MPI_Gather by 0
```

after which execution aborts.

Example 2. In a gather application in which process 0 is the intended root some processes mistakenly specified the last process as root:

```
> mpirun -np 3 example2
VERIFY MPI_GATHER(2): root 2 inconsistent with root 0 of 0
```

Example 3. A classical mistake with an irregular collective: In an MPI_Alltoallv call each process sets its *i*th receive count proportional to *i and* its send count proportional to *i*. This is (of course!) wrong (either send or receive count should have been proportional to the rank of the process). The verification interface was motivated by precisely this situation:

```
> mpirun -np 3 example3
VERIFY MPI_ALLTOALLV(0): sendsize[1]=4 > expected recvsize(1)[0]=0
VERIFY MPI_ALLTOALLV(2): sendsize[0]=0 < expected recvsize(0)[2]=8
```

Example 4. For MPI_Reduce_scatter over intercommunicators, the MPI standard requires that the amount of data scattered by each process group matches the amount of data reduced by the other group. In a test application each of the local groups contributed an amount of data proportional to the size of the group, and scattered these uniformly over the other group. This could be typical – and is wrong if the two groups are not of the same size. The verification interface detects this:

```
>mpirun -np 5 example4
VERIFY MPI_REDUCE_SCATTER (INTERCOMM): scattersize=2 != reducesize=3
```

Example 5. In MPI-IO a *fileview* is used to individually mask out parts of a file for all file accesses performed by a process. The mask (*filetype*), and also the elementary datatype (*etype*), are represented by an MPI datatype. However, they must comply to a number of constraints, eg. that displacements must be monotonically non-decreasing, that elements of the datatype must not overlap if the file is opened with write-access, that the filetype is derived from (or identical to) the etype, that the extent of possible holes in the filetype are multiples of the etype's extent, and that no I/O operations are pending. These are all constant time, local checks (see Section 2.1), nevertheless not always performed by MPI implementations, but caught by the verification interface (in MPI/SX also caught in normal operation):

```
> mpirun -np 4 fview_leaves
VERIFY MPI_File_set_view(0): Filetype is not derived from etype
```

Example 6. File access with shared filepointers requires that all processes (that have opened the file collectively) have identical fileviews in place. The verification interface checks this condition on each access, both collective and non-collective. In the implementation for MPI/SX the overhead for this check is small as the NEC MPI-IO implementation uses *listless I/O* [10] which performs caching of remote fileviews.

```
> mpirun -np 2 shared_fview

VERIFY MPI_FILE_WRITE_SHARED(1): local fileview differs from fileview of (0)
VERIFY MPI_FILE_WRITE_SHARED(0): local fileview differs from fileview of (1)
```

7 Conclusion

We summarized the extensive consistency and correctness checks that can be performed with the *collective semantics verification interface* provided with the NEC MPI implementations. We believe that such a library can be useful for application programmers in early stages of the development process. Partly because of limitations of MPI (many MPI objects are not "first class citizens" and the functions to query them are sometimes limited), which makes the necessary access to internals difficult or not possible, the interface described here is not portable to other MPI implementations. As witnessed by other projects [4, 5, 9] implementing a verification interface in a completely portable manner using only MPI calls is very tedious. For every concrete MPI implementation, given access to information just below the surface, a verifier library as discussed here can be implemented with a modest effort.

References

1. M. Gołebiewski, H. Ritzdorf, J. L. Träff, and F. Zimmermann. The MPI/SX implementation of MPI for NEC's SX-6 and other NEC platforms. *NEC Research & Development*, 44(1):69–74, 2003.
2. W. Gropp. Runtime checking of datatype signatures in MPI. In *Recent Advances in Parallel Virtual Machine and Message Passing Interface. 7th European PVM/MPI Users' Group Meeting*, volume 1908 of *Lecture Notes in Computer Science*, pages 160–167, 2000.
3. W. Gropp, S. Huss-Lederman, A. Lumsdaine, E. Lusk, B. Nitzberg, W. Saphir, and M. Snir. *MPI – The Complete Reference*, volume 2, The MPI Extensions. MIT Press, 1998.
4. B. Krammer, K. Bidmon, M. S. Müller, and M. M. Resch. MARMOT: An MPI analysis and checking tool. In *Parallel Computing (ParCo)*, 2003.
5. G. R. Luecke, H. Chen, J. Coyle, J. Hoekstra, M. Kraeva, and Y. Zou. MPI-CHECK: a tool for checking fortran 90 MPI programs. *Concurrency and Computation: Practice and Experience*, 15(2):93–100, 2003.
6. M. Snir, S. Otto, S. Huss-Lederman, D. Walker, and J. Dongarra. *MPI – The Complete Reference*, volume 1, The MPI Core. MIT Press, second edition, 1998.
7. J. L. Träff. SMP-aware message passing programming. In *Eigth International Workshop on High-level Parallel Programming Models and Supportive Environments (HIPS03), International Parallel and Distributed Processing Symposium (IPDPS 2003)*, pages 56–65, 2003.
8. J. L. Träff, H. Ritzdorf, and R. Hempel. The implementation of MPI-2 one-sided communication for the NEC SX-5. In *Supercomputing*, 2000. http://www.sc2000.org/proceedings/techpapr/index.htm#01.
9. J. S. Vetter and B. R. de Supinski. Dynamic software testing of MPI applications with Umpire. In *Supercomputing (SC)*, 2000. http://www.sc2000.org/proceedings/techpapr/index.htm#01.
10. J. Worringen, J. L. Träff, and H. Ritzdorf. Fast parallel non-contiguous file access. In *Supercomputing*, 2003. http://www.sc-conference.org/sc2003/tech_papers.php.

A   Verification Performed for Collective MPI Functions

MPI function	Non-local semantic checks
MPI_Barrier	Call consistency
MPI_Bcast	Call consistency, datasize match
MPI_Gather	Call, root consistency, datasize match
MPI_Gatherv	Call, root consistency, datasize match
MPI_Scatter	Call, root consistency, datasize match
MPI_Scatterv	Call, root consistency, datasize match
MPI_Allgather	Call, MPI_IN_PLACE consistency, datasize match
MPI_Allgatherv	Call, MPI_IN_PLACE consistency, datasize match
MPI_Alltoall	Call consistency, datasize match
MPI_Alltoallv	Call consistency, datasize match
MPI_Alltoallw	Call consistency, datasize match
MPI_Reduce	Call, root, op consistency, datasize match
MPI_Allreduce	Call, MPI_IN_PLACE, op consistency, datasize match
MPI_Reduce_scatter	Call, MPI_IN_PLACE, op consistency, datasize match
MPI_Scan	Call, op consistency, datasize match
MPI_Exscan	Call, op consistency, datasize match
MPI_Comm_dup	Call consistency
MPI_Comm_create	Call consistency, group consistency
MPI_Comm_split	Call consistency
MPI_Intercomm_merge	Call, high/low consistency
MPI_Intercomm_create	Call, local leader, tag consistency
MPI_Cart_create	Call consistency, dims consistency
MPI_Cart_map	Call consistency, dims consistency
MPI_Graph_create	Call consistency, graph consistency
MPI_Graph_map	Call consistency, graph consistency
MPI_Comm_spawn{_multiple}	Call, root consistency
MPI_Comm_accept	Call, root consistency
MPI_Comm_connect	Call, root consistency
MPI_Comm_disconnect	Call consistency
MPI_Win_create	Info consistency
MPI_Win_fence	Assertion consistency (MPI_MODE_NOPRECEDE, MPI_MODE_NOSUCCEED)
MPI_File_open	Call, info consistency
MPI_File_set_view	Call, info consistency, pending operations, type correctness
MPI_File_set_size	Call consistency, pending operations
MPI_File_preallocate	Call consistency, pending operations
MPI_File_set_atomicity	Call consistency
MPI_File_seek_shared	Call consistency, offset & mode match
MPI_File_{write|read}_shared	Fileview consistency
MPI_File_{write|read}_ordered	Call consistency, fileview consistency
MPI_File_{write|read}_ordered_begin	Call consistency, pending operations, fileview consistency
MPI_File_{write|read}_ordered_end	Call consistency, operation match
MPI_File_{write|read}{_at}_all	Call consistency
MPI_File_{write|read}{_at}_all_begin	Call consistency, pending operations
MPI_File_{write|read}{_at}_all_end	Call consistency, operation match

Fast Tuning of Intra-cluster Collective Communications

Luiz Angelo Barchet-Estefanel* and Grégory Mounié

Laboratoire ID - IMAG, Project APACHE**
51, Avenue Jean Kuntzmann, F38330 Montbonnot St. Martin, France
{Luiz-Angelo.Estefanel,Gregory.Mounie}@imag.fr

Abstract. Recent works try to optimise collective communication in grid systems focusing mostly on the optimisation of communications among different clusters. We believe that intra-cluster collective communications should also be optimised, as a way to improve the overall efficiency and to allow the construction of multi-level collective operations. Indeed, inside homogeneous clusters, a simple optimisation approach rely on the comparison from different implementation strategies, through their communication models. In this paper we evaluate this approach, comparing different implementation strategies with their predicted performances. As a result, we are able to choose the communication strategy that better adapts to each network environment.

1 Introduction

The optimisation of collective communications in grids is a complex task because the inherent heterogeneity of the network forbids the use of general solutions. Indeed, the optimisation cost can be fairly reduced if we consider grids as interconnected islands of homogeneous clusters, if we can identify the network topology.

Most systems only separate inter and intra-cluster communications, optimising communication across wide-area networks, which are usually slower than communication inside LANs. Some examples of this "two-layered" approach include ECO [11], MagPIe [4, 6] and even LAM-MPI 7 [8]. While ECO and MagPIe apply this concept for wide-area networks, LAM-MPI 7 applies it to SMP clusters, where each SMP machine is an island of fast communication. Even though, there is no real restriction on the number of layer and, indeed, the performance of collective communications can still be improved by the use of multi-level communication layers, as observed by [3].

If most works today use the "islands of clusters" approach, to our knowledge none of them tries to optimise the intra-cluster communication. We believe that while inter-cluster communication represents the most important aspect in grid-like environments, intra-cluster optimisation also should be considered, specially

---

* Supported by grant BEX 1364/00-6 from CAPES - Brazil.
** This project is supported by CNRS, INPG, INRIA and UJF.

D. Kranzlmüller et al. (Eds.): EuroPVM/MPI 2004, LNCS 3241, pp. 28–35, 2004.

if the clusters should be structured in multiple layers [3]. In fact, collective communications in local-area networks can still be improved with the use of message segmentation [1, 6] or the use of different communication strategies [12].

In this paper we propose the use of well known techniques for collective communication, that due to the relative homogeneity inside each cluster, may reduce the optimisation cost. Contrarily to [13], we decided to model the performance of different implementation strategies for collective communications and to select, according to the network characteristics, the most adapted implementation technique for each set of parameters (communication pattern, message size, number of processes). Hence, in this paper we illustrate our approach with two examples, the Broadcast and Scatter operations, and we validate our approach by comparing the performance from real communications and the models' predictions.

The rest of this paper is organised as follows: Section 2 presents the definitions and the test environment we will consider along this paper. Section 3 presents the communication models we developed for both Broadcast's and Scatter's most usual implementations. In Section 4 we compare the predictions from the models with experimental results. Finally, Section 5 presents our conclusions, as well as the future directions of the research.

2   System Model and Definitions

In this paper we model collective communications using the *parameterised LogP* model, or simply pLogP [6]. Hence, all along this paper we shall use the same terminology from pLogP's definition, such as $g(m)$ for the gap of a message of size m, L as the communication latency between two nodes, and P as the number of nodes. In the case of message segmentation, the segment size s of the message m is a multiple of the size of the basic datatype to be transmitted, and it splits the initial message m into k segments. Thus, $g(s)$ represents the gap of a segment with size s.

The pLogP parameters used to feed our models were previously obtained with the MPI LogP Benchmark tool [5] using LAM-MPI 6.5.9 [7]. The experiments to obtain pLogP parameters, as well as the practical experiments, were conducted on the **ID/HP icluster-1** from the ID laboratory Cluster Computing Centre[1], with 50 Pentium III machines (850Mhz, 256MB) interconnected by a switched Ethernet 100 Mbps network.

3   Communication Models with pLogP

Due to the limited space, we cannot present models for all collective communication, thus we chose to present the Broadcast and the Scatter operations. Although they are two of the simplest collective communication patterns, practical implementations of MPI usually construct other collective operations, as for example, Barrier, Reduce and Gather, in a very similar way, what makes these two

---

[1] http://www-id.imag.fr/Grappes/

operations a good example for our models accuracy. Further, the optimisation of grid-aware collective communications explores intensively such communication patterns, as for example the AllGather operation in MagPIe, which has three steps: a Gather operation inside each cluster, an AllGatherv among the clusters' roots and a Broadcast to the cluster's members.

3.1 Broadcast

With Broadcast, a single process, called *root,* sends the same message of size m to all other $(P-1)$ processes. Among the classical implementations for broadcast in homogeneous environments we can find flat, binary and binomial trees, as well as chains (or pipelines). It is usual to apply different strategies within these techniques according to the message size, as for example, the use of a *rendezvous* message that prepares the receiver to the incoming of a large message, or the use of non-blocking primitives to improve communication overlap. Based on the models proposed by [6], we developed the communication models for some current techniques and their "flavours", which are presented on Table 1.

We also considered message segmentation [6, 12], which may improve the communication performance under some specific situations. An important aspect, when dealing with message segmentation, is to determine the optimal segment size. Too little messages pay more for their headers than for their content, while too large messages do not explore enough the network bandwidth. Hence, we can use the communication models presented on Table 1 to search the segment size s that minimises the communication time in a given network. Once determined this segment size s, large messages can be split into $\lfloor m/s \rfloor$ segments, while smaller messages will be transmitted without segmentation.

As most of these variations are clearly expensive, we did not consider them on the experiments from Section 4, and focused only in the comparison of the most efficient techniques, the Binomial and the Segmented Chain Broadcasts.

3.2 Scatter

The Scatter operation, which is also called "personalised broadcast", is an operation where the *root* holds $m \times P$ data items that should be equally distributed among the P processes, including itself. It is believed that optimal algorithms for homogeneous networks use flat trees [6], and by this reason, the Flat Tree approach is the *default* Scatter implementation in most MPI implementations. The idea behind a Flat Tree Scatter is that, as each node shall receive a different message, the root shall sends these messages directly to each destination node.

To better explore our approach, we constructed the communication model for other strategies (Table 2) and, in this paper, we compare Flat Scatter and Binomial Scatter in real experiments. In a first look, a Binomial Scatter is not as efficient as the Flat Scatter, because each node receives from the parent node its message as well as the set of messages it shall send to its successors. On the other hand, the cost to send these "combined" messages (where most part is useless to the receiver and should be forwarded again) may be compensate

Table 1. Communication Models for Broadcast.

Implementation Technique	Communication Model
Flat Tree	$(P-1) \times g(m) + L$
Flat Tree Rendezvous	$(P-1) \times g(m) + 2 \times g(1) + 3 \times L$
Segmented Flat Tree	$(P-1) \times (g(s) \times k) + L$
Chain	$(P-1) \times (g(m) + L)$
Chain Rendezvous	$(P-1) \times (g(m) + 2 \times g(1) + 3 \times L)$
Segmented Chain (Pipeline)	$(P-1) \times (g(s) + L) + (g(s) \times (k-1))$
Binary Tree	$\leq \lceil log_2 P \rceil \times (2 \times g(m) + L)$
Binomial Tree	$\lfloor log_2 P \rfloor \times g(m) + \lceil log_2 P \rceil \times L$
Binomial Tree Rendezvous	$\lfloor log_2 P \rfloor \times g(m) + \lceil log_2 P \rceil \times (2 \times g(1) + 3 \times L)$
Segmented Binomial Tree	$\lfloor log_2 P \rfloor \times g(s) \times k + \lceil log_2 P \rceil \times L$

Table 2. Communication Models for Scatter.

Implementation Technique	Communication Model
Flat Tree	$(P-1) \times g(m) + L$
Chain	$\sum_{j=1}^{P-1} g(j \times m) + (P-1) \times L$
Binomial Tree	$\sum_{j=0}^{\lceil log_2 P \rceil - 1} g(2^j \times m) + \lceil log_2 P \rceil \times L$

by the possibility to execute parallel transmissions. As the trade-off between transmission cost and parallel sends is represented in our models, we can evaluate the advantages of each model according to the clusters' characteristics.

4 Practical Results

4.1 Broadcast

To evaluate the accuracy of our optimisation approach, we measured the completion time of the Binomial and the Segmented Chain Broadcasts, and we compared these results with the model predictions. Through the analysis of Figs. 1(a) and 1(b), we can verify that models' predictions follow closely the real experiments. Indeed, both experiments and models predictions show that the Segmented Chain Broadcast is the most adapted strategy to our network parameter, and consequently, we can rely on the models' predictions to chose the strategy we will apply.

Although models were accurate enough to select the best adapted strategy, a close look at the Fig. 1 still shows some differences between model's predictions and the real results. We can observe that, in the case of the Binomial Broadcast, there is a non expected delay when messages are small. In the case of the Segmented Chain Broadcast, however, the execution time is slightly larger than expected. Actually, we believe that both variations derive from the same problem.

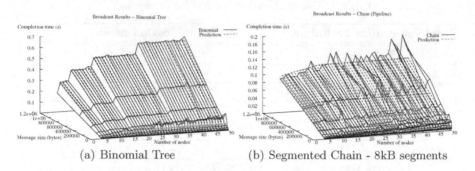

(a) Binomial Tree   (b) Segmented Chain - 8kB segments

Fig. 1. Comparison between models and real results.

Hence, we present in Fig. 2 the comparison of both strategies and their predictions for a fixed number of machines. We can observe that predictions for the Binomial Broadcast fit with enough accuracy the experimental results, except in the case of small messages (less than 128kB). Actually, similar discrepancies were already observed by the LAM-MPI team, and according to [9, 10], they are due to the TCP acknowledgement policy on Linux that may delay the transmission of some small messages even when the TCP_NODELAY socket option is active (actually, only one every n messages is delayed, with n varying from kernel to kernel implementation).

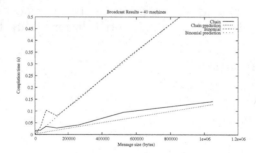

Fig. 2. Comparison between Chain and Binomial Broadcast.

In the case of the Segmented Chain Broadcast, however, this phenomenon affects all message sizes. Because large messages are split into small segments, such segments suffers from the same transmission delays as the Binomial Broadcast with small messages. Further, due to the Chain structure, a delay in one node is propagated until the end of the chain. Nevertheless, the transmission delay for a large message (and by consequence, a large number of segments) does not increases proportionally as it would be expected, but remains constant.

We believe that because these transmission delays are related to the buffering policy from TCP, we believe that the first segments that arrive are delayed by the TCP acknowledge policy, but the successive arrival of the following segments forces the transmission of the remaining segments without any delay.

4.2   Scatter

In the case of Scatter, we compare the experimental results from Flat and Binomial Scatters with the predictions from their models. Due to our network characteristics, our experiments shown that a Binomial Scatter can be more efficient than Flat Scatter, a fact that is not usually explored by traditional MPI implementations. As a Binomial Scatter should balance the cost of combined messages and parallel sends, it might occur, as in our experiments, that its performance outweighs the "simplicity" from the Flat Scatter with considerable gains according to the message size and number of nodes, as shown in Figs. 3(a) and 3(b). In fact, the Flat Tree model is limited by the time the root needs to send successive messages to different nodes (the gap), while the Binomial Tree Scatter depends mostly on the number of nodes, which defines the number of communication steps through the $\lceil log_2 P \rceil \times L$ factor. These results show that the communication models we developes are accurate enough to identify which implementation is the best adapted to a specific environment and a set of parameters (message size, number of nodes).

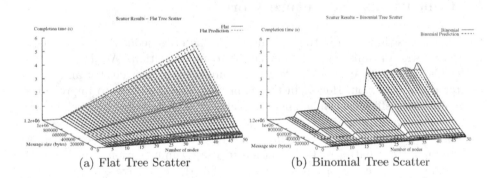

(a) Flat Tree Scatter   (b) Binomial Tree Scatter

Fig. 3. Comparison between models and real results.

Further, although we can observe some delays related to the TCP acknowledgement policy on Linux when messages are small, specially in the Flat Scatter, these variations are less important than those from the Broadcast, as depicted in Fig. 4.

What called our attention, however, was the performance of the Flat Tree Scatter, that outperformed our predictions, while the Binomial Scatter follows the predictions from its model. We think that the multiple transmissions from the Flat Scatter become a "bulk transmission", which forces the communication buffers to transfer the successive messages all together, somehow similarly to the successive sends on the Segmented Chain Broadcast. Hence, we observe that the

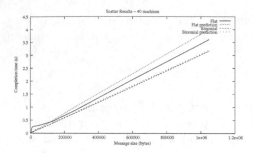

Fig. 4. Comparison between Flat and Binomial Scatter.

pLogP parameters measured by the pLogP benchmark tool are not adapted to such situations, as it considers only individual transmissions, mostly adapted to the Binomial Scatter model.

This behaviour seems to indicate a relationship between the number of successive messages sent by a node and the buffer transmission delay, which are not considered in the pLogP performance model. As this seem a very interesting aspect for the design of accurate communication models, we shall closely investigate and formalise this "multi-message" behaviour in a future work.

5   Conclusions and Future Works

Existing works that explore the optimisation of heterogeneous networks usually focus only the optimisation of inter-cluster communication. We do not agree with this approach, and we suggest to optimise both inter-cluster and intra-cluster communication. Hence, in this paper we described how to improve the communication efficiency on homogeneous cluster through the use of well known implementation strategies.

To compare different implementation strategies, we rely on the modelling of communication patterns. Our decision to use communication models allows a fast and accurate performance prediction for the collective communication strategies, giving the possibility to choose the technique that best adapts to each environment. Additionally, because the intra-cluster communication is based on static techniques, the complexity on the generation of optimal trees is restricted only to the inter-cluster communication.

Nonetheless, as our decisions rely on network models, their accuracy needs to be evaluated. Hence, in this paper we presented two examples that compare the predicted performances and the real results. We shown that the selection of the best communication implementation can be made with the help of the communication models. Even if we found some small variations in the predicted data for small messages, these variations were unable to compromise the final decision, and we could identify the probable origin from these variations. Hence, one of our future works include a deep investigation on the factors that lead to such variations, and in special the relationship between the number of successive messages and the transmission delay, formalising it and proposing extensions to the pLogP model.

In parallel, we will evaluate the accuracy of our models with other network interconnections, specially Ethernet 1Gb and Myrinet, and study how to reflect the presence of multi-processors and multi-networks (division of traffic) in our models. Our research will also include the automatic discovery of the network topology and the construction of optimised inter-cluster trees that work together with efficient intra-cluster communication.

References

1. Barnett, R., Payne, D., van de Geijn, R., Watts, J.: Broadcasting on meshes with wormhole routing. Journal of Parallel and Distributed Computing, Vol 35, No 2,Academic Press (1996) 111-122
2. Bhat, P., Raharendra, C., Prasanna, V.: Efficient Collective Communication in Distributed Heterogeneous Systems. Journal of Parallel and Distributed Computing, No. 63, Elsevier Science. (2003) 251-263
3. Karonis, N. T., Foster, I., Supinski, B., Gropp, W., Lusk, E., Lacour, S.: A Multi-level Approach to Topology-Aware Collective Operations in Computational Grids. Technical report ANL/MCS-P948-0402, Mathematics and Computer Science Division, Argonne National Laboratory (2002).
4. Kielmann, T., Hofman, R., Bal, H., Plaat, A., Bhocdjang, R.: MagPIe: MPI's Collective Communication Operations for Clustered Wide Area Systems. In: Proceeding Seventh ACM SIGPLAN Symposium on Principles and Practice of Parallel Programming, Atlanta, GA, ACM Press. (1999) 131-140
5. Kielmann, T., Bal, H., Verstoep, K.: Fast Measurement of LogP Parameters for Message Passing Platforms. In: 4th Workshop on Runtime Systems for Parallel Programming (RTSPP), Cancun, Mexico, Springer-Verlag, LNCS Vol. 1800 (2000) 1176-1183. http://www.cs.vu.nl/albatross/
6. Kielman, T., Bal, E., Gorlatch, S., Verstoep, K, Hofman, R.: Network Performance-aware Collective Communication for Clustered Wide Area Systems. Parallel Computing, Vol. 27, No. 11, Elsevier Science. (2001) 1431-1456
7. LAM-MPI Team, LAM/MPI Version 6.5.9, http://www.lam-mpi.org/ (2003)
8. LAM-MPI Team, LAM/MPI Version 7, http://www.lam-mpi.org/ (2004)
9. LAM-MPI Team, Performance Issues with LAM/MPI on Linux 2.2.x, http://www.lam-mpi.org/linux/ (2001)
10. Loncaric, J.: Linux TCP Patches to improve acknowledgement policy. http://research.nianet.org/~josip/LinuxTCP-patches.html (2000)
11. Lowekamp, B.: Discovery and Application of Network Information. PhD Thesis, Carnegie Mellon University. (2000)
12. Thakur, R., Gropp, W.: Improving the Performance of Collective Operations in MPICH. In: Proceedings of the Euro PVM/MPI 2003. Springer-Verlag, LNCS Vol. 2840. (2003) 257-267
13. Vadhiyar, S., Fagg, G., Dongarra, J.: Automatically Tuned Collective Communications. In: Proceedings of Supercomputing 2000, Dallas TX. IEEE Computer Society (2000)

More Efficient Reduction Algorithms
for Non-Power-of-Two Number of Processors
in Message-Passing Parallel Systems

Rolf Rabenseifner[1] and Jesper Larsson Träff[2]

[1] High-Performance Computing-Center (HLRS), University of Stuttgart
Allmandring 30, D-70550 Stuttgart, Germany
rabenseifner@hlrs.de
www.hlrs.de/people/rabenseifner/
[2] C&C Research Laboratories, NEC Europe Ltd.
Rathausallee 10, D-53757 Sankt Augustin, Germany
traff@ccrl-nece.de

Abstract. We present improved algorithms for global reduction operations for message-passing systems. Each of p processors has a vector of m data items, and we want to compute the element-wise "sum" under a given, associative function of the p vectors. The result, which is also a vector of m items, is to be stored at either a given *root* processor (MPI_Reduce), or all p processors (MPI_Allreduce). A further constraint is that for each data item and each processor the result must be computed in the same order, and with the same bracketing. Both problems can be solved in $O(m + \log_2 p)$ communication and computation time. Such reduction operations are part of MPI (the *Message Passing Interface*), and the algorithms presented here achieve significant improvements over currently implemented algorithms for the important case where p is not a power of 2. Our algorithm requires $\lceil \log_2 p \rceil + 1$ rounds - one round off from optimal - for small vectors. For large vectors twice the number of rounds is needed, but the communication and computation time is less than $3m\beta$ and $3/2m\gamma$, respectively, an improvement from $4m\beta$ and $2m\gamma$ achieved by previous algorithms (with the message transfer time modeled as $\alpha + m\beta$, and reduction-operation execution time as $m\gamma$). For $p = 3 \times 2^n$ and $p = 9 \times 2^n$ and small $m \le b$ for some threshold b, and $p = q2^n$ with small q, our algorithm achieves the optimal $\lceil \log_2 p \rceil$ number of rounds.

1 Introduction and Related Work

Global reduction operations in three different flavors are included in MPI, the *Message Passing Interface* [14]. The MPI_Reduce collective combines element-wise the input vectors of each of p processes with the result vector stored only at a given *root process*. In MPI_Allreduce, all processes receive the result. Finally, in MPI_Reduce_scatter, the result vector is subdivided into p parts with given (not necessarily equal) numbers of elements, which are then scattered over the

D. Kranzlmüller et al. (Eds.): EuroPVM/MPI 2004, LNCS 3241, pp. 36–46, 2004.

processes. The global reduction operations are among the most used MPI collectives. For instance, a 5-year automatic profiling [12, 13] of all users on a Cray T3E has shown that 37% of the total MPI time was spent in MPI_Allreduce and that 25% of all execution time was spent in runs that involved a non-power-of-two number of processes. Thus improvements to these collectives are almost always worth the effort.

The p processes are numbered consecutively with ranks $i = 0, \ldots, p-1$ (we use MPI terminology). Each has an input vector m_i of m units. Operations are binary, associative, and possibly commutative. MPI poses other requirements that are non-trivial in the presence of rounding errors:

1. For MPI_Allreduce all processes must receive the *same* result vector;
2. reduction must be performed in *canonical order* $m_0 + m_1 + \cdots + m_{p-1}$ (if the operation is not commutative);
3. the same reduction order and bracketing for all elements of the result vector is not strictly required, but should be strived for.

For non-commutative operations $a+b$ may be different from $b+a$. In the presence of rounding errors $a + (b + c)$ may differ from $(a + b) + c$ (two different bracketing's). The requirements ensure consistent results when performing reductions on vectors of floating-point numbers.

We consider 1-ported systems, i.e. each process can send and receive a message at the same time. We assume linear communication and computation costs, i.e. the time for exchanging a message of m units is $t = \alpha + m\beta$ and the time for combining two m-vectors is $t = m\gamma$.

Consider first the MPI_Allreduce collective. For $p = 2^n$ (power-of-2), butterfly-like algorithms that for small m are *latency-optimal*, for large m *bandwidth- and work-optimal*, with a smooth transition from latency dominated to bandwidth dominated case as m increases have been known for a long time [5, 16]. For small m the number of communication rounds is $\log_2 p$ (which is optimal [5]; this is what we mean by *latency optimal*) with $m \log_2 p$ elements exchanged/combined per process. For large m the number of communication rounds doubles because of the required, additional allgather phase, but the number of elements exchanged/combined per process is reduced to $2(m-1/p)$ (which is what we mean by *bandwidth- and work-optimal*). These algorithms are simple to implement and practical.

When p is not a power-of-two the situation is different. The optimal number of communication rounds for small m is $\lceil \log_2 p \rceil$, which is achieved by the algorithms in [3, 5]. However, these algorithms assume commutative reduction operations, and furthermore the processes receive data in different order, such that requirements 1 and 2 cannot be met. These algorithms are therefore not suited for MPI. Also the bandwidth- and work-optimal algorithm for large m in [5] suffers from this problem. A repair for (very) small p would be to collect (allgather) the (parts of the) vectors to be combined on all processes, using for instance the optimal (and very practical) allgather algorithm in [6], and then perform the reduction sequentially in the same order on each process.

Algorithms suitable for MPI (i.e., with respect to the requirements 1 to 3) are based on the butterfly idea (for large m). The butterfly algorithm is executed on the largest power-of-two $p' < p$ processes, with an extra communication round before and after the reduction to cater for the processes in excess of p'. Thus the number of rounds is no longer optimal, and if done naively an extra $2m$ is added to the amount of data communicated for some of the processes. Less straightforward implementations of these ideas can be found in [13, 15], which perform well in practice.

The contributions of this paper are to the practically important non-powers-of-two case. First, we give algorithms with a smooth transition from latency to bandwidth dominated case based on a message threshold of b items. Second, we show that for the general case the amount of data to be communicated in the extra rounds can be reduced by more than a factor of 2 from $2m$ to less than m (precisely to $m/2^{n+1}$ if p is factorized in $p = q2^n$ with q an odd number). Finally, for certain number of processes $p = q2^n$ with $q = 3$ and $q = 9$ we give latency- and bandwidth optimal algorithms by combining the butterfly idea with a ring-algorithm over small rings of 3 processes; in practice these ideas may also yield good results for $q = 5, 7, \ldots$, but this is system dependent and must be determined experimentally.

The results carry over to MPI_Reduce with similar improvements for the non-power-of-two case. In this paper we focus on MPI_Allreduce.

Other related work on reduction-to-all can be found in [1–3]. Collective algorithms for wide-area clusters are developed in [7–9], further protocol tuning can be found in [4, 10, 15], especially on shared memory systems in [11]. Compared to [15], the algorithms of this paper furthermore give a smooth transition from latency to bandwidth optimization and higher bandwidth and shorter latency if the number of processes is not a power-of-two.

2   Allreduce for Powers-of-Two

Our algorithms consist of two phases. In the *reduction phase* reduction is performed with the result scattered over subsets of the p processors. In the *routing phase*, which is only necessary if m is larger than a threshold b, the result vector is computed by gathering the partial results over each subset of processes. Essentially, only a different routing phase is needed to adapt the algorithm to MPI_Reduce or MPI_Reduce_scatter. For MPI_Allreduce and MPI_Reduce the routing phase is most easily implemented by reversing the communication pattern of the reduction phase.

It is helpful first to recall briefly the hybrid butterfly algorithm as found in e.g. [16]. For now $p = 2^n$ and a message threshold b is given.

In the *reduction phase* a number of communication and computation *rounds* is performed. Prior to round $z, z = 0, \ldots, n - 1$ with $m/2^z > b$ each process i possesses a vector of size $m/2^z$ containing a block of the partial result $((m_{i_0} + m_{i_0+1}) + \cdots + (m_{i_0+2^z-2} + m_{i_0+2^z-1}))$ where i_0 is obtained by setting the least significant z bits of i to 0. In round z process i sends half of its partial result

Process i:
```
// Reduction phase
m' ← m // current data size
d ← 1 // "distance"
while d < p do
        // round
        select r − 1 neighbors of i // Protocol decision
        if m' > b then
            exchange(m'/r) with r − 1 neighbors
            Push neighbors and data sizes on stack
            m' ← m'/r
        else exchange(m') with r − 1 neighbors
        local reduction of r (partial) vectors of size m'
        d ← d × r
end while
// Routing phase
while stack non-empty
        pop neighbors and problem size off stack
        exchange with neighbors
end while
```

Fig. 1. High-level sketch of the (butterfly) reduction algorithm. For p a power of two a butterfly exchange step is used with $r = 2$. For other cases different exchange/elimination steps can be used as explained in Section 3.

to process $i \oplus 2^z$ ($\oplus$ denotes bitwise exclusive-or, so the operation corresponds to flipping the zth bit of i), and receives the other half of this process' partial result. Both processes then performs a local reduction, which establishes the above invariant above for round $z + 1$. If furthermore the processes are careful about the order of the local reduction (informally, either from left to right or from right to left), it can be maintained that the partial results on all processes in a group have been computed in canonical order, with the same bracketing, such that the requirements 1 to 3 are fulfilled. If in round z the size of the result vector $m/2^z \leq b$ halving is not performed, in which case processes i and $i \oplus 2^z$ will end up with the same partial result for the next and all succeeding rounds. For the routing phase, nothing needs to be done for these rounds, whereas for the preceding rounds where halving was done, the blocks must be combined.

A high-level sketch of the algorithm is given in Figure 1. In Figure 2 and Figure 3 the execution is illustrated for $p = 8$. The longer boxes shows the process groups for each round. The input buffer is divided into 8 segments A-H$_i$ on process i. The figure shows the buffer data after each round: X-Y$_{i-j}$ is the result of the reduction of the segments X to Y from processes i to j.

Following this sketch it is easy to see that the reduction phase as claimed takes $n = \log_2 p$ rounds. For $m/p \geq b$ the amount of data sent and received per process is $\sum_{k=0}^{n-1} m/2^{k+1} = m(1 - 1/p)$. For $m \leq b$ the routing phase is empty so the optimal $\log_2 p$ rounds suffice. For $m > b$ some allgather rounds are necessary, namely one for each reduction round in which $m/2^k > b$. At most, the number of rounds doubles.

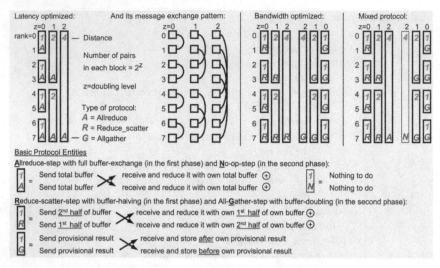

Fig. 2. The butterfly reduction algorithm.

0:	$A\text{-}H_0$	1	$A\text{-}D_{0-1}$	2	$A\text{-}B_{0-3}$	4	$A\text{-}B_{0-7}$	4	$A\text{-}B_{0-7}$	2	$A\text{-}D_{0-7}$	1	$A\text{-}H_{0-7}$
1:	$A\text{-}H_1$	R	$E\text{-}H_{0-1}$		$E\text{-}F_{0-3}$		$E\text{-}F_{0-7}$		$E\text{-}F_{0-7}$		$E\text{-}H_{0-7}$	G	$A\text{-}H_{0-7}$
2:	$A\text{-}H_2$	1	$A\text{-}D_{2-3}$		$C\text{-}D_{0-3}$		$C\text{-}D_{0-7}$		$C\text{-}D_{0-7}$		$A\text{-}D_{0-7}$	1	$A\text{-}H_{0-7}$
3:	$A\text{-}H_3$	R	$E\text{-}H_{2-3}$	R	$G\text{-}H_{0-3}$		$G\text{-}H_{0-7}$		$G\text{-}H_{0-7}$	G	$E\text{-}H_{0-7}$	G	$A\text{-}H_{0-7}$
4:	$A\text{-}H_4$	1	$A\text{-}D_{4-5}$	2	$A\text{-}B_{4-7}$		$A\text{-}B_{0-7}$		$A\text{-}B_{0-7}$	2	$A\text{-}D_{0-7}$	1	$A\text{-}H_{0-7}$
5:	$A\text{-}H_5$	R	$E\text{-}H_{4-5}$		$E\text{-}F_{4-7}$		$E\text{-}F_{0-7}$		$E\text{-}F_{0-7}$		$E\text{-}H_{0-7}$	G	$A\text{-}H_{0-7}$
6:	$A\text{-}H_6$	1	$A\text{-}D_{6-7}$		$C\text{-}D_{4-7}$		$C\text{-}D_{0-7}$		$C\text{-}D_{0-7}$		$A\text{-}D_{0-7}$	1	$A\text{-}H_{0-7}$
7:	$A\text{-}H_7$	R	$E\text{-}H_{6-7}$	R	$G\text{-}H_{4-7}$	A	$G\text{-}H_{0-7}$	N	$G\text{-}H_{0-7}$	G	$E\text{-}H_{0-7}$	G	$A\text{-}H_{0-7}$

Fig. 3. Intermediate results after each protocol step when the threshold b is reached in round 3.

3   The Improvements: Odd Number of Processes

We now present our improvements to the butterfly scheme when p is not a power-of-two. Let for now $p = q2^n$ where 2^n is the largest power of two smaller than p (and q is odd).

For the general case we introduce a more communication efficient way to include data from processes in excess of 2^n into the butterfly algorithm. We call this step *3-2 elimination*. Based on this we give two different algorithms for the general case, both achieving the same bounds. For certain small values of q we show that a ring based algorithm can be used in some rounds of the butterfly algorithm, and for certain values of q results in the optimal number of communication rounds.

3.1   The 3-2 Elimination Step

For $m' > b$ the 3-2 elimination step is used on a group of three processes p_0, p_1, and p_2, to absorb the vector of process p_2 into the partial results of process p_0

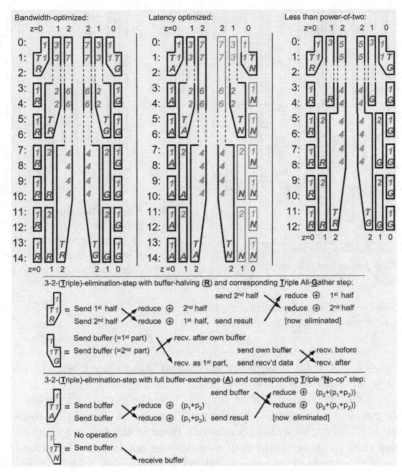

Fig. 4. Overlapping elimination protocol for $p = 15$ and $p = 13$ using 3-2-elimination-steps.

and p_1, which will survive for the following rounds. The step is as follows: process p_2 sends $m'/2$ (upper) elements to process p_1, and simultaneously receives $m'/2$ (lower) elements from process p_1. Process p_1 and p_2 can then perform the reduction operation on their respective part of the vector. Next, process p_0 receives the $m'/2$ (lower) elements of the partial result just computed from process p_2, and sends $m'/2$ (upper) elements to process p_1. Process p_0 and p_1 compute a new partial result from the $m'/2$ elements received.

As can be seen process p_0 and p_1 can finish after two rounds, both with the half of the elements of the result vector $[m'_0 + (m'_1 + m'_2)]$. The total time for process p_0 and p_1 is $2\alpha + \beta m' + \gamma m'$.

Compare this to the trivial solution based on *2-1 elimination*. First process p_2 sends all its m' elements to process p_1 (2-1 elimination), after which process p_0 and p_1 performs a butterfly exchange of $m'/2$ elements. The time for this solution is $2\alpha + 3/2\beta m' + 3/2\gamma m'$.

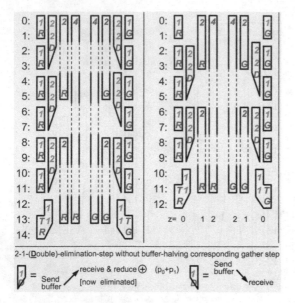

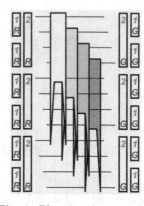

Fig. 6. Plugging in any algorithm for odd number (here 3) of processes after reducing p with the butterfly algorithm (here with two steps) to its odd factor.

Fig. 5. Non-Overlapping elimination protocol for $p = 15$ and $p = 13$ using 3-2- and 2-1-elimination-steps.

For $m' \leq b$, the total buffers are exchanged and reduced, see the protocol entities described in Fig. 4.

The 3-2 elimination step can be plugged into the general algorithm of Figure 1. For $p = 2^n + Q$ with $Q < 2^n$, the total number of elimination steps to be performed is Q. The problem is to schedule these in the butterfly algorithm in such a way that the total number of rounds does not increase by more than 1 for a total of $n + 1 = \lceil \log_2 p \rceil$ rounds. Interestingly we have found two solutions to this problem, which are illustrated in the next subsections.

3.2   Overlapping 3-2 Elimination Protocol

Figure 4 shows the protocol examples with 15 and 13 processes. In general, this protocol schedules 3-2-elimination steps for a group of on $2^z \times 3$ processes in each round z for which the zth bit of p is 1. The 3-2-steps exchange two messages of the same size and are therefore drawn with double width. The first process is not involved in the first message exchange, therefore this part is omitted from the shape in the figure. After each 3-2-step, the third process is eliminated, which is marked with dashed lines in the following rounds. The number of independent pairs or triples in each box is 2^z. As can be seen the protocol does not introduce delays where some processes have to wait for other processes to complete their 3-2 elimination steps of previous rounds, but different groups of processes can simultaneously be at different rounds. Note, that this protocol can be used in general for any number of processes. If p includes a factor 2^n then it starts with n butterfly steps.

3.3 Non-overlapping Elimination Protocol

Figure 5 shows a different protocol that eliminates all excess processes at round $z = 1$. With the combination of one 3-2-elimination-step and pairs of 2-1-elimination-steps any odd number of processes p is thus reduced to its next smaller power-of-two value. Note that for $m > b$ in round $z = 1$ only $m/2$ data are sent in the 2-1-elimination step (instead of m if the 2-1 elimination would have been performed prior to round $z = 0$).

Both the overlapping and the non-overlapping protocol are exchanging the same amount of data and number of messages. For small $m \leq b$ the total time is $t = (1 + \lceil \log_2 p \rceil)\alpha + m(1 + \lceil \log_2 p \rceil)\beta + m\lceil \log_2 p \rceil\gamma$, where the extra round (the α-term) stems from the need to send the final result to the eliminated processes. For large $m > b$ the total time is $t = 2\lceil \log_2 p \rceil\alpha + 2m(1.5 - 1/p')\beta + m(1.5 - 1/p')\gamma$ with $p' = 2^n$ being the largest power of two smaller than p.

This protocol is designed only for odd numbers of processes. For any number of processes it must be combined with the butterfly.

3.4 Small Ring

Let now $p = r^q 2^n$. The idea here is to handle the reduction step for the r^q factor by a ring. For $r - 1$ rounds process i receives data from process $(i-1) \bmod r$ and sends data to process $(i+1) \bmod r$. For $m > b$ each process sends/receives only m/r elements per round, whereas for $m \leq b$ each process sends its full input vector along the ring. After the last step each process sequentially reduces the elements received: the requirements 1 and 2 make it necessary to postpone the local reductions until data from all processes have been received. For $m > b$ each process has m/r elements of the result vector $m_0 + m_1 + \ldots + m_{r-1}$. We note that the butterfly exchange step can be viewed as a 2-ring; the ring algorithm is thus a natural generalization of the butterfly algorithm.

For small $m < b$ and if also $r > 3$ the optimal allgather algorithm of [6] would actually be much preferable; however, the sequential reduction remains a bottleneck, and this idea is therefore only attractive for small p (dependent on the ratio of α and β to γ).

Substituting the ring algorithm for the neighbor exchange step in the algorithm of Figure 1, we can implement the complete reduction phase in $(r-1)q + n$ rounds. This gives a theoretical improvement for $r = 3$ and $q = 1, 2$ to the optimal number of $\lceil \log_2 p \rceil$ rounds. The general algorithm would require $\lceil \log_2 p \rceil + 1$ rounds, one more than optimal, whereas the algorithm with ring steps takes 1 round less. Let for example $p = 12 = 3 \times 2^2$. The ring based algorithm needs $2 + 2 = 4$ rounds, whereas the general algorithm would take $\lceil \log_2 12 \rceil + 1 = 4 + 1 = 5$ rounds.

3.5 Comparison

The time needed for latency-optimized (exchange of full buffers) and bandwidth-optimized (recursive buffer halving or exchange of $1/p$ of the buffer) protocols are:

Table 1. Execution time of the four protocols for odd numbers of processes (p) and different message sizes. The time is displayed as multiples of the message transfer latency α. In each line, the fastest protocol is marked (*).

p	m	Ring, latency-opt.	Elimination, lat.-opt.	Ring, bandwidth-opt.	Elimination, bw-opt.
3	S	2 + 0.2 + 0.02 = 2.22 *	3 + 0.3 + 0.02 = 3.32	4 + 0.13 + 0.01 = 4.14	4 + 0.20 + 0.01 = 4.21
	M	2 + 2.0 + 0.20 = 4.20 *	3 + 3.0 + 0.20 = 6.20	4 + 1.33 + 0.07 = 5.40	4 + 2.00 + 0.10 = 6.10
	L	2 + 20. + 2.00 = 24.0	3 + 30. + 2.00 = 35.0	4 + 13.3 + 0.67 = 18.0 *	4 + 20.0 + 1.00 = 25.0
5	S	3 + 0.4 + 0.04 = 3.44 *	4 + 0.4 + 0.03 = 4.43	7 + 0.16 + 0.01 = 7.17	6 + 0.25 + 0.01 = 6.26
	M	3 + 4.0 + 0.40 = 7.40 *	4 + 4.0 + 0.30 = 8.30	7 + 1.60 + 0.08 = 8.68	6 + 2.50 + 0.13 = 8.63
	L	3 + 40. + 4.00 = 47.0	4 + 40. + 3.00 = 47.0	7 + 16.0 + 0.80 = 23.8 *	6 + 25.0 + 1.25 = 32.3
7	S	3 + 0.6 + 0.06 = 3.66 *	4 + 0.4 + 0.03 = 4.43	9 + 0.17 + 0.01 = 9.18	6 + 0.25 + 0.01 = 6.26
	M	3 + 6.0 + 0.60 = 9.60	4 + 4.0 + 0.30 = 8.30 *	9 + 1.71 + 0.09 = 10.8	6 + 2.50 + 0.13 = 8.63
	L	3 + 60. + 6.00 = 69.0	4 + 40. + 3.00 = 47.0	9 + 17.1 + 0.86 = 27.0 *	6 + 25.0 + 1.25 = 32.3
13	S	4 + 1.2 + 0.12 = 5.32 *	5 + 0.5 + 0.04 = 5.54	16 + 0.19 + 0.01 = 16.2	8 + 0.28 + 0.01 = 8.29
	M	4 + 12. + 1.20 = 17.2	5 + 5.0 + 0.40 = 10.4 *	16 + 1.85 + 0.09 = 18.0	8 + 2.75 + 0.14 = 10.9
	L	4 + 120. + 12.0 = 136.	5 + 50. + 4.00 = 59.0	16 + 18.5 + 0.92 = 35.4 *	8 + 27.5 + 1.38 = 36.9
15	S	4 + 1.4 + 0.14 = 5.54 *	5 + 0.5 + 0.04 = 5.54	18 + 0.19 + 0.01 = 18.2	8 + 0.28 + 0.01 = 8.29
	M	4 + 14. + 1.40 = 19.4	5 + 5.0 + 0.40 = 10.4 *	18 + 1.87 + 0.09 = 20.0	8 + 2.75 + 0.14 = 10.9
	L	4 + 140. + 14.0 = 158.	5 + 50. + 4.00 = 59.0	18 + 18.7 + 0.93 = 37.6	8 + 27.5 + 1.38 = 36.9 *
23	S	5 + 2.2 + 0.22 = 7.42	6 + 0.6 + 0.05 = 6.65 *	27 + 0.19 + 0.01 = 27.2	10 + 0.29 + 0.01 = 10.3
	M	5 + 22. + 2.20 = 29.2	6 + 6.0 + 0.50 = 12.5 *	27 + 1.91 + 0.10 = 29.0	10 + 2.88 + 0.14 = 13.0
	L	5 + 220. + 22.0 = 247.	6 + 60. + 5.00 = 71.0	27 + 19.1 + 0.96 = 47.1	10 + 28.8 + 1.44 = 40.2 *
	XL	5 + 2200. + 220. = 2425	6 + 600. + 50.0 = 656.	27 + 191. + 9.60 = 228. *	10 + 288. + 14.4 = 312.
63	S	6 + 6.2 + 0.62 = 12.8	7 + 0.6 + 0.06 = 7.66 *	68 + 0.19 + 0.01 = 68.2	12 + 0.29 + 0.01 = 12.3
	M	6 + 62. + 6.20 = 74.2	7 + 6.0 + 0.60 = 13.6 *	68 + 1.97 + 0.10 = 70.1	12 + 2.94 + 0.15 = 15.1
	L	6 + 620. + 62.0 = 688.	7 + 60. + 6.00 = 73.0	68 + 19.7 + 0.98 = 88.7	12 + 29.4 + 1.47 = 42.9
	XL	6 + 6200. + 620. = 6826	7 + 600. + 60.0 = 667.	68 + 197. + 9.80 = 275. *	12 + 294. + 14.7 = 321.

message size m: S: $\beta m = 0.1\alpha$, $\gamma m = 0.01\alpha$;   L: $\beta m = 10\,\alpha$, $\gamma m = 1.00\alpha$;

M: $\beta m = 1.0\alpha$, $\gamma m = 0.10\alpha$;   XL: $\beta m = 100\alpha$, $\gamma m = 10.0\alpha$;

$$t_{ring,lat-opt.} = \alpha\lceil\log_2 p\rceil \qquad\qquad +\beta m(p-1) \qquad\qquad +\gamma m(p-1)$$
$$t_{elim.,lat-opt.} = \alpha(\lceil\log_2 p\rceil + 1) \qquad +\beta m(\lceil\log_2 p\rceil + 1) \quad +\gamma m(\lceil\log_2 p\rceil)$$
$$t_{ring,bw-opt.} = \alpha(\lceil\log_2 p\rceil + p - 1) +\beta m(2(1 - 1/p)) \qquad +\gamma m(1 - 1/p)$$
$$t_{elim.,bw-opt.} = \alpha(2\lceil\log_2 p\rceil) \qquad\qquad +\beta m(2(1.5 - 1/p')) +\gamma m(1.5 - 1/p')$$

with $p' = 2^{\lfloor\log_2 p\rfloor}$. Table 1 compares the 4 algorithms for four cases based on different rations $\beta m/\alpha$ and $\gamma m/\alpha$, and for several numbers of processes p. The fastest protocol is marked in each line. Note, that this table does not necessarily gives the optimal values for the elimination protocols because they may be achieved by using some internal steps with buffer halving and the further steps without buffer halving. One can see that each algorithm has a usage range, where it is significantly faster than the other protocols.

3.6   Putting the Pieces Together

The 3-2-elimination step and the ring exchange were two alternative exchange patterns that could be plugged into the high-level algorithm of Figure 1 for non-powers-of-two, see also Fig 6. The number of processes $p = 2^n q_1 q_2 ... q_h$ is factorized in a) 2^n for the butterfly protocol, b) small odd numbers $q_1, ... q_{h-1}$ for the ring protocol, and c) finally an odd number q_h for the 3-2-elimination or 2-1-elimination protocol. For given p it is of course essential that each process i at each round z can determine efficiently (i.e., in *constant time*) what protocol is to be used. This amounts to determining a) exchange step (butterfly, 3-2-elimination, 2-1-elimination, ring), b) neighboring process(es), and c) whether the process will be active for the following rounds. We did not give the details; however, for all protocols outlined in the paper this is indeed the case, but as shortcut, Table 1 is now used for the odd factors q_i and vector size reduced by $1/2^n$ if the butterfly protocol uses buffer halving due to long vectors.

4   Conclusion and Open Problems

We presented an improved algorithm for the MPI_Allreduce collective for the important case where the number of participating processes (p) is not a power of two, i.e., $p = 2^n q$ with odd q and $n \geq 0$. For general non-powers-of-two and small vectors, our algorithm requires $\lceil \log_2 p \rceil + 1$ rounds - one round off from optimal. For **large vectors** twice the number of rounds is needed, but the communication and computation time is less than $(1+1/2^{n+1})(2m\beta + m\gamma)$, i.e., an improvement from $2(2m\beta + m\gamma)$ achieved by previous algorithms [15], e.g., with $p = 24$ or 40, the execution time can be reduced by 47%. For **small vectors** and small q our algorithm achieves the optimal $\lceil \log_2 p \rceil$ number of rounds.

The main open problem is whether a latency optimal allreduce algorithm under the MPI constraint 1- 3 with $\lceil \log_2 p \rceil$ rounds is possible for any number of processes. We are not aware of results to the contrary.

References

1. M. Barnett, S. Gupta, D. Payne, L. Shuler, R. van de Gejin, and J. Watts, *Interprocessor collective communication library (InterCom)*, in Proceedings of Supercomputing '94, Nov. 1994.
2. A. Bar-Noy, J. Bruck, C.-T. Ho, S. Kipnis, and B. Schieber. Computing global combine operations in the multiport postal model. *IEEE Transactions on Parallel and Distributed Systems*, 6(8):896–900, 1995.
3. A. Bar-Noy, S. Kipnis, and B. Schieber. An optimal algorithm for computing census functions in message-passing systems. *Parallel Processing Letters*, 3(1):19–23, 1993.
4. E. K. Blum, X. Wang, and P. Leung. Architectures and message-passing algorithms for cluster computing: Design and performance. *Parallel Computing* 26 (2000) 313–332.
5. J. Bruck and C.-T. Ho. Efficient global combine operations in multi-port message-passing systems. *Parallel Processing Letters*, 3(4):335–346, 1993.
6. J. Bruck, C.-T. Ho, S. Kipnis, E. Upfal, and D. Weathersby. Efficient algorithms for all-to-all communications in multiport message-passing systems. *IEEE Transactions on Parallel and Distributed Systems*, 8(11):1143–1156, Nov. 1997.
7. E. Gabriel, M. Resch, and R. Rühle, *Implementing MPI with optimized algorithms for metacomputing*, in Proceedings of the MPIDC'99, Atlanta, USA, pp 31–41.
8. N. Karonis, B. de Supinski, I. Foster, W. Gropp, E. Lusk, and J. Bresnahan, *Exploiting hierarchy in parallel computer networks to optimize collective operation performance*, in Proceedings of the 14th International Parallel and Distributed Processing Symposium (IPDPS '00), 2000, pp 377–384.
9. T. Kielmann, R. F. H. Hofman, H. E. Bal, A. Plaat, R. A. F. Bhoedjang, *MPI's reduction operations in clustered wide area systems*, in Proceedings of the MPIDC'99, Atlanta, USA, March 1999, pp 43–52.

10. A. D. Knies, F. Ray Barriuso, W. J. H., G. B. Adams III, *SLICC: A low latency interface for collective communications*, in Proceedings of the 1994 conference on Supercomputing, Washington, D.C., Nov. 14–18, 1994, pp 89–96.
11. Howard Pritchard, Jeff Nicholson, and Jim Schwarzmeier, *Optimizing MPI Collectives for the Cray X1*, in Proceeding of the CUG 2004 conference, Knoxville, Tennessee, USA, May, 17-21, 2004 (personal communication).
12. R. Rabenseifner, *Automatic MPI counter profiling of all users: First results on a CRAY T3E 900-512*, Proceedings of the Message Passing Interface Developer's and User's Conference 1999 (MPIDC'99), Atlanta, USA, March 1999, pp 77–85.
13. R. Rabenseifner. Optimization of collective reduction operations. In *M. Bubak et al. (Eds.): International Conference on Computational Science (ICCS 2004)*, volume 3036 of *Lecture Notes in Computer Science*, pages 1–9, 2004.
14. M. Snir, S. Otto, S. Huss-Lederman, D. Walker, and J. Dongarra. *MPI – The Complete Reference*, volume 1, The MPI Core. MIT Press, second edition, 1998.
15. R. Thakur and W. D. Gropp. Improving the performance of collective operations in MPICH. In *Recent Advances in Parallel Virtual Machine and Message Passing Interface. 10th European PVM/MPI Users' Group Meeting*, volume 2840 of *Lecture Notes in Computer Science*, pages 257–267, 2003.
16. R. van de Geijn. On global combine operations. *Journal of Parallel and Distributed Computing*, 22:324–328, 1994.

Zero-Copy MPI Derived Datatype Communication over InfiniBand*

Gopalakrishnan Santhanaraman, Jiesheng Wu, and Dhabaleswar K. Panda

Department of Computer Science and Engineering
The Ohio State University
{santhana,wuj,panda}@cse.ohio-state.edu

Abstract. This paper presents a new scheme, Send Gather Receive
Scatter (SGRS), to perform zero-copy datatype communication over InfiniBand. This scheme leverages the gather/scatter feature provided by
InfiniBand channel semantics. It takes advantage of the capability of
processing non-contiguity on both send and receive sides in the Send
Gather and Receive Scatter operations. In this paper, we describe the
design, implementation and evaluation of this new scheme. Compared
to the existing Multi-W zero-copy datatype scheme, the SGRS scheme
can overcome the drawbacks of low network utilization and high startup
costs. Our experimental results show significant improvement in both
point-to-point and collective datatype communication. The latency of a
vector datatype can be reduced by up to 62% and the bandwidth can
be increased by up to 400%. The Alltoall collective benchmark shows a
performance benefit of up to 23% reduction in latency.

1   Introduction

The MPI (Message Passing Interface) Standard [3] has evolved as a *de facto*
parallel programming model for distributed memory systems. As one of its most
important features, MPI provides a powerful and general way of describing arbitrary collections of data in memory in a compact fashion. The MPI standard also
provides run time support to create and manage such MPI derived datatypes.
MPI derived datatypes are expected to become a key aid in application development.

In principle, there are two main goals in providing derived datatypes in MPI.
First, several MPI applications such as (de)composition of multi-dimensional
data volumes [1,4] and finite-element codes [2] often need to exchange data with
algorithm-related layouts between two processes. In the NAS benchmarks such
as MG, LU, BT, and SP, non-contiguous data communication has been found to
be dominant [10]. Second, MPI derived datatypes provide opportunities for MPI
implementations to optimize datatype communication. Therefore, applications

---

* This research is supported in part by Department of Energy's Grant #DE-FC02-
  01ER25506, and National Science Foundation's grants #CNS-0204429, and #CCR-
  0311542.

D. Kranzlmüller et al. (Eds.): EuroPVM/MPI 2004, LNCS 3241, pp. 47–56, 2004.

developed with datatype can achieve portable performance over different MPI applications with optimized datatype communication.

In practice, however, the poor performance of many MPI implementations with derived datatypes [2, 5] becomes a barrier to using derived datatypes. A programmer often prefers packing and unpacking noncontiguous data manually even with considerable effort. Recently, a significant amount of research work have concentrated on improving datatype communication in MPI implementations, including 1) *Improved datatype processing system* [5, 13], 2) *Optimized packing and unpacking procedures* [2, 5], and 3) *Taking advantage of network features to improve noncontiguous data communication* [17].

In this paper, we focus on improving non-contiguous data communication by taking advantage of InfiniBand features. We focus on zero-copy datatype communication over InfiniBand. Zero copy communication protocols are of increased importance because they improve memory performance and also have reduced host cpu involvement in moving data. Our previous work [17] used multiple RDMA writes, *Multi-W*, as an effective solution to achieve zero-copy datatype communication. In this paper we look at an alternate way of achieving zero-copy datatype communication using the *send/receive* semantics with the *gather/scatter* feature provided by InfiniBand. We call this scheme *SGRS* (Send Gather Receive Scatter) in the rest of this paper. This scheme can overcome two main drawbacks in the Multi-W scheme: *low network utilization* and *high startup cost*. We have implemented and evaluated our proposed SGRS scheme in MVAPICH, an MPI implementation over InfiniBand [12, 9].

The rest of the paper is organized as follows. We first give a brief overview of InfiniBand and MVAPICH in Section 2. Section 3 provides the motivation for the SGRS scheme. Section 4 describes the basic approach, the design issues involved and the implementation details. The performance results are presented in Section 5. Section 6 presents related work. We draw our conclusions and possible future work in Section 7.

2   Background

In this section we provide an overview of the Send Gather/Recv Scatter feature in InfiniBand Architecture and MVAPICH.

2.1   Send Gather/Recv Scatter in InfiniBand

The InfiniBand Architecture (IBA) [6] defines a System Area Network (SAN) for interconnecting processing nodes and I/O nodes. It supports both channel and memory semantics. In channel semantics, send/receive operations are used for communication. In memory semantics, RDMA write and RDMA read operations are used instead. In channel semantics, the sender can gather data from multiple locations in one operation. Similarly, the receiver can receive data into multiple locations. In memory semantics, non-contiguity is allowed only in one side. RDMA write can gather multiple data segments together and write all data into a contiguous buffer on the remote node in one single operation. RDMA read can scatter data into multiple local buffers from a contiguous buffer on the remote node.

2.2 Overview of MVAPICH

MVAPICH is a high performance implementation of MPI over InfiniBand. Its design is based on MPICH [15] and MVICH [8]. The Eager protocol is used to transfer small and control messages. The Rendezvous protocol is used to transfer large messages. Datatype communication in the current MVAPICH is directly derived from MPICH and MVICH without any change. Basically, the generic packing and unpacking scheme is used inside the MPI implementation. When sending a datatype message, the sender first packs the data into a contiguous buffer and follows the contiguous path. On the receiver side, it first receives data into a contiguous buffer and then unpacks data into the user buffers. In the rest of this paper, we refer to this scheme as *Generic scheme*.

3 Motivating Case Study for the Proposed SGRS Scheme

Consider a case study involving the transfer of multiple columns in a two dimensional $M \times N$ integer array from one process to another. There are two possible zero-copy schemes. The first one uses multiple RDMA writes, one per row. The second one uses Send Gather/Receive Scatter. We compare these two schemes over the VAPI layer, which is an InfiniBand API provided by Mellanox [11]. The first scheme posts a list of RDMA write descriptors. Each descriptor writes one contiguous block in each row. The second scheme posts multiple Send Gather descriptors and Receiver Scatter descriptors. Each descriptor has 50 blocks from 50 different rows (50 is the maximum number of segments supported in one descriptor in the current version of Mellnox SDK). We will henceforth refer to these two schemes as "Multi-W" and "SGRS" in the plots. In the first test, we consider a 64×4096 integer array. The number of columns varies from 8 to 2048. The total message size varies from 2 KBytes to 512 KBytes accordingly. The bandwidth test is used for evaluation. As shown in Figure 1, the SGRS scheme consistently outperforms the Multi-W scheme.

In the second test, the number of blocks varies from 4 to 64. The total message size we studied is 128 KBytes, 256 KBytes, and 512 KBytes. Figure 2 shows the bandwidth results with different number of blocks and different message sizes. When the number of blocks is small, both Multi-W and SGRS schemes perform comparably. This is because the block size is relatively large. The network utilization in the Multi-W is still high. As the number of segments increase we observe a significant fall in bandwidth for the Multi-W scheme whereas the fall in bandwidth is negligible for the SGRS scheme. There are two reasons. First, the network utilization becomes lower when the block size decreases (i.e. the number of blocks increases) in the Multi-W scheme. However, in the SGRS scheme, the multiple blocks in one send or receive descriptor are considered as one message. Second, the total startup costs in the Multi-W scheme increases with the

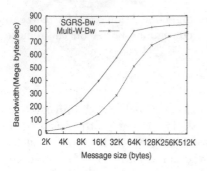
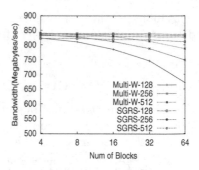

Fig. 1. Bandwidth Comparison over VAPI with 64 Blocks.

Fig. 2. Bandwidth Comparison over VAPI with Different Number of Blocks.

increase of the number of blocks because each block is treated as an individual message in the Multi-W scheme and hence the startup cost is associated with each block. From these two examples, it can be observed that the SGRS scheme can overcome the two drawbacks in the Multi-W by increasing network utilization and reducing startup costs. These potential benefits motivate us to design MPI datatype communication using the SGRS scheme described in detail in Section 4.

4   Proposed SGRS (Send Gather/Recv Scatter) Approach

In this section we first describe the SGRS scheme. Then we discuss the design and implementation issues and finally look at some optimizations to this scheme.

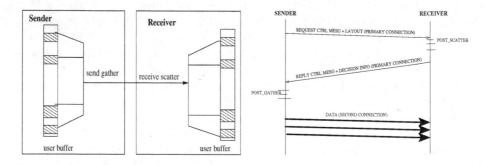

Fig. 3. Basic Idea of the SGRS Scheme.

Fig. 4. SGRS Protocol.

4.1   Basic Idea

The basic idea behind the SGRS scheme is to use the scatter/gather feature associated with the send receive mechanism to achieve zero copy communication. Using this feature we can send/receive multiple data blocks as a single message by posting a send gather descriptor at source and a receive scatter descriptor at destination. Figure 3 illustrates this approach. The SGRS scheme can handle non-contiguity on both sides. As mentioned in Section 2, RDMA Write Gather or RDMA Read Scatter handles non-contiguity only on one side. Hence, to achieve zero-copy datatype communication based on RDMA operations, the Multi-W scheme is needed [17].

4.2   Design and Implementation Issues

Communication Protocol. The SGRS scheme is deployed in Rendezvous protocol to transfer large datatype messages. For small datatype messages, the Generic scheme is used. As shown in Figure 4, the sender first sends the Rendezvous start message with the data layout information out. Second, the receiver receives the above message and figures out how to match the sender's layout with its own layout. Then, the receiver sends the layout matching decision to the sender. After receiving the reply message, the sender posts send gather descriptors. It is possible that the sender may break one block into multiple blocks to meet the layout matching decision. There are four main design issues: Secondary connection, Layout exchange, Posting descriptors and Registration.

Secondary Connection. The SGRS scheme needs a second connection to transmit the non-contiguous data. This need arises because it is possible in the existing MVAPICH design to prepost some receive descriptors on the main connection as a part of its flow control mechanism. These descriptors could unwittingly match with the gather-scatter descriptors associated with the non-contiguous transfer. One possible issue with the extra connection is scalability. In our design, there are no buffers/resources for the second connection. The HCA usually can support a large number of connections. Hence the extra connection does not hurt the scalability.

Layout Exchange. The MPI datatype has only local semantics. To enable zero-copy communication, both sides should have an agreement on how to send and receive data. In our design, the sender first sends its layout information to the receiver in the Rendezvous start message as shown in Figure 4. Then the receiver finds a solution to match these layouts. This decision information is also sent back to the sender for posting send gather descriptors. To reduce the overhead for transferring datatype layout information, a layout caching mechanism is desirable [7]. Implementation details of this cache mechanism in MVAPICH can be found in [17]. In Section 5, we evaluate the effectiveness of this cache mechanism.

Posting Descriptors. There are three issues in posting descriptors. First, if the number of blocks in the datatype message is larger than the maximum allowable gather/scatter limit, the message has to be chopped into multiple gather/scatter descriptors. Second, the number of posted send descriptors and the number of posted receive descriptors must be equal. Third, for each pair of matched send and receive descriptors, the data length must be same. This basically needs a negotiation phase. Both these issues can be handled by taking advantage of the Rendezvous start and reply message in the Rendezvous protocol. In our design, the receiver makes the matching decision taking into account the layouts as well as scatter-gather limit. Both the sender and the receiver post their descriptors with the guidance of the matching decision.

User Buffer Registration. To send data from and receive data into user buffer directly, the user buffers need to be registered. Given a non-contiguous datatype we can register each contiguous block one by one. We could also register the whole region which covers all blocks and gaps between blocks. Both attempts have their drawbacks [16]. In [16], *Optimistic Group Registration(OGR)* has been proposed to make a tradeoff between the number of registration and deregistration operations and the total size of registered space to achieve efficient memory registration on datatype message buffers.

5   Performance Evaluation

In this section we evaluate the performance of our SGRS scheme with the Multi-W zero copy scheme as well as the generic scheme in MVAPICH. We do latency, bandwidth and CPU overhead tests using a vector datatype to demonstrate the effectiveness of our scheme. Then we show the potential benefits that can be observed for collective communication such as MPI_Alltoall that are built on top of point to point communication. Further we investigate the impact of layout caching for our design.

5.1   Experimental Testbed

A cluster of 8 SuperMicro SUPER X5DL8-GG nodes, each with dual Intel Xeon 3.0 GHz processors, 512 KB L2 cache, PCI-X 64-bit 133 MHz bus, and connected to Mellanox InfiniHost MT23108 DualPort 4x HCAs. The nodes are connected using the Mellanox InfiniScale 24 port switch MTS 2400. The kernel version used is Linux 2.4.22smp. The InfiniHost SDK version is 3.0.1 and HCA firmware version is 3.0.1. The Front Side Bus (FSB) runs at 533MHz. The physical memory is 1 GB of PC2100 DDR-SDRAM memory.

5.2   Vector Latency and Bandwidth Tests

In this benchmark, increasing number of columns in a two dimensional M*4096 integer array are transferred between two processes. These columns can be represented by a vector datatype. Figure 5 compares the ping-pong latency in the MPI implementation using the two zero-copy schemes. We set up two cases for the number of rows (M) in this array: one is 64 and one is 128. The number of columns varies from 4 to 2048, the corresponding message size varies from 2 KBytes to 512 KBytes. We also compare it with the latency of the contiguous transfer which serves as the lower bound. We observe that the SGRS scheme reduces the latency by up to 61% compared to that of the Multi-W scheme. Figure 6 shows the bandwidth results. The improvement factor over the Multi-W scheme varies from 1.12 to 4.0.

In both latency and bandwidth tests, it can also be observed that when the block size is smaller, the improvement of the SGRS scheme over the Multi-W scheme is higher. This is because the improved network utilization in the SGRS scheme is more significant when the block size is small. When the block size is large enough, RDMA operations on each block can achieve good network utilization as well. Both schemes perform comparably.Compared to the Generic scheme, the latency results of the SGRS scheme are better in cases when the block size is larger than 512 bytes. When the message size is small and the block size is small, the Generic scheme performs the best. This is because the memory copy cost is not substantial and the Generic scheme can achieve better network utilization. The bandwidth results of the SGRS scheme are always better than the Generic scheme.

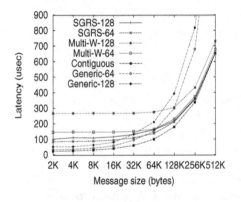 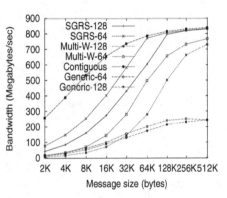

Fig. 5. MPI Level Vector Latency.   **Fig. 6.** MPI Level Vector Bandwidth.

5.3   CPU Overhead Tests

In this section we measure the CPU overhead involved for the two schemes. Figures 7 and 8 compare the CPU overheads associated at the sender side and receiver side, respectively. The SGRS scheme has lower CPU involvement on the sender side as compared to Multi-W scheme. However on the receiver side the SGRS scheme has an additional overhead as compared to practically close to zero overhead incase of Multi-W scheme.

5.4   Performance of MPI_Alltoall

Collective datatype communication can benefit from high performance point-to-point datatype communication provided in our implementation. We designed a test to evaluate MPI_Alltoall performance with derived datatypes. We use the same vector datatype we had used for our earlier evaluation.

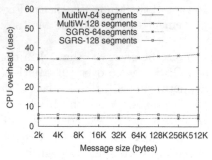

Fig. 7. Sender side CPU overhead.

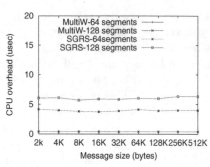

Fig. 8. Receiver side CPU overhead.

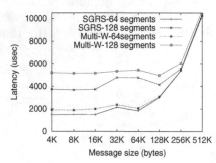

Fig. 9. MPI_Alltoall Latency.

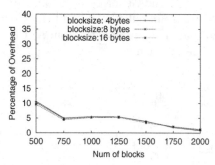

Fig. 10. Overhead of Transferring Layout Information.

Figure 9 shows the MPI_Alltoall latency performance of the various schemes on 8 nodes. We study the Alltoall latency over the message range 4K-512K. We ran these experiments for two different numbers of blocks: 64 and 128. We observe that the SGRS scheme outperforms the Multi-W scheme consistently. The gap widens as the number of blocks increases. This is because the startup costs in the Multi-W scheme increase with the increase of the number of blocks. In addition, given a message size, the network utilization decreases with the increase of the number of blocks in the Multi-W scheme.

5.5   Impact of Layout Caching

In both the Multi-W and SGRS schemes, the layout has to be exchanged between the sender and receiver before data communication. In this test, we studied the overhead of transferring the layout information. We consider a synthetic benchmark where this effect might be prominent. In our benchmark, we need to transfer the two leading diagonals of a square matrix between two processes. These diagonal elements are actually small blocks rather than single elements. Hence, the layout information is complex and we need considerable layout size to describe it. As the size of the matrix increases, the number of non-contiguous blocks correspondingly increases as well as the layout description. Figure 10 shows the percentage of overhead that is incurred in transferring this layout information when there is no layout cache as compared with the case that has a layout cache. For smaller message sizes, we can see a benefit of 10 percent and this keeps diminishing as the message size increases. Another aspect here is that even though for small messages the layout size is comparable with message size, since the layout is transferred in a contiguous manner, it takes a lesser fraction of time to transfer this as compared to the non-contiguous message of comparable size. Since the cost associated in maintaining this cache is virtually zero, for message sizes in this range we can benefit from layout caching.

6   Related Work

Many researchers have been working on improving MPI datatype communication. Research in datatype processing system includes [5, 13]. Research in optimizing packing and unpacking procedures includes [2, 5]. The closest work to ours is the work [17, 14] to take advantage of network features to improve noncontiguous data communication. In [14], the use of InfiniBand features to transfer non-contiguous data is discussed in the context of ARMCI which is a one sided communication library. In [17], Wu et al. have systematically studied two main types of approach for MPI datatype communication (*Pack/Unpack-based approaches* and *Copy-Reduced approaches*) over InfiniBand. The Multi-W scheme has been proposed to achieve zero-copy datatype communication.

7   Conclusions and Future Work

In this paper we presented a new zero-copy scheme to efficiently implement datatype communication over InfiniBand. The proposed scheme, *SGRS*, leverages the Send Gather/Recv Scatter feature of InfiniBand to improve the datatype communication performance. The experimental results we achieved show that this scheme outperforms the existing Multi-W zero-copy scheme in all cases for both point to point as well as collective operations. Compared to the Generic scheme, for many cases, the SGRS reduces the latency by 62%, and increases the bandwidth by 400%. In the cases where the total datatype message size is small and the contiguous block sizes are relatively small, packing/unpacking based schemes [17] perform better. But beyond a particular "cutoff" point, the zero-copy scheme performs better. The SGRS scheme pushes this cutoff point to a relatively smaller value compared to the Multi-W scheme. As part of future work, we would like to compare this scheme with other schemes and evaluate this scheme at the application level.A combination of this scheme with other schemes can be incorporated to choose the best scheme for a given datatype message adaptively.

References

1. Mike Ashworth. A Report on Further Progress in the Development of Codes for the CS2. In *Deliverable D.4.1.b F. Carbonnell (Eds), GPMIMD2 ESPRIT Project, EU DGIII, Brussels*, 1996.

2. Surendra Byna, Xian-He Sun, William Gropp, and Rajeev Thakur. Improving the Performance of MPI Derived Datatypes by Optimizing Memory-Access Cost. In *Proceedings of the IEEE International Conference on Cluster Computing*, 2003.
3. Message Passing Interface Forum. MPI: A message-passing interface standard. *The International Journal of Supercomputer Applications and High Performance Computing*, 8(3–4), 1994.
4. B. Fryxell, K. Olson, P. Ricker, F. X. Timmes, M. Zingale, D. Q. Lamb, P. Mac-Neice, R. Rosner, and H. Tufo. FLASH: An Adaptive Mesh Hydrodynamics Code for Modelling Astro physical Thermonuclear Flashes. *Astrophysical Journal Suppliment*, 131:273, 2000.
5. William Gropp, Ewing Lusk, and Deborah Swider. Improving the Performance of MPI Derived Datatypes. In *MPIDC*, 1999.
6. InfiniBand Trade Association. InfiniBand Architecture Specification, Release 1.0, October 24, 2000.
7. J. L. Träff, H. Ritzdorf and R. Hempel. The Implementation of MPI–2 One-sided Communication for the NEC SX. In *Proceedings of Supercomputing*, 2000.
8. Lawrence Berkeley National Laboratory. MVICH: MPI for Virtual Interface Architecture, August 2001.
9. Jiuxing Liu, Jiesheng Wu, Sushmitha P. Kini, Pete Wyckoff, and Dhabaleswar K. Panda. High Performance RDMA-Based MPI Implementation over InfiniBand. In *17th Annual ACM International Conference on Supercomputing*, June 2003.
10. Qingda Lu, Jiesheng Wu, Dhabaleswar K. Panda, and P. Sadayappan. Employing MPI Derived Datatypes to the NAS Benchmarks: A Case Study . Technical Report OSU-CISRC-02/04-TR10, Dept. of Computer and Information Science, The Ohio State University, Feb. 2004.
11. Mellanox Technologies. Mellanox InifniBand Technologies. http://www.mellanox.com.
12. Network-Based Computing Laboratory. MVAPICH: MPI for InfiniBand on VAPI Layer. http://nowlab.cis.ohio-state.edu/projects/mpi-iba/index.html.
13. Robert Ross, Neill Miller, and William Gropp. Implementing Fast and Reusable Datatype Processing. In *EuroPVM/MPI*, Oct. 2003.
14. V. Tipparaju, G. Santhanaraman, J. Nieplocha, and D. K. Panda. Host-Assisted Zero-Copy Remote Memory Access Communication on InfiniBand. In *IPDPS '04*, April 2004.
15. W. Gropp and E. Lusk and N. Doss and A. Skjellum. A High-Performance, Portable Implementation of the MPI, Message Passing Interface Standard.
16. Jiesheng Wu, Pete Wyckoff, and Dhabaleswar K. Panda. Supporting Efficient Noncontiguous Access in PVFS over InfiniB and. In *Proceedings of the IEEE International Conference on Cluster Computing*, 2003.
17. Jiesheng Wu, Pete Wyckoff, and Dhabaleswar K. Panda. High Performance Implementation of MPI Datatype Communication over InfiniBand. In *International Parallel and Distributed Processing Symposium (IPDPS '04)*, April 2004.

Minimizing Synchronization Overhead in the Implementation of MPI One-Sided Communication

Rajeev Thakur, William D. Gropp, and Brian Toonen

Mathematics and Computer Science Division
Argonne National Laboratory
Argonne, IL 60439, USA

Abstract. The one-sided communication operations in MPI are intended to provide the convenience of directly accessing remote memory and the potential for higher performance than regular point-to-point communication. Our performance measurements with three MPI implementations (IBM MPI, Sun MPI, and LAM) indicate, however, that one-sided communication can perform much worse than point-to-point communication if the associated synchronization calls are not implemented efficiently. In this paper, we describe our efforts to minimize the overhead of synchronization in our implementation of one-sided communication in MPICH-2. We describe our optimizations for all three synchronization mechanisms defined in MPI: fence, post-start-complete-wait, and lock-unlock. Our performance results demonstrate that, for short messages, MPICH-2 performs six times faster than LAM for fence synchronization and 50% faster for post-start-complete-wait synchronization, and it performs more than twice as fast as Sun MPI for all three synchronization methods.

1   Introduction

MPI defines one-sided communication operations that allow users to directly access the memory of a remote process [9]. One-sided communication both is convenient to use and has the potential to deliver higher performance than regular point-to-point (two-sided) communication, particularly on networks that support one-sided communication natively, such as InfiniBand and Myrinet. On networks that support only two-sided communication, such as TCP, it is harder for one-sided communication to do better than point-to-point communication. Nonetheless, a good implementation should strive to deliver performance as close as possible to that of point-to-point communication.

One-sided communication in MPI requires the use of one of three synchronization mechanisms: fence, post-start-complete-wait, or lock-unlock. The synchronization mechanism defines the time at which the user can initiate one-sided communication and the time when the operations are guaranteed to be completed. The true cost of one-sided communication, therefore, must include the

D. Kranzlmüller et al. (Eds.): EuroPVM/MPI 2004, LNCS 3241, pp. 57–67, 2004.

time taken for synchronization. An unoptimized implementation of the synchronization functions may perform more communication and synchronization than necessary (such as a barrier), which can adversely affect performance, particularly for short and medium-sized messages.

We measured the performance of three MPI implementations, IBM MPI, Sun MPI, and LAM [8], for a test program that performs nearest-neighbor ghost-area exchange, a communication pattern common in many scientific applications such as PDE simulations. We wrote four versions of this program: using point-to-point communication (isend/irecv) and using one-sided communication with fence, post-start-complete-wait, and lock-unlock synchronization. We measured the time taken for a single communication step (each process exchanges data with its four neighbors) by doing the step a number of times and calculating the average. Figure 1 shows a snippet of the fence version of the program, and Figure 2 shows the performance results.

```
for (i=0; i<ntimes; i++) {
    MPI_Win_fence(MPI_MODE_NOPRECEDE, win);
    for (j=0; j<nbrs; j++) {
        MPI_Put(sbuf + j*n, n, MPI_INT, nbr[j], j, n, MPI_INT, win);
    }
    MPI_Win_fence(MPI_MODE_NOSTORE | MPI_MODE_NOPUT | MPI_MODE_NOSUCCEED, win);
}
```

Fig. 1. Fence version of the test.

With IBM MPI on an SP, one-sided communication is almost two orders of magnitude slower than point-to-point (pt2pt) for short messages and remains significantly slower until messages get larger than 256 KB. With Sun MPI on a shared-memory SMP, all three one-sided versions are about six times slower than the point-to-point version for short messages. With LAM on a Linux cluster connected with fast ethernet, for short messages, post-start-complete-wait (pscw) is about three times slower than point-to-point, and fence is about 18 times slower than point-to-point[1]. As shown in Figure 1, we pass appropriate assert values to MPI_Win_fence so that the MPI implementation can optimize the function. Since LAM does not support asserts, we commented them out when using LAM. At least some of the poor performance of LAM with fence can be attributed to not taking advantage of asserts.

We observed similar results for runs with different numbers of processes on all three implementations. Clearly, the overhead associated with synchronization significantly affects the performance of these implementations. Other researchers [4] have found similarly high overheads in their experiments with four MPI implementations: NEC, Hitachi, Sun, and LAM.

Our goal in the design and implementation of one-sided communication in our MPI implementation, MPICH-2, has been to minimize the amount of additional communication and synchronization needed to implement the semantics defined by the synchronization functions. We particularly avoid using a barrier anywhere.

---

[1] LAM does not support lock-unlock synchronization.

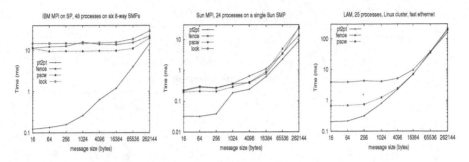

Fig. 2. Performance of IBM MPI, Sun MPI, and LAM for a nearest-neighbor ghost-area exchange test.

As a result, we are able to achieve much higher performance than do other MPI implementations. We describe our optimizations and our implementation in this paper.

2   Related Work

One-sided communication as a programming paradigm was made popular initially by the SHMEM library on the Cray T3D and T3E [6], the BSP library [5], and the Global Arrays library [12]. After the MPI-2 Forum defined an interface for one-sided communication in MPI, several vendors and a few research groups implemented it, but, as far as we know, none of these implementations specifically optimizes the synchronization overhead. For example, the implementations of one-sided communication for Sun MPI by Booth and Mourão [3] and for the NEC SX-5 by Träff et al. [13] use a barrier to implement fence synchronization. Other efforts at implementing MPI one-sided communication include the implementation for InfiniBand networks by Jiang et al. [7], for a Windows implementation of MPI (WMPI) by Mourão and Silva [11], for the Fujitsu VPP5000 vector machine by Asai et al. [1], and for the SCI interconnect by Worringen et al. [14]. Mourão and Booth [10] describe issues in implementing one-sided communication in an MPI implementation that uses multiple protocols, such as TCP and shared memory.

3   One-Sided Communication in MPI

In MPI, the memory that a process allows other processes to access via one-sided communication is called a *window*. Processes specify their local windows to other processes by calling the collective function MPI_Win_create. The three functions for one-sided communication are MPI_Put (remote write), MPI_Get (remote read), and MPI_Accumulate (remote update). They are nonblocking functions: They initiate but not necessarily complete the one-sided operation. These three functions are not sufficient by themselves because one needs to know when

```
Process 0                      Process 1
MPI_Win_fence(win)             MPI_Win_fence(win)
MPI_Put(1)                     MPI_Put(0)
MPI_Get(1)                     MPI_Get(0)
MPI_Win_fence(win)             MPI_Win_fence(win)
```

a. Fence synchronization

```
Process 0                Process 1              Process 2
                         MPI_Win_post(0,2)
MPI_Win_start(1)                                MPI_Win_start(1)
MPI_Put(1)                                      MPI_Put(1)
MPI_Get(1)                                      MPI_Get(1)
MPI_Win_complete(1)                             MPI_Win_complete(1)
                         MPI_Win_wait(0,2)
```

b. Post-start-complete-wait synchronization

```
Process 0                  Process 1                Process 2
MPI_Win_create(&win)       MPI_Win_create(&win)     MPI_Win_create(&win)
MPI_Win_lock(shared,1)                              MPI_Win_lock(shared,1)
MPI_Put(1)                                          MPI_Put(1)
MPI_Get(1)                                          MPI_Get(1)
MPI_Win_unlock(1)                                   MPI_Win_unlock(1)
MPI_Win_free(&win)         MPI_Win_free(&win)       MPI_Win_free(&win)
```

c. Lock-unlock synchronization

Fig. 3. The three synchronization mechanisms for one-sided communication in MPI.

a one-sided operation can be initiated (that is, when the remote memory is ready to be read or written) and when a one-sided operation is guaranteed to be completed. To specify these semantics, MPI defines three different synchronization mechanisms.

Fence. Figure 3a illustrates the fence method of synchronization (without the syntax). MPI_Win_fence is collective over the communicator associated with the window object. A process may issue one-sided operations after the first call to MPI_Win_fence returns. The next fence completes any one-sided operations that this process issued after the preceding fence, as well as the one-sided operations other processes issued that had this process as the target. The drawback of the fence method is that if only small subsets of processes are actually communicating with each other, the collectiveness of the fence function over the entire communicator results in unnecessary synchronization overhead.

Post-Start-Complete-Wait. To avoid the drawback of fence, MPI defines a second mode of synchronization in which only subsets of processes need to synchronize, as shown in Figure 3b. A process that wishes to expose its local window to remote accesses calls MPI_Win_post, which takes as argument an MPI_Group object that specifies the set of processes that will access the window. A process that wishes to perform one-sided communication calls MPI_Win_start, which also takes as argument an MPI_Group object that specifies the set of processes that will be the target of one-sided operations from this process. After issuing all the one-sided operations, the origin process calls MPI_Win_complete to complete the operations at the origin. The target calls MPI_Win_wait to complete the operations at the target.

Lock-Unlock. In this synchronization method, the origin process calls MPI_Win_lock to obtain either shared or exclusive access to the window on the target, as shown in Figure 3c. After issuing the one-sided operations, it calls MPI_Win_unlock. The target does not make any synchronization call. When MPI_Win_unlock returns, the one-sided operations are guaranteed to be completed at the origin and the target. MPI_Win_lock is not required to block until the lock is acquired, except when the origin and target are one and the same process.

4   Implementing MPI One-Sided Communication

Our current implementation of one-sided communication in MPICH-2 is layered on the same lower-level communication abstraction we use for point-to-point communication, called CH3 [2]. CH3 uses a two-sided communication model in which the sending side sends packets followed optionally by data, and the receiving side explicitly posts receives for packets and, optionally, data. The content and interpretation of the packets are decided by the upper layers. We have simply added new packet types for one-sided communication. So far, CH3 has been implemented on top of TCP and shared memory, and therefore our implementation of one-sided communication runs on TCP and shared memory. Implementations of CH3 on other networks are in progress.

For all three synchronization methods, we do almost nothing in the first synchronization call; do nothing in the calls to put, get, or accumulate other than queuing up the requests locally; and instead do everything in the second synchronization call. This approach allows the first synchronization call to return immediately without blocking, reduces or eliminates the need for extra communication in the second synchronization call, and offers the potential for communication operations to be aggregated and scheduled efficiently as in BSP [5]. We describe our implementation below.

4.1   Fence

An implementation of fence synchronization must take into account the following semantics: A one-sided operation cannot access a process's window until that process has called fence, and the next fence on a process cannot return until all processes that need to access that process's window have completed doing so.

A naïve implementation of fence synchronization could be as follows. At the first fence, all processes do a barrier so that everyone knows that everyone else has called fence. Puts, gets, and accumulates can be implemented either as blocking or nonblocking operations. In the second fence, after all the one-sided operations have been completed, all processes again do a barrier to ensure that no process leaves the fence before other processes have finished accessing its window. This method requires two barriers, which can be quite expensive.

In our implementation, we avoid the two barriers completely. In the first call to fence, we do nothing. For the puts, gets, and accumulates that follow, we simply queue them up locally and do nothing else, with the exception that any one-sided operation whose target is the origin process itself is performed immediately by doing a simple memory copy or local accumulate. In the second fence, each process goes through its list of queued one-sided operations and determines, for every other process i, whether any of the one-sided operations have i as the target. This information is stored in an array, such that a 1 in the ith location of the array means that one or more one-sided operations are targeted to process i, and a 0 means no one-sided operations are targeted to that process. All processes now do a reduce-scatter sum operation on this array (as in MPI_Reduce_scatter). As a result, each process now knows how many processes will be performing one-sided operations on its window, and this number is stored in a counter in the MPI_Win object. Each process is now free to perform the data transfer for its one-sided operations; it needs only to ensure that the window counter at the target gets decremented when all the one-sided operations from this process to that target have been completed.

A put is performed by sending a put packet containing the address, count, and datatype information for the target. If the datatype is a derived datatype, an encapsulated version of the derived datatype is sent next. Then follows the actual data. The MPI progress engine on the target receives the packet and derived datatype, if any, and then directly receives the data into the correct memory locations. No rendezvous protocol is needed for the data transfer, because the origin has already been authorized to write to the target window. Gets and accumulates are implemented similarly.

For the last one-sided operation, the origin process sets a field in the packet header indicating that it is the last operation. The target therefore knows to decrement its window counter after this operation has completed at the target. When the counter reaches 0, it indicates that all remote processes that need to access the target's window have completed their operations, and the target can therefore return from the second fence. This scheme of decrementing the counter only on the last operation assumes that data delivery is ordered, which is a valid assumption for the networks we currently support. On networks that do not guarantee ordered delivery, a simple sequence-numbering scheme can be added to achieve the same effect.

We have thus eliminated the need for a barrier in the first fence and replaced the barrier at the *end* of the second fence by a reduce-scatter at the *beginning* of the second fence before any data transfer. After that, all processes can do their communication independently and return when they are done.

4.2   Post-Start-Complete-Wait

An implementation of post-start-complete-wait synchronization must take into account the following semantics: A one-sided operation cannot access a process's window until that process has called MPI_Win_post, and a process cannot return from MPI_Win_wait until all processes that need to access that process's window have completed doing so and called MPI_Win_complete.

A naïve implementation of this synchronization could be as follows. MPI_Win _start blocks until it receives a message from all processes in the target group indicating that they have called MPI_Win_post. Puts, gets, and accumulates can be implemented as either blocking or nonblocking functions. MPI_Win_complete waits until all one-sided operations initiated by that process have completed locally and then sends a done message to each target process. MPI_Win_wait on the target blocks until it receives the done message from each origin process. Clearly, this method involves a great deal of synchronization.

We have eliminated most of this synchronization in our implementation as follows. In MPI_Win_post, if the assert MPI_MODE_NOCHECK is not specified, the process sends a zero-byte message to each process in the origin group to indicate that MPI_Win_post has been called. It also sets the counter in the window object to the size of this group. As in the fence case, this counter will get decremented by the completion of the last one-sided operation from each origin process. MPI_Win_wait simply blocks and invokes the progress engine until this counter reaches zero.

On the origin side, we do nothing in MPI_Win_start. All the one-sided operations following MPI_Win_start are simply queued up locally as in the fence case. In MPI_Win_complete, the process first waits to receive the zero-byte messages from the processes in the target group. It then performs all the one-sided operations exactly as in the fence case. The last one-sided operation has a field set in its packet that causes the target to decrement its counter on completion of the operation. If an origin process has no one-sided operations destined to a target that was part of the group passed to MPI_Win_start, it still needs to send a packet to that target for decrementing the target's counter. MPI_Win_complete returns when all its operations have locally completed.

Thus the only synchronization in this implementation is the wait at the beginning of MPI_Win_complete for a zero-byte message from the processes in the target group, and this too can be eliminated if the user specifies the assert MPI_MODE_NOCHECK to MPI_Win_post and MPI_Win_start (similar to MPI_Rsend).

4.3   Lock-Unlock

Implementing lock-unlock synchronization when the window memory is not directly accessible by all origin processes requires the use of an asynchronous agent at the target to cause progress to occur, because one cannot assume that the user program at the target will call any MPI functions that will cause progress periodically.

Our design for the implementation of lock-unlock synchronization involves the use of a thread that periodically wakes up and invokes the MPI progress engine if it finds that no other MPI function has invoked the progress engine within some time interval. If the progress engine had been invoked by other calls to MPI, the thread does nothing. This thread is created only when MPI_Win_create is called and if the user did not pass an info object to MPI_Win_create with the key no_locks set to true (indicating that he will not be using lock-unlock synchronization). In MPI_Win_lock, we do nothing but queue up the lock request locally and return immediately. The one-sided operations are also queued up locally. All the work is done in MPI_Win_unlock.

For the general case where there are multiple one-sided operations, we implement MPI_Win_unlock as follows. The origin sends a "lock-request" packet to the target and waits for a "lock-granted" reply. When the target receives the lock request, it either grants the lock by sending a lock-granted reply to the origin or queues up the lock request if it conflicts with the existing lock on the window. When the origin receives the lock-granted reply, it performs the one-sided operations exactly as in the other synchronization modes. The last one-sided operation, indicated by a field in the packet header, causes the target to release the lock on the window after the operation has completed. Therefore, no separate unlock request needs to be sent from origin to target.

The semantics specify that MPI_Win_unlock cannot return until the one-sided operations are completed at both origin and target. Therefore, if the lock is a shared lock and none of the operations is a get, the target sends an acknowledgment to the origin after the last operation has completed. If any one of the operations is a get, we reorder the operations and perform the get last. Since the origin must wait to receive data, no additional acknowledgment is needed. This approach assumes that data transfer in the network is ordered. If not, an acknowledgment is needed even if the last operation is a get. If the lock is an exclusive lock, no acknowledgment is needed even if none of the operations is a get, because the exclusive lock on the window prevents another process from accessing the data before the operations have completed.

Optimization for Single Operations. If the lock-unlock is for a single short operation and predefined datatype at the target, we send the put/accumulate data or get information along with the lock-request packet itself. If the target can grant the lock, it performs the specified operation right away. If not, it queues up the lock request along with the data or information and performs the operation when the lock can be granted. Except in the case of get operations, MPI_Win_unlock blocks until it receives an acknowledgment from the target that the operation has completed. This acknowledgment is needed even if the lock is an exclusive lock because the origin does not know whether the lock has been granted.

Similar optimizations are possible for multiple one-sided operations, but at the cost of additional queuing/buffering at the target.

5   Performance Results

To study the performance of our implementation, we use the same ghost-area exchange program described in Section 1. Figure 4 shows the performance of the test program with MPICH-2 on a Linux cluster with fast ethernet and on a Sun

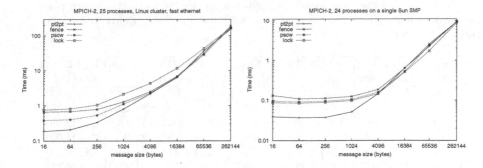

Fig. 4. Performance of MPICH-2 on a Linux cluster and Sun SMP.

SMP with shared memory[2]. We see that the time taken by the point-to-point version with MPICH-2 is about the same as with other MPI implementations in Figure 2, but the time taken by the one-sided versions is much lower. To compare the performance of MPICH-2 with other MPI implementations, we calculated the ratio of the time with point-to-point communication to the time with one-sided communication and tabulated the results in Table 1[3]. For short messages, MPICH-2 is almost six times faster than LAM for the fence version and about 50% faster for the post-start-complete-wait version. Compared with Sun MPI for short messages, MPICH-2 is more than twice as fast for all three synchronization methods. The difference narrows for large message sizes where the synchronization overheads are less of an issue, but MPICH-2 still performs better than both LAM and Sun MPI. In some cases, we see that the ratio is less than one, which means that one-sided communication is actually faster than point-to-point communication. We attribute this to the waiting time for the receive to be called in the rendezvous protocol used in point-to-point communication for large messages.

6   Conclusions and Future Work

This paper shows that an optimized implementation of the synchronization functions significantly improves the performance of MPI one-sided communication. Nonetheless, several opportunities exist for improving the performance further, and we plan to explore them. For example, in the case of lock-unlock synchronization for a single put or accumulate, we can improve the performance substantially by not having the origin process wait for an acknowledgment from the target at the end of the unlock. This optimization, however, breaks the semantics of unlock, which state that when the unlock returns, the operation is complete at

---

[2] Since the MPICH-2 progress engine is not yet fully thread safe, we ran the lock-unlock test without a separate thread for making progress. The MPI calls on the main thread made progress.

[3] Since MPICH-2 does not run on an IBM SP yet, we could not compare with IBM MPI.

Table 1. Ratio of the time with one-sided communication to the time with point-to-point communication on the Linux cluster (left) and Sun SMP (right) (the smaller the ratio, the better).

Size	LAM		MPICH-2		Size	Sun MPI			MPICH-2		
(bytes)	fence	pscw	fence	pscw	(bytes)	fence	pscw	lock	fence	pscw	lock
16	18.9	3.3	3.5	2.03	16	6.9	6.0	6.5	3.4	2.45	2.24
64	18.1	3.2	3.28	1.94	64	9.2	6.5	8.8	2.94	2.47	2.30
256	13.6	2.4	2.35	1.60	256	7.0	5.3	6.7	3.0	2.55	2.38
1K	5.1	1.52	1.59	1.39	1K	2.0	1.55	1.93	2.43	2.06	1.92
16K	1.40	1.02	1.08	1.05	16K	1.79	1.44	1.26	0.99	0.82	0.79
64K	1.03	0.95	0.85	0.78	64K	1.48	2.16	1.54	1.13	1.06	0.77
256K	1.30	1.21	1.22	1.08	256K	2.76	2.59	1.51	0.99	1.01	0.94

both origin and target. We plan to explore the possibility of allowing the user to pass an assert or info key to select weaker semantics that do not require the operation to be completed at the target when unlock returns. We plan to extend our implementation to work efficiently on networks that have native support for remote-memory access, such as InfiniBand and Myrinet. We also plan to cache derived datatypes at the target so that they need not be communicated each time and aggregate short messages and communicate them as a single message instead of communicating them separately.

Acknowledgments

This work was supported by the Mathematical, Information, and Computational Sciences Division subprogram of the Office of Advanced Scientific Computing Research, Office of Science, U.S. Department of Energy, under Contract W-31-109-ENG-38. We thank Chris Bischof for giving us access to the Sun SMP machines at the University of Aachen and Dieter an May for helping us in running our tests on those machines. We also thank Don Frederick for giving us access to the IBM SP at the San Diego Supercomputer Center.

References

1. Noboru Asai, Thomas Kentemich, and Pierre Lagier. MPI-2 implementation on Fujitsu generic message passing kernel. In *Proceedings of SC99: High Performance Networking and Computing*, November 1999.
2. David Ashton, William Gropp, Rajeev Thakur, and Brian Toonen. The CH3 design for a simple implementation of ADI-3 for MPICH-2 with a TCP-based implementation. Technical Report ANL/MCS-P1156-0504, Mathematics and Computer Science Division, Argonne National Laboratory, May 2004.
3. S. Booth and E. Mourão. Single sided MPI implementations for SUN MPI. In *Proceedings of SC2000: High Performance Networking and Computing*, November 2000.

4. Edgar Gabriel, Graham E. Fagg, and Jack J. Dongarra. Evaluating the performance of MPI-2 dynamic communicators and one-sided communication. In Jack Dongarra, Domenico Laforenza, and Salvatore Orlando, editors, *Recent Advances in Parallel Virtual Machine and Message Passing Interface, 10th European PVM/MPI Users' Group Meeting*, pages 88–97. Lecture Notes in Computer Science 2840, Springer, September 2003.
5. J. M. D. Hill, B. McColl, D. C. Stefanescu, M. W. Goudreau, K. Lang, S. B. Rao, T. Suel, T. Tsantilas, and R. H. Bisseling. BSPlib: The BSP programming library. *Parallel Computing*, 24(14):1947–1980, December 1998.
6. Cray Research Inc. Cray T3E C and C++ optimization guide, 1994.
7. Weihang Jiang, Jiuxing Liu, Hyun-Wook Jin, Dhabaleswar K. Panda, William Gropp, and Rajeev Thakur. High performance MPI-2 one-sided communication over InfiniBand. In *Proc. of 4th IEEE/ACM Int'l Symp. on Cluster Computing and the Grid*, April 2004.
8. LAM/MPI Parallel Computing. http://www.lam-mpi.org.
9. Message Passing Interface Forum. MPI-2: Extensions to the Message-Passing Interface, July 1997. http://www.mpi-forum.org/docs/docs.html.
10. Elson Mourão and Stephen Booth. Single sided communications in multi-protocol MPI. In Jack Dongarra, Peter Kacsuk, and Norbert Podhorszki, editors, *Recent Advances in Parallel Virutal Machine and Message Passing Interface, 7th European PVM/MPI Users' Group Meeting*, pages 176–183. Lecture Notes in Computer Science 1908, Springer, September 2000.
11. Fernando Elson Mourão and João Gabriel Silva. Implementing MPI's one-sided communications for WMPI. In Jack Dongarra, Emilio Luque, and Tomàs Margalef, editors, *Recent Advances in Parallel Virtual Machine and Message Passing Interface, 6th European PVM/MPI Users' Group Meeting*, pages 231–238. Lecture Notes in Comp. Science 1697, Springer, Sept. 1999.
12. Jaroslaw Nieplocha, Robert J. Harrison, and Richard J. Littlefield. Global Arrays: A non-uniform-memory-access programming model for high-performance computers. *The Journal of Supercomputing*, 10(2):169–189, 1996.
13. Jesper Larsson Träff, Hubert Ritzdorf, and Rolf Hempel. The implementation of MPI-2 one-sided communication for the NEC SX-5. In *Proceedings of SC2000: High Performance Networking and Computing*, November 2000.
14. Joachim Worringen, Andreas Gäer, and Frank Reker. Exploiting transparent remote memory access for non-contiguous and one-sided-communication. In *Proceedings of the 2002 Workshop on Communication Architecture for Clusters (CAC)*, April 2002.

Efficient Implementation
of MPI-2 Passive One-Sided Communication
on InfiniBand Clusters*

Weihang Jiang[1], Jiuxing Liu[1], Hyun-Wook Jin[1], Dhabaleswar K. Panda[1],
Darius Buntinas[2], Rajeev Thakur[2], and William D. Gropp[2]

[1] Department of Computer Science and Engineering
The Ohio State University, Columbus, OH 43210
{jiangw,liuj,jinhy,panda}@cse.ohio-state.edu
[2] Mathematics and Computer Science Division
Argonne National Laboratory, Argonne, IL 60439
{buntinas,thakur,gropp}@mcs.anl.gov

Abstract. In this paper we compare various design alternatives for synchronization in MPI-2 passive one-sided communication on InfiniBand clusters. We discuss several requirements for synchronization in passive one-sided communication. Based on these requirements, we present four design alternatives, which can be classified into two categories: thread-based and atomic operation-based. In thread-based designs, synchronization is achieved with the help of extra threads. In atomic operation-based designs, we exploit InfiniBand atomic operations such as Compare-and-Swap and Fetch-and-Add. Our performance evaluation results show that the atomic operation-based design can require less synchronization overhead, achieve better concurrency, and consume fewer computing resources compared with the thread based design.

1 Introduction

MPI has been the de facto standard in high-performance computing for writing parallel applications. As an extension to MPI, MPI-2 [11] introduces several new features. One important new feature is one-sided communication. In the traditional two-sided communication, both parties must perform matching communication operations (e.g., a *send* and a *receive*). In one-sided operation, a matching operation is not required from the remote party. All parameters for the operation, such as source and destination buffers, are provided by the initiator

---

* This research is supported by Department of Energy's grant #DE-FC02-01ER25506, National Science Foundation's grants #CNS-0204429 and #CCR-0311542, and Post-doctoral Fellowship Program of Korea Science & Engineering Foundation(KOSEF). This work is also supported by the Mathematical, Information, and Computational Sciences Division subprogram of the Office of Advanced Scientific Computing Research, Office of Science, U.S. Department of Energy, under Contract W-31-109-ENG-38.

D. Kranzlmüller et al. (Eds.): EuroPVM/MPI 2004, LNCS 3241, pp. 68–76, 2004.

of the operation. One-sided operations can support more flexible communication patterns and improve performance in certain applications.

In MPI-2 one-sided communication, the process that initiates the operation is called the *origin* process, and the process being accessed is called the *target* process. The memory area in the target process that can be accessed is called the *window*. MPI-2 requires explicit synchronization to guarantee the completion of data communication operations on windows. MPI-2 supports two synchronization modes: active and passive. In active mode, the target is actively involved in synchronization, whereas in passive mode, the target is not explicitly involved in synchronization.

We have previously implemented MPI-2 active mode one-sided communication in [7] using RDMA-based communication over InfiniBand. In this paper we extend this implementation to allow passive one-sided communication. The main challenge in extending active mode to passive mode is designing an efficient synchronization mechanism that allows one-sided operations to be performed at the target node independently of what the target process is doing.

Several design challenges are involved in implementing efficient MPI-2 passive one-sided communications. First, the implementation must be able to make independent progress in passive one-sided communications. Next, concurrent communications must be handled efficiently. Another issue is efficient implementation of shared and exclusive locks. A related issue is how to implement MPI_Win_lock() in a nonblocking fashion. Further, since modern network interfaces, such as InfiniBand, offer RDMA-based operations, these operations should be used in the most efficient manner.

In this paper we take on these challenges. We implement and evaluate four design alternatives for passive-mode synchronization: dedicated thread, event-driven thread, Test-and-Set, and MCS based designs.

All these designs have been incorporated in our MPI-2 implementation over InfiniBand, MVAPICH2 [14] [9], which is based on MPICH2 developed by Argonne National Laboratory [1].

The paper is organized as follows. In Section 2, we briefly describe Infini-Band. In Sections 3–5, we present design issues. In Section 6, we evaluate the performance of our various designs. In Section 7, we discuss related work. In Section 8, we draw conclusions and discuss future work.

2   InfiniBand and Atomic Operations

The InfiniBand Architecture is an industry standard for high-performance interconnects between processing nodes and I/O nodes [5]. Host channel adapters (HCAs) connect nodes to the InfiniBand fabric. InfiniBand provides both channel and memory semantics. In channel semantics, send and receive are used for communication. In memory semantics, InfiniBand supports remote direct memory access (RDMA) operations.

Another emerging feature of InfiniBand is remote atomic operation. This allows us to efficiently implement the synchronization algorithms designed for a shared-memory environment to a distributed private-memory environment. In-finiBand supports two 64-bit atomic operations: Compare-and-Swap and Fetch-and-Add. The Compare-and-Swap operation reads a 64-bit content from the memory of a remote process, compares the content with the *compare_value* parameter of this atomic operation, and puts the value of the *swap_value* parameter into the remote memory if the two compared values are the same. The Fetch-and-Add operation reads data from the remote memory, performs an addition operation between the data and the *add_value* parameter of this atomic operation, and updates the result to the remote memory. Both the Compare-and-Swap and Fetch-and-Add operations bring back the old value of the variable in the remote memory. As the name "atomic operation" suggests, Compare-and-Swap and Fetch-and-Add operations are handled atomically, and more important, they are processed by the processor on the HCAs, without CPU involvement at the remote side.

3   Design Issues in Passive One-Sided Communication

In this section we discuss challenges in implementing efficient MPI-2 passive one-sided communication.

Synchronization Performance: When the contention for synchronization functions is low, low synchronization overhead is important. When the contention is high, low synchronization delay is desirable. In both cases, the number of messages exchanged for synchronization functions should be small. **Independent Progress:** In MPI-2 passive one-sided communication, the target process does not make any MPI calls to cooperate with the origin process for communication or synchronization. Therefore, the implementation cannot rely on the progress engine being called by MPI functions, the strategy commonly used for two-sided communication. **Concurrent Communication:** In MPI-2, one-sided communication and two-sided communication may happen concurrently. One-sided communication from different origin processes to disjoint windows at a target process may also happen concurrently. Hence we need to handle this situation in our design. **Nonblocking MPI_Win_lock():** When MPI_Win_lock() is called, if the lock at the target process is held by other processes, the lock cannot be acquired until the lock is released. The nonblocking MPI_Win_lock() allows the current process to continue without waiting for the lock. However, the lock must be acquired before the first communication operation takes effect. **Shared Locks and Exclusive Locks:** In MPI-2, both shared locks and exclusive locks are supported. **RDMA for Data Transfer:** Modern network interfaces, such as InfiniBand, offer RDMA-based operations, which can be used for high-performance data transfer in one-sided communication [7]. Therefore, synchronization mechanisms must work correctly with RDMA-based communication.

In following sections, we address these challenges and describe thread-based and atomic operation-based designs for MPI-2 passive mode synchronization.

4   Thread-Based Designs

Thread-based design is widely used in MPI implementations to support one-sided communication. In our MPI-2 implementation over InfiniBand, we use RDMA operations to transfer data, while we still use a thread running at the target process to handle special cases such as noncontiguous data transfer and the accumulate function. Therefore, we are interested in whether we can use this thread to handle passive synchronization efficiently.

4.1   Dedicated Thread-Based Design

In a dedicated thread-based design, an assisting thread runs at the target side in a dedicated manner and handles all passive synchronization requests from the origin processes. In order to achieve low latency, the thread is always active and uses polling to process communication. The characteristics of the thread guarantee the independent progress of the synchronization process. Before using RDMA operations to transfer data, we need to acquire the lock. This is done by sending a control message and getting an acknowledgment back. In order to implement MPI_Win_lock() in a nonblocking manner, after sending the lock request, the origin process continues its work, buffering any one sided operations until the lock is acquired. Since in a thread-based design, the target process knows whether the request message is for a shared lock or an exclusive lock, the target process can coordinate between shared-lock requests and exclusive-lock requests to support both of them. In this implementation, passive synchronization messages and two-sided messages are handled by different threads. Eventually the MPICH2 internal progress engine will be thread safe, until then, we need to provide our own mutual exclusion mechanisms to serialize access by the two threads.

4.2   Event Driven Based Design

A problem in the dedicated thread-based design is that since the dedicated thread keeps running at a target process, CPU cycles are consumed by the thread even if there is no passive one-sided communication. In this section, we introduce an event-driven based design to solve this problem.

InfiniBand provides both channel and memory semantics. By using channel semantics, one process can generate a signal at a remote process. Also, InfiniBand HCA supports event handlers, which can be driven by such a signal.

Combining these features, we propose an event driven-based design to reduce the CPU utilization. Initially, a predefined event handler that can resume a blocked thread is registered to an HCA. When the assisting thread is created, the process posts a receive operation and blocks the thread. If a process wants to communicate with a remote process, it first posts a send operation matching with the prepost receive operation at the remote party, to generate a signal. Then the event wakes up the thread. The remaining steps are similar to the dedicated thread-based design. Finally, before the thread is blocked again, another receive operation is posted.

5   Atomic Operation-Based Design

In Section 2, we described hardware-level remote atomic operations in Infini-Band. These give us the opportunity to exploit some well-known algorithms proposed for shared-memory synchronization. In this section, we present two designs based on the Test-and-Set and MCS lock algorithms.

5.1   Test-and-Set-Based Design

In the Test-and-Set algorithm, a flag is used to indicate whether the lock is held. To acquire a lock, a processor tries to change the flag from false to true by executing a Test-and-Set instruction. The processor releases the lock by changing the flag back to true. To implement MPI-2 synchronization functions using the Test-and-Set lock algorithm, we can use the atomic operation Compare-and-Swap.

Since the atomic operations are handled by HCA and the MPI library at the target process is not involved, the progress of passive synchronization is independent of the progress of the target process. Once the Compare-and-Swap for acquiring a lock succeeds, the process can start using RDMA to transfer data. To implement nonblocking MPI_Win_lock, a process issues the first Compare-and-Swap operation in MPI_Win_lock(), without waiting for it to complete. The waiting is delayed until the first communication operation.

We can easily extend the Test-and-Set algorithm to support both shared lock and exclusive lock, by checking the value returned by Compare-and-Swap operation. Details can be found in [6].

One drawback of the Test-and-Set-based design is the high network traffic caused by repeated issue of Compare-and-Swap operations. Using an exponential back-off mechanism can alleviate this problem.

5.2   MCS-Based Design

The MCS algorithm is proposed as a scalable synchronization algorithm for shared-memory multiprocessors [10]. The main idea of MCS is to maintain a distributed queue for processes competing for the lock. Scalability is achieved by avoiding spinning on remote memory, and by decreasing the lock synchronization delay.

For each window, each origin process maintains three data structures – *flag*, *previous*, and *next* – and each target process maintains one data structure, called *lock*. When origin process A requests a lock on the target process, it swaps its process id with the value of *lock*. Then origin process B requests the same lock by swapping. Based on the value swapped back, origin process B knows that origin process A is queued before it. Thus, it updates the value of *next* in origin process A. When origin process A releases the lock, based on the value of *next*, origin process A updates the value of *flag* in the origin process B. Finally, when the origin process B releases the lock, it resets *lock* to NULL at the target process. The atomic Swap operation is used to update the *lock* value atomically, and RDMA Write is used to update the *next* and *flag* values.

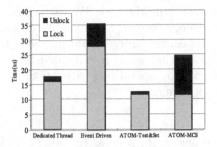

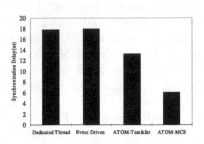

Fig. 1. Synchronization Overhead.   **Fig. 2.** Synchronization Delay.

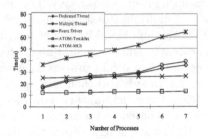

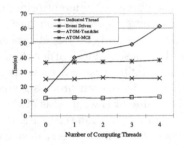

Fig. 3. Concurrency.   **Fig. 4.** Computing Thread.

6 Performance Evaluation

The performance of MPI-2 passive synchronization functions can be evaluated with respect to the following metrics: (1) synchronization overhead: time spent on synchronization functions, (2) synchronization delay: time required after one origin process releases a lock on a remote window and another origin process acquires the same lock, (3) concurrency: the capability to handle multiple concurrent passive synchronization functions, (4) message complexity: the number of messages exchanged for synchronization functions, and (5) CPU utilization: the CPU cycles involved in the synchronization process.

Our experimental testbed consists of a cluster of eight SuperMicro SUPER X5DL8-GG nodes, each with dual Intel Xeon 3.0 GHz processors, PCI-X 64-bit 133 MHz bus, and connected to Mellanox InfiniHost MT23108 DualPort 4x HCAs. Detailed configuration can be found in [6].

6.1 Synchronization Overhead

We begin with a simple approach to measure synchronization overhead. In this test, one process calls only MPI-2 passive synchronization functions (MPI_Win_lock and MPI_Win_unlock) on a window at the other process for multiple iterations. We then report the time taken for each iteration. Figure 1 shows the synchronization overhead for all four designs. We also report separately the time spent acquiring the lock and the time spent releasing the lock. We can see that the Test-and-Set-based design shows the best performance, approximately 12.83 μs. Releasing a lock is faster for this design because processes do not need to wait for the completion of unlock. We also see that using a dedicated thread can achieve better performance than using atomic operations with MCS. The event-driven approach shows the worst performance.

6.2   Synchronization Delay

Synchronization delay is the delay between one origin process releasing a lock on a remote window and another origin process acquiring the same lock. It is an important performance metric for a lock algorithm, especially when the competition between different origin processes for a given lock is heavy. The test for measuring synchronization delay consists of multiple iterations, using two origin processes and one target process. In the even-numbered iterations, origin process 1 requests a lock earlier than origin process 2, and in the odd-numbered iterations, origin process 2 requests a lock earlier than origin process 1. After acquiring the lock, each process holds the lock for time E and then releases the lock. The value of E we used is always longer than the synchronization overhead of all designs. As we can see in Figure 2, the designs based on atomic operations outperform the thread-based designs. The MCS-based design shows the best synchronization delay because locks can be transferred to the next process by using a single message. In all other designs, at least a roundtrip time is required.

6.3   Concurrency

For some MPI-2 applications, the target process may have a large volume of data to be accessed by multiple origin processes. One way to improve the application performance is to use multiple windows and let different origin processes concurrently access the data in different windows. To evaluate how different designs handle concurrent accesses, we used a test with multiple origin processes and one target process. In the target process, multiple windows are created, and the number of windows is equal to the number of origin processes. In each iteration, each origin process calls only MPI_Win_lock and MPI_Win_unlock on the corresponding target window. We then report the average time spent on each iteration. Figure 3 shows that for Test-and-Set-based design and MCS-based design, the time spent on synchronization functions does not change. This result indicates that they can handle concurrent accesses efficiently. However, for thread-based designs, the time increases when the number of origin processes increases.

Further, we can see that even using multiple threads, we cannot achieve better concurrency. The reason is that as long as all the threads are sharing the same progress engine, the time spent on the progress engine cannot be overlapped, because of the mutexes controlling access to the progress engine.

6.4   CPU Utilization

In an MPI-2 application, each process may have multiple computing threads running. We evaluated the performance of synchronization functions under this scenario. Our test uses two processes: one target process and one origin process. The target process spawns several computing threads, and the origin process calls synchronization functions MPI_Win_lock and MPI_Win_lock on a window at the target process for multiple iterations. We then measured the time for each iteration. Figure 4 shows that for both atomic-based designs, the time remains almost unchanged, while for the dedicated thread-based design, the time increases with an increase in the number of computing threads. For the event-driven based design, since the assisting thread is awakened by a signal, the time almost remains constant, too.

6.5   Discussion

From the performance results we can see that, in general, atomic operation-based designs outperform thread-based designs. By taking advantage of atomic operations in InfiniBand, we can achieve better synchronization overhead and synchronization delay. Atomic operation-based designs can also achieve better concurrency and independent communication progress. One possible drawback of the atomic operation-based design is that the number of messages to acquire a lock increases when there is high contention for the lock. For the Test-and-Set-based design, this problem can be solved by using exponential backoff. Detailed evaluation of the message complexity of the proposed schemes together with the exponential backoff is included in [6]. For the MCS-based design, the problem stems from the lack of an atomic Swap operation in the current InfiniBand implementation.

7   Related Work

Most implementations of MPI-2 passive one-sided communication are implemented based on the thread-based design [3], [12], [13], [17], [15], [2]. Work in [16], [4] describe nonthreaded implementations of MPI-2 passive one-sided communication, where the progress engine takes charge of performing synchronization. This design does not satisfy the requirements we have defined in Section 3. The latest version of LAM-MPI [8] supports a part of MPI-2 functions that do not include the passive one-sided communication.

8   Conclusions and Future Work

In this paper, we analyzed issues and concerns related to designing a high-performance MPI-2 passive synchronization mechanisms on InfiniBand clusters. We proposed, implemented, and evaluated two thread-based designs (i.e., dedicated thread and event-driven blocking thread-based designs) and two atomic operation-based designs (i.e., Test-and-Set and MCS-based design). We demonstrated that by taking advantage of InfiniBand atomic operations, we can achieve efficient synchronization and deliver good performance.

In the future, we plan to use real applications to study the impact of synchronization in MPI-2 passive communication. We also plan to investigate how to handle datatype efficiently in MPI-2 one-sided communication.

References

1. Argonne National Laboratory. MPICH2. http://www.mcs.anl.gov/mpi/mpich2/.
2. N. Asai, T. Kentemich, and P. Lagier. MPI-2 Implementation on Fujitsu Generic Message Passing Kernel. In *SC*, 1999.
3. S. Booth and F. E. Mourao. Single Sided MPI Implementations for SUN MPI. In *SC*, 2000.
4. M. Golebiewski and J. L. Traff. MPI-2 One-Sided Communications on a Giganet SMP Cluster. In *EuroPVM/MPI*, 2001.
5. InfiniBand Trade Association. InfiniBand Architecture Specification, Release 1.0, October 24 2000.
6. W. Jiang, J. Liu, H.-W. Jin, D. K. Panda, D. Buntinas, R. Thakur, and W. Gropp. Efficient Implementation of MPI-2 Passive One-Sided Communication over InfiniBand Clusters. Technical Report OSU-CISRC-5/04-TR34, May 2004.
7. W. Jiang, J. Liu, H.-W. Jin, D. K. Panda, W. Gropp, and R. Thakur. High Performance MPI-2 One-Sided Communication over InfiniBand. In *IEEE/ACM CCGrid*, 2004.
8. LAM Team, Indiana University. LAM 7.0.4.
9. J. Liu, W. Jiang, P. Wyckoff, D. K. Panda, D. Ashton, D. Buntinas, W. Gropp, and B. Toonen. Design and Implementation of MPICH2 over InfiniBand with RDMA Support. In *IPDPS*, April 2004.
10. J. M. Mellor-Crummey and M. L. Scott. Algorithms for Scalable Synchronization on Shared-Memory Multiprocessors. In *ACM Trans. on Computer System*, 1991.
11. Message Passing Interface Forum. MPI-2: A Message Passing Interface Standard. *High Performance Computing Applications*, 12(1–2):1–299, 1998.
12. E. Mourao and S. Booth. Single Sided Communications in Multi-Protocol MPI. In *EuroPVM/MPI*, 2000.
13. F. E. Mourao and J. G. Silva. Implementing MPI's One-Sided Communications for WMPI. In *EuroPVM/MPI*, September 1999.
14. Network-Based Computing Laboratory. MVAPICH2: MPI-2 for InfiniBand on VAPI Layer. http://nowlab.cis.ohio-state.edu/projects/mpi-iba/index.html, January 2003.
15. M. Schulz. Efficient Coherency and Synchronization Management in SCI based DSM systems. In *SCI-Europe, Conference Stream of Euro-Par*, 2000.
16. J. Traff, H. Ritzdorf, and R. Hempel. The Implementation of MPI-2 One-Sided Communication for the NEC SX. In *SC*, 2000.
17. J. Worringen, A. Gaer, and F. Reker. Exploiting Transparent Remote Memory Access for Non-Contiguous and One-Sided-Communication. In *CAC*, April 2002.

Providing Efficient I/O Redundancy in MPI Environments[*]

Willam D. Gropp, Robert Ross, and Neill Miller

Mathematics and Computer Science Division
Argonne National Laboratory
Argonne, IL 60439
{gropp,rross,neillm}@mcs.anl.gov

Abstract. Highly parallel applications often use either highly parallel file systems or large numbers of independent disks. Either approach can provide the high data rates necessary for parallel applications. However, the failure of a single disk or server can render the data useless. Conventional techniques, such as those based on applying erasure correcting codes to each file write, are prohibitively expensive for massively parallel scientific applications because of the granularity of access at which the codes are applied. In this paper we demonstrate a scalable method for recovering from single disk failures that is optimized for typical scientific data sets. This approach exploits coarser-grained (but precise) semantics to reduce the overhead of constructing recovery data and makes use of parallel computation (proportional to the data size and independent of number of processors) to construct data. Experiments are presented showing the efficiency of this approach on a cluster with independent disks, and a technique is described for hiding the creation of redundant data within the MPI-IO implementation.

1 Introduction

The scale of today's systems and applications requires very high performance I/O systems. Because single-disk performance has not improved at an adequate rate, current I/O systems employ hundreds of disks in order to obtain the needed aggregate performance. This approach effectively solves the performance problem, but it brings with it a new problem: increased chance of component failure. General approaches for maintaining redundant data have been applied to locally attached disk storage systems (e.g. RAID [7]). These approaches, however, work at a very fine granularity. Applying this type of approach in a parallel file system or across the local disks in a cluster, where disks are distributed and latencies are higher, imposes unacceptable overhead on the system.

Fortunately, two characteristics of computational science applications and storage systems provide opportunities for more efficient approaches. First, the

---

[*] This work was supported by the Mathematical, Information, and Computational Sciences Division subprogram of the Office of Advanced Scientific Computing Research, Office of Science, U.S. Department of Energy, under Contract W-31-109-Eng-38.

D. Kranzlmüller et al. (Eds.): EuroPVM/MPI 2004, LNCS 3241, pp. 77–86, 2004.

applications that generate large amounts of data typically have phases of I/O where much data is written at once, such as in checkpointing. A partially written checkpoint isn't particularly useful, so generally one or more previous checkpoints are preserved in case a failure occurs during the checkpointing process. Thus, creating redundant data for a checkpoint once the application is finished writing it adequately covers the failure cases in which the application is interested. Second, we can generally detect a failed disk. Failures where the location of lost data is known are categorized in coding theory as *erasures*. This category of data loss is more easily corrected than the general case of potentially corrupted data located anywhere in the system; that is, more algorithms are applicable to the problem of recovering from failures in this environment.

This paper describes *lazy redundancy*, a technique for matching the generation of recovery data to the needs of high-performance parallel scientific applications and the systems on which these applications run. The lazy redundancy technique embodies two main principles: aggregating the creation of error-correcting data at explicit points, and leveraging the resources of clients and storage devices in creating error correcting data or recovering from failures. Specifically we describe a method for efficient creation of the data necessary to recover from a single erasure failure that is applied only when processes reach an I/O synchronization point (e.g., `MPI_File_close` or `MPI_File_sync`). Because all processes have reached this point, we may use the processes to perform the recovery data calculation. Moreover, MPI collectives may be used to make communication as efficient as possible. This approach may be applied both when applications write out individual files to local disks and when a parallel file system is used as the backing store. Further, the approach can be generalized to provide redundancy in the presence of more than one failure through the application of well-known algorithms (such as [1] for two erasures (still using only XOR operations) or the more general approaches based on Reed-Solomon codes [9]).

A great deal of work has been done on recovery from erasures in file systems. Much of this work focuses on relatively small numbers of separate disks and on preserving data after each `write` operation, as required by POSIX semantics. Work on large arrays of disks has often focused on the multiple-erasure case (e.g., [4, 1, 9]). These approaches can be adapted to the techniques in this paper. Other work has exploited the semantics of operations to provide improved performance. Ligon [6] demonstrated recovery of lost data in a PVFS file system using serial recovery algorithms to create error recovery data after application completion and to reconstruct data after a single server failure. Pillai and Lauria studied the implementation of redundancy schemes in PVFS, maintaining the existing PVFS semantics and switching between mirroring and parity based on write size [8]. Ladin et al. discussed *lazy replication* in the context of distributed services [5].

In Section 2 we describe the lazy redundancy technique and our implementation. In Section 3 we demonstrate the technique in a cluster with separate local disks. In Section 4 we summarize the contributions of this work and point to future research directions.

2   Implementing Lazy Redundancy

Let there be p "storage devices," where a storage device may be a disk local to a processor or a server in a parallel file system. Here we assume that an erasure failure affects exactly one storage device. Files are striped across all storage devices in a round-robin fashion. We denote the piece of a stripe residing on one storage device as a *data block*, or *block*. Each data block is of some fixed size (e.g., a natural block size for the disk or parallel file system server, such as 64 Kbytes). A file consists of a set of blocks $A_{i,j}$, where storage device j has blocks $i = 0, 1, \ldots$. The top diagram in Figure 1 shows the layout of these blocks on storage devices in a round-robin fashion. We wish to compute data P_j, called the *parity blocks*, stored on device j, that allows the reconstruction of $A_{i,m}$ if any one storage device m is lost.

The particular algorithms for calculating these parity blocks and placing these blocks on storage devices in this implementation of the lazy redundancy technique are based closely on the RAID5 [7] work.

For simplicity, we assume that $0 \le i < p$. If $i \ge p$, this approach may be applied iteratively, starting with a new parity block stored on the first storage device. Let $a \oplus b$ represent a bitwise exclusive OR operation. Define

$$P_j = \left(\bigoplus_{0 \le k < j} A_{j-1,k} \right) \oplus \left(\bigoplus_{j < k < p} A_{j,k} \right) \tag{1}$$

Note that the parity block P_j does not involve any data from device j and that, because of the property of exclusive OR, for any m, all of the data blocks $A_{i,m}$ can be recomputed by using the data $A_{i,k}$ for $k \ne m$ and the parity blocks P_k for $k \ne m$.

These parity blocks are evenly distributed among the storage devices to best balance the I/O load; and, just as in RAID5, the extra space required by the parity blocks is just $1/p$, where there are p disk storage devices. The middle diagram of Figure 1 illustrates the relationship between data blocks and parity blocks, including the placement of P_j so that it is not on a device storing one of the data blocks used in its calculation. The bottom diagram of Figure 1 points out that the parity blocks are not necessarily stored in the same file(s) as the data blocks; they could be stored in a separate file or files or stored in blocks associated with the file.

The lazy redundancy technique differs from RAID5 in three ways: when error correcting data (parity blocks) are computed, how the parity blocks are computed, and how data is reconstructed in the event of an erasure. In the RAID5 approach parity blocks are recomputed on each individual write. This results in a great deal of I/O to parity blocks and at a very fine granularity ($1/p$ the size of the original data). In the lazy redundancy scheme, parity blocks are computed only at explicit synchronization points. This avoids the overhead of computing the parity blocks on small I/O operations and of writing the corresponding even smaller parity blocks, instead aggregating both the calculation and I/O into fewer, larger operations.

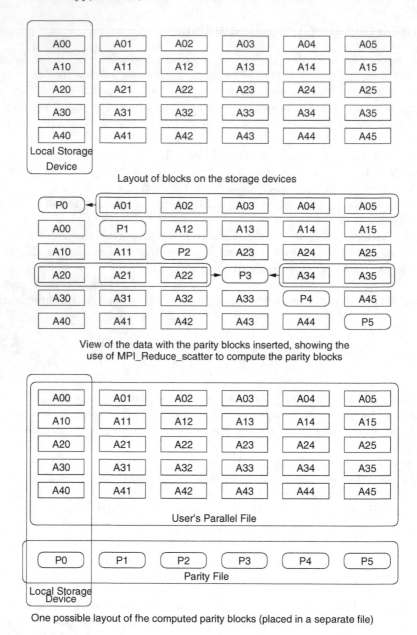

Layout of blocks on the storage devices

View of the data with the parity blocks inserted, showing the
use of MPI_Reduce_scatter to compute the parity blocks

One possible layout of the computed parity blocks (placed in a separate file)

Fig. 1. Real and logical layout of the data blocks $A_{i,j}$ and parity blocks P_j.

The second difference is in the computation of the parity blocks themselves,
in particular the communication cost associated with combining data from each
storage device. We leverage the fact that all processes have reached a synchronization point, so we can calculate parity blocks collectively (in the MPI sense).

```
/* Create the datatype */
b = 65536; /* Bytes per block */
MPI_Comm_rank( comm, &j );
MPI_Comm_size( comm, &p );
blklens[0] = j*b;
blklens[1] = (p-1-j)*b;
displs[0]  = 0;
displs[1]  = (j+1)*b;
MPI_Type_indexed( 2, blklens, displs, MPI_BYTE, &rtype );
MPI_Type_commit( &rtype );

/* Clear the part of the buffer corresponding to our process */
for (i=0; i<b; i++)
   buf[j*b + i] = 0;
MPI_File_open( MPI_COMM_SELF, filename, ..., &fh );
MPI_File_read( fh, buf, 1, rtype, MPI_STATUS_IGNORE );
MPI_File_close( &fh );

for (i=0; i<p; i++)
   recvcounts[i] = b;
MPI_Reduce_scatter( buf, rbuf, recvcounts, MPI_BYTE, MPI_BXOR, comm );
MPI_File_open( parityfile, ..., MPI_COMM_SELF, fh );
MPI_File_write( fh, rbuf, b, MPI_BYTE, MPI_STATUS_IGNORE );
MPI_File_close( &fh );
```

Fig. 2. Pseudocode for computing the parity blocks and distributing them among the processes. This code assumes that each process in the communicator comm has an associated independent disk, rather than using a parallel file system.

We can use algorithms that have been optimized for this sort of calculation, as opposed to performing many independent operations. Because there is no knowledge of a group of processes collectively operating on the file in the general RAID5 model, this type of optimization is impossible.

Equation 1 can be implemented in MPI by using the MPI_Reduce_scatter operation, using the predefined combiner operation MPI_BXOR. The reduce-scatter operation in MPI is functionally equivalent to an MPI_Reduce followed by an MPI_Scatter. However, the combined operation can often be implemented more efficiently than using the two separate operations. The time to create the parity block can be estimated as

$$T_{parity} = T_{read}(b) + T_{write}(b/p) + T_c(b/p) \qquad (2)$$

Here, $T_{read}(n)$ and $T_{write}(n)$ are the times to read and write n bytes respectively. These are single operations involving large blocks that, with some care, can be aligned with the blocks on the storage device itself, leading to optimal I/O transfers. The third term is the cost of the MPI_Reduce_scatter used to compute the parity blocks (this assumes an optimal implementation of the reduce-scatter operation, such as that described in [10], whose cost is proportional to the data size for sufficiently large data). Pseudo-code implementing the computation of the parity blocks is shown in Figure 2.

The use of MPI_Reduce_scatter allows this code to exploit any optimizations in the implementation of this operation that may be included within the MPI library [10].

```
/* Create the Datatype rtype as in the parity construction code */
/* Create a communicator ordered in the same way as the original job */
MPI_Comm_split( incomm, 0, oldrank, &comm );
MPI_Comm_rank( comm, &j );
if (j != failedRank) {
    MPI_File_open( MPI_COMM_SELF, filename, ..., &fh );
    MPI_File_read( fh, buf, 1, rtype, MPI_STATUS_IGNORE );
    MPI_File_close( &fh );
    MPI_File_open( MPI_COMM_SELF, parityfile, ..., &fh );
    MPI_File_read( fh, &buf[j*b], b, MPI_BYTE, MPI_STATUS_IGNORE );
    MPI_File_close( &fh );
}
else {
    for (i=0; i<p*b; i++) buf[i] = 0;
}
MPI_Reduce( buf, outbuf, p*b, MPI_BYTE, MPI_BXOR, failedRank, comm );
if (j == failedRank) {
    MPI_File_open( MPI_COMM_SELF, filename, ..., &fh );
    MPI_File_write( fh, outbuf, p*b, MPI_BYTE, MPI_STATUS_IGNORE );
    MPI_File_close( &fh );
}
```

Fig. 3. Pseudocode to reconstruct a block. MPI I/O was used to keep the MPI flavor and to use a single library for all operations in the examples. The read operation used is nothing more than a Unix-style **readv** with a two element **iovec**.

Finally, a similar approach to the one used in computing parity blocks is also applied to reconstruction. Specifically, to reconstruct the data blocks for a failed device, we exploit the properties of exclusive OR. If the rank of the failed device is f, then the data blocks of the failed storage device can be recovered using by Equation 3.

$$
A_{j,f} = P_j \oplus \left(\bigoplus_{\substack{0 \le k < j \\ k \ne f}} A_{j-1,k} \right) \oplus \left(\bigoplus_{\substack{j < k < p \\ k \ne f}} A_{j,k} \right) \tag{3}
$$

Reconstruction of a block can be implemented with the routine MPI_Reduce; this combines data from all processes to a single process, as shown in Figure 3.

This combination of aggregation of calculations and use of optimized collectives results in very efficient parity calculations.

3   Experiments

In this section, we demonstrate the effectiveness of our approach on a cluster where each node has a separate disk. The parallel application writes data files, one per process, to the local disk. This is a common approach for parallel applications, particularly those running on systems that do not provide an effective parallel file system.

We do not compare this approach to a traditional RAID-5 approach for several reasons. First, we do not have a comparable system available. Second, building such a system would be an endeavor of larger scope than the work presented here. Third, the performance of a RAID-5 system would depend heavily on the pattern of access of the processes, while the lazy redundancy approach does not.

We consider the following scenario. Each process in an MPI program writes data to a local disk (that is, a disk on the same processor node as the MPI process). This collection of files represents the output from the MPI program. If one node fails for any reason (such as a failure in the network card, fan, or power supply, not only the disk), the data from that disk is no longer accessible. With the lazy redundancy approach described in the previous section, it is possible to recover from this situation. If the parity blocks have been computed as described in Section 2, the user can restore the "lost" data by running an MPI program on the remaining $p-1$ nodes and on one new node. In these experiments we assume that failure detection is handled by some other system software component.

Our experiment simulates this scenario by performing the following steps on a Linux cluster [3] where each node has a local disk. Two underlying communication methods implemented in MPICH2 were used: TCP and an implementation of the GASNet [2] protocol on top of the GM based Myrinet interconnect. We would expect the Myrinet results to be superior when communication is necessary, such as in the process of computing parity or restoring from an erasure. Each step is timed so that the relative cost of each step can be computed:

1. All processes generate and write out data to their local disks.
2. All processes read the data needed to compute the parity blocks.
3. The parity blocks are computed and distributed to each participating process.
4. The parity blocks are written to local disk in a separate file.
5. One process is selected as the one with the failed data (whose data and parity data is removed), and the reconstruction algorithm is executed on this storage device, recreating the missing data.

The results of running this experiment on 4, 16, and more storage devices (and the same number of MPI processes) are shown in Table 1. In each example, the total size of the data scales with p (there are $p-1$ blocks of the same size on each storage device). Thus, the best performance is linear scaling of the time with respect to the number of processes. Hardware problems precluded running Myrinet experiments at 128 processes.

In this experiment, each process generates a separate data file with the name `datafile.`*rank*, where *rank* is the process rank according to MPI_Comm_rank. The parity data, once computed and gathered, is written to local disk alongside the data file as a "hidden" file with the name `.LZY.datafile.`*rank*. The size of the parity data remains fixed for all tests, regardless of the increasing number of participating processes and data sizes. To achieve the larger data sizes used, we applied the parity computation iteratively a fixed number of times over the varying size of the data composed of data blocks.

The results in Table 1 show the scalability of the approach. For the tests on TCP, the performance scales with the data size to 64 processes. For Myrinet, the times scale better than linearly; we believe that this is due to the serial overheads;

Table 1. Average time (in seconds) to create a file, create the parity blocks using lazy redundancy, and to restore a file on 4 to 128 storage devices. The step to create the parity block is broken down into the three substeps shown. Parity data size is fixed at 512 Kbytes per process.

	TCP/IP				Myrinet			
	4	16	64	128	4	16	32	64
Data per Client (KB)	1536	7680	32256	65024	1536	7680	16128	32256
Create File	0.040	0.193	0.810	1.632	0.041	0.155	0.319	0.645
Create Parity								
Read Blocks	0.030	0.160	0.656	1.741	0.023	0.123	0.270	0.506
Compute Parity	0.288	1.812	8.270	62.71	0.378	0.463	0.594	1.307
Write Parity	0.013	0.013	0.013	0.013	0.013	0.013	0.013	0.013
Restore Erasure	0.307	1.310	5.490	15.06	0.092	0.425	1.212	1.696

in fact, the time for the compute parity step is dominated by a constant overhead of about 0.36 seconds.

Note that the time to create the file is the time seen by the application; this does not necessarily include the time to write the data to disk (the data may be in a write cache).

One benefit of using this approach is that only a very small amount of additional data is necessary to provide fault tolerance. However, calculating this data does become more expensive as the number of clients increases, as we see in the TCP case at 128 processes. In this particular case we may be seeing the impact of memory exhaustion for large reduce-scatter operations (which may need to be broken up in the implementation). However, applying the redundancy scheme to smaller sets of storage devices would also alleviate the problem, at the cost of somewhat larger storage requirements. For example, we could consider the 128 processes as two groups of 64 for the purposes of redundant data calculation, doubling the amount of parity data but allowing for the reduce-scatters to proceed in parallel.

From the data gathered, it is apparent that a major part of the cost in these examples is reading the file. If either the user-level I/O library (such as an MPI-I/O implementation or a higher-level library) or a parallel file system performed the "Compute Parity" step when sufficient data was available, rather than rereading all the data as in our tests, then this cost would disappear.

4   Conclusions and Future Work

In this work we have presented the lazy redundancy technique for providing erasure correction in an efficient manner on highly parallel systems. The technique exploits collective computation, I/O, and message passing to best leverage the resources available in the system. While tested in an independent file environment, the same approach is equally usable for data stored on a parallel file system. Further, with user-defined reduction operations, this approach can be extended to more complex erasure-correcting codes, such as Reed-Solomon, that would handle larger numbers of failures than the algorithm shown in this work. An additional optimization would leverage accumulate operations during collective writes to avoid the read I/O step of recovery data calculation. This would, of course, impose additional complexity and overhead in the write phase, and that tradeoff warrants further study.

Although we have presented this work in the context of independent disks, the technique is equally suitable to the parallel file system environment. In order to implement lazy redundancy transparently in this environment, some augmentation is necessary to the parallel file system. We intend to implement this technique as a component of the PVFS2 file system [11]. Our approach will be to transparently perform the lazy redundancy operations within the PVFS2-specific portion of the ROMIO MPI-IO implementation, hiding the details of this operation from the user.

Three capabilities will be added to PVFS2 to facilitate this work: a mechanism for allowing clients to obtain the mapping of file data blocks to servers, an interface allowing clients to perform I/O operations when the file system is only partially available (and be notified of what servers have failed), and a distribution scheme that reserves space for parity data as part of the file object, while keeping parity data out of the file byte stream itself.

References

1. Mario Blaum, Jim Brady, Jehoshua Bruck, Jai Menon, and Alexander Vardy. The EVENODD code and its generalization: An efficient scheme for tolerating multiple disk failures in RAID architectures. In Hai Jin, Toni Cortes, and Rajkumar Buyya, editors, *High Performance Mass Storage and Parallel I/O: Technologies and Applications*, chapter 14, pages 187–208. IEEE Computer Society Press and Wiley, New York, NY, 2001.

2. Dan Bonachea. Gasnet specification, v1.1. Technical Report CSD-02-1207, University of California, Berkeley, October 2002.

3. R. Evard, N. Desai, J. P. Navarro, and D. Nurmi. Clusters as large-scale development facilities. In *Proceedings of IEEE International Conference on Cluster Computing (CLUSTER02)*, pages 54–66, 2002.

4. Lisa Hellerstein, Garth A. Gibson, Richard M. Karp, Randy H. Katz, and David A. Patterson. Coding techniques for handling failures in large disk arrays. *Algorithmica*, 12(2/3):182–208, 1994.

5. Rivka Ladin, Barbara Liskov, and Liuba Shrira. Lazy replication: Exploiting the semantics of distributed services. In *IEEE Computer Society Technical Committee on Operating Systems and Application Environments*, volume 4, pages 4–7. IEEE Computer Society, 1990.

6. Walt Ligon. Private communication. 2002.

7. David A. Patterson, Garth Gibson, and Randy H. Katz. A case for redundant arrays of inexpensive disks (RAID). In ACM, editor, *Proceedings of Association for Computing Machinery Special Interest Group on Management of Data: 1988 Annual Conference, Chicago, Illinois, June 1–3*, pages 109–116, New York, NY 10036, USA, 1988. ACM Press.

8. Manoj Pillai and Mario Lauria. A high performance redundancy scheme for cluster file systems. In *Proceedings of the 2003 IEEE International Conference on Cluster Computing*, Kowloon, Hong Kong, December 2003.

9. James S. Plank. A tutorial on reed-solomon coding for fault-tolerance in RAID-like systems. *Software – Practice and Experience*, 27(9):995–1012, September 1997.

10. Rajeev Thakur and William Gropp. Improving the performance of collective operations in MPICH. In Jack Dongarra, Domenico Laforenza, and Salvatore Orlando, editors, *Recent Advances in Parallel Virtual Machine and Message Passing Interface*, number LNCS2840 in Lecture Notes in Computer Science, pages 257–267. Springer Verlag, 2003. 10th European PVM/MPI User's Group Meeting, Venice, Italy.

11. The PVFS2 parallel file system. http://www.pvfs.org/pvfs2/.

The Impact of File Systems
on MPI-IO Scalability*

Rob Latham, Rob Ross, and Rajeev Thakur

Argonne National Laboratory, Argonne, IL 60439, USA
{robl,rross,thakur}@mcs.anl.gov

Abstract. As the number of nodes in cluster systems continues to grow, leveraging scalable algorithms in all aspects of such systems becomes key to maintaining performance. While scalable algorithms have been applied successfully in some areas of parallel I/O, many operations are still performed in an uncoordinated manner. In this work we consider, in three file system scenarios, the possibilities for applying scalable algorithms to the many operations that make up the MPI-IO interface. From this evaluation we extract a set of file system characteristics that aid in developing scalable MPI-IO implementations.

1 Introduction

The MPI-IO interface [10] provides many opportunities for optimizing access to underlying storage. Most of these opportunities arise from the interface's ability to express noncontiguous accesses, the collective nature of many operations, and the precise but somewhat relaxed consistency model. Significant research has used these features to improve the scalability of MPI-IO data operations. Implementations use two-phase [13], data sieving [14], and data shipping [11], among others, to efficiently handle I/O needs when many nodes are involved.

On the other hand, little attention has been paid to the remaining operations, which we will call the *management operations*. MPI-IO semantics provide opportunities for scalable versions of open, close, resize, and other such operations. Unfortunately, the underlying file system API can limit the implementation's ability to exploit these opportunities just as it does in the case of the I/O operations.

We first discuss the opportunities provided by MPI-IO and the potential contributions that the parallel file system can make toward an efficient, scalable MPI-IO implementation. We then focus specifically on the issue of providing scalable management operations in MPI-IO, using the PVFS2 parallel file system as an example of appropriate support. We also examine the scalability of common MPI-IO management operations in practice on a collection of underlying file systems.

---

* This work was supported by the Mathematical, Information, and Computational Sciences Division subprogram of the Office of Advanced Scientific Computing Research, Office of Science, U.S. Department of Energy, under Contract W-31-109-Eng-38.

D. Kranzlmüller et al. (Eds.): EuroPVM/MPI 2004, LNCS 3241, pp. 87–96, 2004.

1.1   MPI-IO Opportunities

Implementations can take advantage of three aspects of the MPI-IO specification to maximize scalability: semantics, noncontiguous I/O, and collective functions.

MPI-IO provides more relaxed consistency semantics than the traditional POSIX [6] interface provides. These semantics are relaxed on two fronts: in terms of the scope of the consistency (just the processes in the communicator) and the points in time at which views from different processes are synchronized. Under the default MPI-IO semantics, simultaneous writes to the same region yield an undefined result. Further, writes from one process are not immediately visible to another. Active buffering with threads [9], for example, takes advantage of MPI-IO consistency semantics to hide latency of write operations.

Additionally, MPI datatypes may be used to describe noncontiguous regions both in file and in memory, providing an important building block for efficient access for scientific applications. Several groups have implemented support for efficient noncontiguous I/O, including listless I/O [15], data sieving [14], list I/O [2], and datatype I/O [4].

MPI-IO also affords many opportunities for scalable implementations through collective operations. These collective functions enable the implementation to use scalable communication routines and to reorganize how operations are presented to the file system. The focus of optimizations of collective MPI-IO routines to this point has been on read and write operations. Optimizations such as two-phase have had a significant impact on the performance of collective I/O, particularly at large scale.

1.2   MPI-IO with POSIX and NFS

POSIX is not the ideal underlying interface for MPI-IO for three reasons. First, the readv, writev, and lio_listio calls are not efficient building blocks for noncontiguous I/O. The readv and writev calls only allow describing noncontiguous regions in memory, while these often occur in the file as well. The lio_listio API does allow for multiple file regions, but the language for describing them through the API is verbose, leading to descriptions larger than the data itself. Data sieving [14] is not so much an optimization as a workaround for these shortcomings; it is often more efficient to read an entire region containing the data of interest, discarding much of that data, than to construct and use a noncontiguous request with the POSIX noncontiguous functions.

The POSIX stateful model is also problematic. The open, read, write, close model of access requires that all processes desiring to access files directly perform many system calls. The file descriptor returned by open has meaning to just one client. Hence, file descriptors cannot be shared among all processors. Each client must make the open system call. In a large parallel program, opening a file on a parallel file system can put a large strain on the servers as thousands of clients simultaneously call open.

NFS provides an interesting contrast to POSIX. Clients access NFS file systems using the same functions as POSIX and must deal with same issues with file descriptors. Although the API is the same, however, the consistency semantics are quite different and impose an additional set of problems. Because clients cache data aggressively and without synchronization among clients, it is difficult to predict when writes from one client will be visible to another. Metadata caching further complicates parallel I/O: when one process modifies the file size or file attributes, it is difficult to know when those modifications will be visible to the other processes. The NFS consistency semantics work well in the serial environment for which they were designed, but they are a poor fit for parallel I/O.

MPI-IO implementations can function on top of a wide variety of file systems, but the underlying file system can greatly help the implementation achieve real scalability, particularly if it addresses the problems outlined above.

1.3   Parallel File System Building Blocks

A parallel file system can provide three fundamentals to aid in a scalable MPI-IO implementation:

- Efficient noncontiguous I/O support
- Consistency semantics closely matching the MPI-IO model
- Client-independent references to files

Efficient noncontiguous I/O support in the file system has been a focus of a great deal of recent research [15]. The datatype I/O concept [2] in particular provides an efficient infrastructure for noncontiguous I/O in MPI-IO, with similar concepts seen in the View I/O [7] work.

One of the most significant influences on performance of a file system, both in data and metadata operations, is the consistency semantics implemented by the file system. For example, the POSIX consistency semantics require essentially sequential consistency of operations. Enforcing these semantics can result in high overhead and reduced parallelism. The NFS consistency semantics, on the other hand, require no additional overhead because there is no guarantee of consistency between clients. Thus, the consistency semantics drive caching policies, dictating how clients can cache data and when they must synchronize.

The "nonconflicting write" semantics of PVFS and PVFS2 are an ideal building block from the MPI-IO perspective. A write operation is nonconflicting with another write operation if no part of the two operations overlap (interleaved requests can still be nonconflicting). If two processes perform nonconflicting write operations, then all other processes will see the data from the writers after their respective writes have completed. If two clients write data to the same region of a file (i.e., a conflicting write), the result is undefined. The file system counts on the MPI-IO implementation handling any additional consistency requirements.

While relaxed data consistency can improve scalability, *metadata* consistency semantics have an impact on scalable optimizations as well. Some mechanism has to ensure that all clients have a consistent view of metadata from the MPI-IO perspective. For file systems such as NFS, all clients end up performing the same

Table 1. MPI-IO Management Operations (all clients call).

Function	Collective?	NFS	POSIX	PVFS2
MPI_File_get_size	no	$O(n)$	$O(n)$	$O(n)$
MPI_File_seek	no	$O(n)$	$O(n)$	—
MPI_File_delete	no	$O(1)$	$O(1)$	$O(1)$
MPI_File_open	yes	$O(n)$	$O(n)$	$O(1)$
MPI_File_close	yes	$O(n)$	$O(n)$	$O(1)$
MPI_File_sync	yes	$O(n)$	$O(n)$	$O(1)$
MPI_File_set_size	yes	$O(n)$	$O(1)$	$O(1)$
MPI_File_preallocate	yes	$O(1)$	$O(1)$	$O(1)$
MPI_File_set_info	yes	$O(n)$	$O(n)$	$O(1)$
MPI_File_set_view	yes	$O(n)$	$O(n)$	$O(1)$
MPI_File_get_position_shared	no	—	—	—
MPI_File_seek_shared	yes	—	—	—

operations because the MPI-IO implementation cannot control caching, limiting our ability to implement scalable operations. Thus, it is important to have not just relaxed consistency semantics but *appropriate* consistency semantics, and the right hooks to control the data and metadata caches.

A third mechanism that parallel file systems can use to achieve high performance is client-independent references to files. As opposed to the POSIX file descriptor, which has meaning only to one client, these references can be used by any client to refer to a file. By sharing these file references among all clients, programs place fewer demands on the file system. As we will see, this feature can significantly improve the performance of MPI-IO management operations.

2   MPI-IO Management Operations

We can roughly split the MPI-IO operations into two groups: operations that read or write data, and operations that do not. We call this second group MPI-IO management operations. Table 1 lists the management operations that interact with the file system (calls such as MPI_File_get_position generally require no corresponding file system operations, using cached data instead). The point of the table is to help shed light on the options for creating scalable implementations of the MPI-IO functions.

Some functions, such as MPI_File_get_size, are not collective and thus cannot be optimized – every process wanting the file size would need to make the call, or the application programmer would need to synchronize, make a single MPI-IO call, and then broadcast the result. Little can be done in this case. Functions such as MPI_File_delete fall into the same category, except that it is generally assumed that the application programmer will perform synchronization and make only one call, since calling this function many times would likely result in success on one process and failure on all others.

Next is a core set of collective management functions that are often used in MPI-IO applications, including MPI_File_open and MPI_File_close among others. The stateful nature of the POSIX and NFS APIs requires that open and close file system operations be performed on all processes. Likewise, the fsync operation that is the interface for synchronizing data in the POSIX and NFS APIs flushes changes only on the local node, requiring a file system operation per node (generally implemented by calling fsync on each process for simplicity). The NFS metadata caching makes relying on NFS file sizes problematic, so ROMIO chooses to call ftruncate on all processes. This situation could perhaps be avoided by maintaining file size data within ROMIO rather than relying on the file system, but implementing such a feature would touch many other calls.

For example, the file truncation function MPI_File_set_size is a collective operation that may be made scalable even under the POSIX API. A single POSIX ftruncate operation may be performed and the result broadcast to remaining processes (see Figure 1). We can use MPI-IO semantics here to further improve scalability: MPI_File_set_size is treated like a write operation, so the caller must synchronize client calls if that is desired. We note, however, that this cannot be done in the NFS environment, where metadata is not kept consistent between nodes; in that case we must perform the truncate on all nodes.

```
if (rank == 0) {
        /* perform the truncate on one node */
        ret = ftruncate(fd, size);
        MPI_Bcast(&ret, 1, MPI_INT, 0, comm);
} else {
        /* the result is broadcast to the other processors */
        MPI_Bcast(&ret, 1, MPI_INT, 0, comm);
}
/* at this point, all processors know the status of the ftruncate
 * call, even though only one processor actually sent the request */
```

Fig. 1. Scalable MPI_File_set_size (pseudocode).

The MPI_File_sync function offers a slightly different example. It, too, is a collective operation but has the added property that no outstanding write operations should be in progress on any process. Figure 2 demonstrates one possible scalable implementation. By using MPI_Reduce (or MPI_Gather), we can ensure that all processes have performed their write operations before one process initiates the flush. This scheme assumes there will be no client-side caching of data. In the POSIX and NFS environments, where local caching occurs and the fsync flushes only local buffers, calls must be performed by all clients to ensure that changes on all nodes make it out to disk.

The MPI_File_set_info and related calls are interesting because they may or may not require file system operations, depending on the hints passed in and supported by the file system. For example, setting the MPIO_DIRECT_READ option on file systems that support it (e.g., XFS) would require a file system call from each process.

```
MPI_Reduce(&dummy1, &dummy1, 1, MPI_INT, MPI_SUM, 0, comm);

if (rank == 0) {
        ret = fsync(fd);
        MPI_Bcast(&ret, 1, MPI_INT, 0, comm);
} else {
        MPI_Bcast(&ret, 1, MPI_INT, 0, comm);
}
/* at this point, all processors know the status of the fsync call,
 * even though only one processor actually sent the request */
```

Fig. 2. Scalable `MPI_File_sync` (pseudocode). We do not want to sync until we know all other processors have finished writing. The call to `MPI_Reduce` ensures that all processes have completed any outstanding write operations. In this example, rank 0 will not call fsync until all other processors have sent rank 0 an integer.

Moreover, because no commonly used file systems support shared file pointers, the implementation in ROMIO uses a shared file approach to store the pointer. There may be more scalable options for implementing this support; this is an open research area.

3   File System Support: PVFS2

The new PVFS2 parallel file system [12] is a good example of providing efficient building blocks for scalable MPI-IO. It is no accident that PVFS2 is well suited for MPI-IO: it was expressly designed with such a goal in mind.

We took advantage of several PVFS2 features to optimize our MPI-IO implementation. Naturally, these features resemble the points laid out in Section 1.3:

- support for arbitrary noncontiguous I/O patterns
- consistency semantics well-suited for MPI-IO
- client-independent handles
- no client-side cache

Support for noncontiguous access in PVFS2, similar to the datatype I/O prototype in PVFS1 ([4],[3]), provides the necessary API for efficient independent I/O. Nonconflicting write consistency semantics, which leave the results of byte-overlapped concurrent writes undefined, provide sufficient consistency for building the nonatomic MPI-IO semantics.

Opaque, client-independent file references allow open operations to be performed scalably, and the stateless nature of the file system means that close operations are also trivial (a single synchronize operation is performed to flush all changes to disk).

For example, only one process in a parallel program has to perform actual PVFS2 function calls to create files (see Figure 3). One process performs a lookup, creating the file if it does not exist. The PVFS2 server responds with a reference to the file system object. The client then broadcasts the result to the other clients. `MPI_File_set_size` is another win: one client resizes the file and then broadcasts the result to the other clients. Servers experience less load

```
/* PVFS2 is stateless: clients perform a 'lookup' operation
 * to convert a path into a handle.  This handle can then
 * be passed around to all clients.  */

if (rank == 0) {
        ret = PVFS_sys_lookup(fs_id, path_name,
                credentials, &response, PVFS2_LOOKUP_LINK_FOLLOW);
        if (ret == ENOENT) {
                ret = PVFS_sys_create(name, parent, attribs,
                        credentials, NULL, &response);
        }
}
MPI_Bcast(&ret, 1, MPI_INT, 0, comm);
MPI_Bcast(&response, 1, MPI_INT, 0, comm);

/* now all processors know if the lookup succeeded, and
 * if it did, the handle for the entity   */
```

Fig. 3. Scalable open for PVFS2 (heavily simplified). In a real implementation, one could create an MPI datatype to describe the handle and error code and perform just one `MPI_Bcast`. We perform two for simplicity.

because only one request comes in. The same approach is used for `MPI_File_sync` (Figure 2). Only one process has to ask the file system to flush data. The result is the same: all I/O servers write out their caches, but they have to handle only one request each to do so.

4   Results

To evaluate MPI-IO implementations, we performed experiments on the Jazz cluster at Argonne National Laboratory [8], the ALC cluster at Lawrence Livermore National Laboratory [1], and the DataStar cluster at NPACI/SDSC [5]. Jazz users have access to two clusterwide file systems: NFS-exported GFS volumes and PVFS (version 1). Additionally, we temporarily deployed PVFS2 across a subset of available compute nodes for testing purposes. The ALC cluster has a Lustre file system and the DataStar cluster has GPFS: we have included them for reference, even though we do not discuss their design elsewhere in this paper. In these tests, PVFS2 ran with 8 I/O servers, one of which also acted as a metadata server. For fairness in these tests, the PVFS2 servers used TCP over Fast Ethernet. We used a CVS version of MPICH2 from mid-April 2004, including the ROMIO MPI-IO implementation (also from mid-April).

In the first experiment, we created 1,000 files in an empty directory with `MPI_File_open` and computed the average time per create. Table 2 summarizes the results. PVFS2 – the only stateless file system in the table – achieves consistent open times as the number of clients increased. Additionally, the average time to create a file on PVFS2 is an order of magnitude faster than the time it takes to do so on any of the other file systems. From a scalability standpoint, the NFS+GFS file system performs remarkably well. File creation may be relatively expensive, but as the number of clients increases, the cost to open a file remains virtually constant. PVFS1 demonstrates poor scalability with the number of clients, because PVFS1 each client must open the file to get a file descriptor (PVFS1 is stateful in this regard). With more clients, the metadata server has to handle increasingly large numbers of requests. Thus, the metadata server becomes a serialization point for these clients, and the average time per request goes up. While Lustre and GPFS both outperform PVFS1, they too must perform an open on each client. The time per operation increases as the number of clients and the demand placed on the file system increases significantly.

Table 2. Results: A Comparison of several cluster file systems(milliseconds).

No. of Clients	NFS+GFS	Lustre	GPFS	PVFS1	S-PVFS1	PVFS2
			Create			
1	3.368	8.585	16.38	41.78	-	18.82
4	178.1	51.62	29.77	221.8	-	18.73
8	191.6	56.68	45.80	292.4	-	22.02
16	176.6	67.03	280.2	241.1	-	20.66
25	183.0	146.8	312.0	2157	-	19.05
50	204.1	141.5	400.6	2447	-	24.73
75	212.7	231.2	475.3	3612	-	24.82
100	206.8	322.2	563.9	1560	-	28.14
128	204.0	463.3	665.8	1585	-	32.94
			Resize			
1	0.252	1.70	7.0	1.26	1.37	0.818
4	3.59	05.39	14.4	2.23	1.54	0.823
8	1.75	13.49	36.0	3.25	1.44	0.946
16	14.88	29.7	36.6	2.75	1.86	0.944
25	36.5	66.0	35.7	25.0	4.07	0.953
50	1960	113	39.5	16.2	2.02	1.11
75	2310	179	43.5	15.0	2.62	1.26
100	4710	233	40.5	19.1	3.10	1.46
128	2820	254	42.4	18.6	3.38	1.07

In the second experiment, we opened a file and then timed how long it took to perform 100 calls to `MPI_File_set_size` on one file with a random `size` parameter (ranging between 0 and `RAND_MAX`. We then computed the average time for one resize operation. Table 2 summarizes our results. The file-based locking of GFS clearly hurts resize performance. The GFS lock manager becomes the serialization point for the resize requests from the clients, and performance degrades drastically as the number of clients increases. The NFS client-side caches mean we must resize the file on each client to ensure consistency, so we cannot use the scalable techniques outlined earlier. Again, as with the create test, we see Lustre's performance getting worse as the number of clients increases. GPFS performance appears virtually independent of the number of clients; it would be interesting to know what approach GPFS takes to achieving scalable operations for this case. The PVFS column shows performance without the scalable algorithm from Figure 1. We modified the ROMIO PVFS1 resize routine to use the more scalable approach(the S-PVFS1 column) . Using the algorithm shown in Figure 1, both PVFS1 and PVFS2 both show consistent performance. The small increase in time as the number of clients increases can be attributed to the increased synchronization time at the reduce and the time taken to broadcast the results to a larger number of processors.

5   Conclusions and Future Directions

Many opportunities exist for optimizing MPI-IO operations, even in general purpose file systems. Some of these opportunities have been heavily leveraged, in particular those for collective I/O. Others require additional support from the file system. In this work we have described the collection of MPI-IO operations and categorized these based on the ability of the MPI-IO implementor to optimize them given specific underlying interfaces. We have pointed out some characteristics of file system APIs and semantics that more effectively serve as the basis for MPI-IO implementations: support for noncontiguous I/O, better consistency semantics, client-independent file references, a stateless I/O model, and a caching model that allows a single file system operation to sync to storage (in the case of PVFS2, no client-side caching at all). By building parallel file systems with these characteristics in mind, MPI-IO implementations can leverage MPI collective communication and achieve good performance from the parallel file system even as the number of clients increases.

In future work, we will examine scalable support for the shared file pointer and atomic access modes of the MPI-IO interface. While inherently less scalable than private file pointers and the more relaxed default semantics, these are important components in need of optimization. It is not clear whether file system support for shared file pointers and the more strict atomic data mode is warranted or if this support should be provided at the MPI-IO layer. Additionally, we continue to examine options for more aggressively exploiting the I/O semantics through client-side caching at the MPI-IO layer.

Acknowledgments

We gratefully acknowledge use of "Jazz," a 350-node computing cluster operated by the Mathematics and Computer Science Division at Argonne National Laboratory as part of its Laboratory Computing Resource Center. Jianwei Li at Northwestern University and Tyce McLarty at Lawrence Livermore National Laboratory contributed benchmark data. We thank them for their efforts.

References

1. ALC, the ASCI Linux Cluster. http://www.llnl.gov/linux/alc/.
2. Avery Ching, Alok Choudhary, Kenin Coloma, Wei keng Liao, Robert Ross, and William Gropp. Noncontiguous I/O accesses through MPI-IO. In *Proceedings of the Third IEEE/ACM International Symposium on Cluster Computing and the Grid*, pages 104–111, Tokyo, Japan, May 2003. IEEE Computer Society Press.

3. Avery Ching, Alok Choudhary, Wei keng Liao, Robert Ross, and William Gropp. Noncontiguous I/O through PVFS. In *Proceedings of the 2002 IEEE International Conference on Cluster Computing*, September 2002.
4. A. Ching, A. Choudhary, W. Liao, R. Ross, and W. Gropp. Efficient structured data access in parallel file systems. In *Proceedings of Cluster 2003*, Hong Kong, November 2003.
5. IBM DataStar Cluster. http://www.npaci.edu/DataStar/.
6. IEEE/ANSI Std. 1003.1. Portable operating system interface (POSIX)–part 1: System application program interface (API) [C language], 1996 edition.
7. Florin Isaila and Walter F. Tichy. View I/O: Improving the performance of non-contiguous I/O. In *Proceedings of IEEE Cluster Computing Conference, Hong Kong*, December 2003.
8. LCRC, the Argonne National Laboratory Computing Project. http://www.lcrc.anl.gov.
9. Xiasong Ma, Marianne Winslett, Jonghyun Lee, and Shengke Yu. Improving MPI IO output performance with active buffering plus threads. In *Proceedings of the International Parallel and Distributed Processing Symposium*. IEEE Computer Society Press, April 2003.
10. MPI-2: Extensions to the message-passing interface. The MPI Forum, July 1997.
11. Jean-Pierre Prost, Richard Treumann, Richard Hedges, Bin Jia, and Alice Koniges. MPI-IO GPFS, an optimized implementation of MPI-IO on top of GPFS. In *Proceedings of Supercomputing 2001*, November 2001.
12. The Parallel Virtual File System, version 2. http://www.pvfs.org/pvfs2.
13. Rajeev Thakur and Alok Choudhary. An Extended Two-Phase Method for Accessing Sections of Out-of-Core Arrays. *Scientific Programming*, 5(4):301–317, Winter 1996.
14. Rajeev Thakur, William Gropp, and Ewing Lusk. A case for using MPI's derived datatypes to improve I/O performance. In *Proceedings of SC98: High Performance Networking and Computing*. ACM Press, November 1998.
15. Joachim Worringen, Jesper Larson Traff, and Hubert Ritzdorf. Fast parallel non-contiguous file access. In *Proceedings of SC2003: High Performance Networking and Computing*, Phoenix, AZ, November 2003. IEEE Computer Society Press.

Open MPI: Goals, Concept, and Design of a Next Generation MPI Implementation

Edgar Gabriel[1], Graham E. Fagg[1], George Bosilca[1],
Thara Angskun[1], Jack J. Dongarra[1], Jeffrey M. Squyres[2],
Vishal Sahay[2], Prabhanjan Kambadur[2], Brian Barrett[2],
Andrew Lumsdaine[2], Ralph H. Castain[3], David J. Daniel[3],
Richard L. Graham[3], and Timothy S. Woodall[3]

[1] Innovative Computing Laboratory, University of Tennessee
{egabriel,fagg,bosilca,anskun,dongarra}@cs.utk.edu
[2] Open System Laboratory, Indiana University
{jsquyres,vsahay,pkambadu,brbarret,lums}@osl.iu.edu
[3] Advanced Computing Laboratory, Los Alamos National Lab
{rhc,ddd,rlgraham,twoodall}@lanl.gov

Abstract. A large number of MPI implementations are currently available, each of which emphasize different aspects of high-performance computing or are intended to solve a specific research problem. The result is a myriad of incompatible MPI implementations, all of which require separate installation, and the combination of which present significant logistical challenges for end users. Building upon prior research, and influenced by experience gained from the code bases of the LAM/MPI, LA-MPI, and FT-MPI projects, *Open MPI* is an all-new, production-quality MPI-2 implementation that is fundamentally centered around component concepts. Open MPI provides a unique combination of novel features previously unavailable in an open-source, production-quality implementation of MPI. Its component architecture provides both a stable platform for third-party research as well as enabling the run-time composition of independent software add-ons. This paper presents a high-level overview the goals, design, and implementation of Open MPI.

1 Introduction

The evolution of parallel computer architectures has recently created new trends and challenges for both parallel application developers and end users. Systems comprised of tens of thousands of processors are available today; hundred-thousand processor systems are expected within the next few years. Monolithic high-performance computers are steadily being replaced by clusters of PCs and workstations because of their more attractive price/performance ratio. However, such clusters provide a less integrated environment and therefore have different (and often inferior) I/O behavior than the previous architectures. Grid and meta-computing efforts yield a further increase in the number of processors available to parallel applications, as well as an increase in the physical distances between computational elements.

D. Kranzlmüller et al. (Eds.): EuroPVM/MPI 2004, LNCS 3241, pp. 97–104, 2004.

These trends lead to new challenges for MPI implementations. An MPI application utilizing thousands of processors faces many scalability issues that can dramatically impact the overall performance of any parallel application. Such issues include (but are not limited to): process control, resource exhaustion, latency awareness and management, fault tolerance, and optimized collective operations for common communication patterns.

Network layer transmission errors—which have been considered highly improbable for moderate-sized clusters—cannot be ignored when dealing with large-scale computations [4]. Additionally, the probability that a parallel application will encounter a process failure during its run increases with the number of processors that it uses. If the application is to survive a process failure without having to restart from the beginning, it either must regularly write checkpoint files (and restart the application from the last consistent checkpoint [1, 8]) or the application itself must be able to adaptively handle process failures during runtime [3]. All of these issues are current, relevant research topics. Indeed, some have been addressed at various levels by different projects. However, no MPI implementation is currently capable of addressing all of them comprehensively.

This directly implies that a new MPI implementation is necessary: one that is capable of providing a framework to address important issues in emerging networks and architectures. Building upon prior research, and influenced by experience gained from the code bases of the LAM/MPI [9], LA-MPI [4], and FT-MPI [3] projects, *Open MPI* is an all-new, production-quality MPI-2 implementation. Open MPI provides a unique combination of novel features previously unavailable in an open-source, production-quality implementation of MPI. Its component architecture provides both a stable platform for cutting-edge third-party research as well as enabling the run-time composition of independent software add-ons.

1.1   Goals of Open MPI

While all participating institutions have significant experience in implementing MPI, Open MPI represents more than a simple merger of LAM/MPI, LA-MPI and FT-MPI. Although influenced by previous code bases, Open MPI is an all-new implementation of the Message Passing Interface. Focusing on production-quality performance, the software implements the full MPI-1.2 [6] and MPI-2 [7] specifications and fully supports concurrent, multi-threaded applications (i.e., MPI_THREAD_MULTIPLE).

To efficiently support a wide range of parallel machines, high performance "drivers" for all established interconnects are currently being developed. These include TCP/IP, shared memory, Myrinet, Quadrics, and Infiniband. Support for more devices will likely be added based on user, market, and research requirements. For network transmission errors, Open MPI provides optional features for checking data integrity. By utilizing message fragmentation and striping over multiple (potentially heterogeneous) network devices, Open MPI is capable of both maximizing the achievable bandwidth to applications and providing the ability to dynamically handle the loss of network devices when nodes

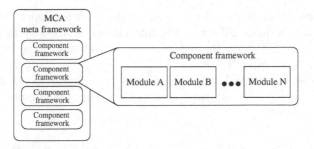

Fig. 1. Three main functional areas of Open MPI: the MCA, its component frameworks, and the modules in each framework.

are equipped with multiple network interfaces. Thus, the handling of network failovers is completely transparent to the application.

The runtime environment of Open MPI will provide basic services to start and manage parallel applications in interactive and non-interactive environments. Where possible, existing run-time environments will be leveraged to provide the necessary services; a portable run-time environment based on user-level daemons will be used where such services are not already available.

2 The Architecture of Open MPI

The Open MPI design is centered around the MPI Component Architecture (MCA). While component programming is widely used in industry, it is only recently gaining acceptance in the high performance computing community [2, 9]. As shown in Fig. 1, Open MPI is comprised of three main functional areas:

- MCA: The backbone component architecture that provides management services for all other layers;
- Component frameworks: Each major functional area in Open MPI has a corresponding back-end component framework, which manages modules;
- Modules: Self-contained software units that export well-defined interfaces that can be deployed and composed with other modules at run-time.

The MCA manages the component frameworks and provides services to them, such as the ability to accept run-time parameters from higher-level abstractions (e.g., mpirun) and pass them down through the component framework to individual modules. The MCA also finds components at build-time and invokes their corresponding hooks for configuration, building, and installation.

Each component framework is dedicated to a single task, such as providing parallel job control or performing MPI collective operations. Upon demand, a framework will discover, load, use, and unload modules. Each framework has different policies and usage scenarios; some will only use one module at a time while others will use all available modules simultaneously.

Modules are self-contained software units that can configure, build, and install themselves. Modules adhere to the interface prescribed by the component framework that they belong to, and provide requested services to higher-level tiers and other parts of MPI.

The following is a partial list of component frameworks in Open MPI (MPI functionality is described; run-time environment support components are not covered in this paper):

- Point-to-point Transport Layer (PTL): a PTL module corresponds to a particular network protocol and device. Mainly responsible for the "wire protocols" of moving bytes between MPI processes, PTL modules have no knowledge of MPI semantics. Multiple PTL modules can be used in a single process, allowing the use of multiple (potentially heterogeneous) networks. PTL modules supporting TCP/IP, shared memory, Quadrics elan4, Infiniband and Myrinet will be available in the first Open MPI release.
- Point-to-point Management Layer (PML): the primary function of the PML is to provide message fragmentation, scheduling, and re-assembly service between the MPI layer and all available PTL modules. More details to the PML and the PTL modules can be found at [11].
- Collective Communication (COLL): the back-end of MPI collective operations, supporting both intra- and intercommunicator functionality. Two collective modules are planned at the current stage: a basic module implementing linear and logarithmic algorithms and a module using hierarchical algorithms similar to the ones used in the MagPIe project [5].
- Process Topology (TOPO): Cartesian and graph mapping functionality for intracommunicators. Cluster-based and Grid-based computing may benefit from topology-aware communicators, allowing the MPI to optimize communications based on locality.
- Reduction Operations: the back-end functions for MPI's intrinsic reduction operations (e.g., MPI_SUM). Modules can exploit specialized instruction sets for optimized performance on target platforms.
- Parallel I/O: I/O modules implement parallel file and device access. Many MPI implementations use ROMIO [10], but other packages may be adapted for native use (e.g., cluster- and parallel-based filesystems).

The wide variety of framework types allows third party developers to use Open MPI as a research platform, a deployment vehicle for commercial products, or even a comparison mechanism for different algorithms and techniques.

The component architecture in Open MPI offers several advantages for end-users and library developers. First, it enables the usage of multiple components within a single MPI process. For example, a process can use several network device drivers (PTL modules) simultaneously. Second, it provides a convenient possibility to use third party software, supporting both source code and binary distributions. Third, it provides a fine-grained, run-time, user-controlled component selection mechanism.

2.1 Module Lifecycle

Although every framework is different, the COLL framework provides an illustrative example of the usage and lifecycle of a module in an MPI process:

1. During MPI_INIT, the COLL framework finds all available modules. Modules may have been statically linked into the MPI library or be shared library modules located in well-known locations.
2. All COLL modules are queried to see if they want to run in the process. Modules may choose not to run; for example, an Infiniband-based module may choose not to run if there are no Infiniband NICs available. A list is made of all modules who choose to run – the list of "available" modules.
3. As each communicator is created (including MPI_COMM_WORLD and MPI_COMM_SELF), each available module is queried to see if wants to be used on the new communicator. Modules may decline to be used; e.g., a shared memory module will only allow itself to be used if all processes in the communicator are on the same physical node. The highest priority module that accepted is selected to be used for that communicator.
4. Once a module has been selected, it is initialized. The module typically allocates any resources and potentially pre-computes information that will be used when collective operations are invoked.
5. When an MPI collective function is invoked on that communicator, the module's corresponding back-end function is invoked to perform the operation.
6. The final phase in the COLL module's lifecycle occurs when that communicator is destroyed. This typically entails freeing resources and any pre-computed information associated with the communicator being destroyed.

3 Implementation Details

Two aspects of Open MPI's design are discussed: its object-oriented approach and the mechanisms for module management.

3.1 Object Oriented Approach

Open MPI is implemented using a simple C-language object-oriented system with single inheritance and reference counting-based memory management using a retain/release model. An "object" consists of a structure and a singly-instantiated "class" descriptor. The first element of the structure must be a pointer to the parent class's structure.

Macros are used to effect C++-like semantics (e.g., new, construct, destruct, delete). The experience with various software projects based on C++ and the according compilation problems on some platforms has encouraged us to take this approach instead of using C++ directly.

Upon construction, an object's reference count is set to one. When the object is retained, its reference count is incremented; when it is released, its reference count is decreased. When the reference count reaches zero, the class's destructor (and its parents' destructor) is run and the memory is freed.

3.2 Module Discovery and Management

Open MPI offers three different mechanisms for adding a module to the MPI library (and therefore to user applications):

- During the configuration of Open MPI, a script traverses the build tree and generates a list of modules found. These modules will be configured, compiled, and linked statically into the MPI library.
- Similarly, modules discovered during configuration can also be compiled as shared libraries that are installed and then re-discovered at run-time.
- Third party library developers who do not want to provide the source code of their modules can configure and compile their modules independently of Open MPI and distribute the resulting shared library in binary form. Users can install this module into the appropriate directory where Open MPI can discover it at run-time.

At run-time, Open MPI first "discovers" all modules that were statically linked into the MPI library. It then searches several directories (e.g., $HOME/ompi/, ${INSTALLDIR}/lib/ompi/, etc.) to find available modules, and sorts them by framework type. To simplify run-time discovery, shared library modules have a specific file naming scheme indicating both their MCA component framework type and their module name.

Modules are identified by their name and version number. This enables the MCA to manage different versions of the same component, ensuring that the modules used in one MPI process are the same—both in name and version number–as the modules used in a peer MPI process. Given this flexibility, Open MPI provides multiple mechanisms both to choose a given module and to pass run-time parameters to modules: command line arguments to mpirun, environment variables, text files, and MPI attributes (e.g., on communicators).

4   Performance Results

A performance comparison of Open MPI's point-to-point methodology to other, public MPI libraries can be found in [11]. As a sample of Open MPI's performance in this paper, a snapshot of the development code was used to run the Pallas benchmarks (v2.2.1) for MPI_Bcast and MPI_Alltoall. The algorithms used for these functions in Open MPI's basic COLL module were derived from their corresponding implementations in LAM/MPI v6.5.9, a monolithic MPI implementation (i.e., not based on components). The collective operations are based on standard linear/logarithmic algorithms using MPI's point-to-point message passing for data movement. Although Open MPI's code is not yet complete, measuring its performance against the same algorithms in monolithic architecture provides a basic comparison to ensure that the design and implementation are sound.

The performance measurements were executed on a cluster of 2.4 GHz dual processor Intel Xeon machines connected via fast Ethernet. The results shown in Fig. 2 indicate that the performance of the collective operations using the Open

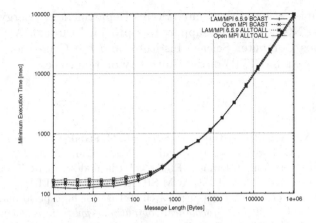

Fig. 2. Performance comparison for MPI_BCAST and MPI_ALLTOALL operations in Open MPI and in LAM/MPI v6.5.9.

MPI approach is identical for large message sizes to its LAM/MPI counterpart. For short messages, there is currently a slight overhead for Open MPI compared to LAM/MPI. This is due to point-to-point latency optimizations in LAM/MPI not yet included in Open MPI; these optimizations will be included in the release of Open MPI. The graph shows, however, that the design and overall approach is sound, and simply needs optimization.

5 Summary

Open MPI is a new implementation of the MPI standard. It provides functionality that has not previously been available in any single, production-quality MPI implementation, including support for all of MPI-2, multiple concurrent user threads, and multiple options for handling process and network failures. The Open MPI group is furthermore working on establishing a proper legal framework, which enbales third party developers to contribute source code to the project.

The first full release of Open MPI is planned for the 2004 Supercomputing Conference. An initial beta release supporting most of the described functionality and an initial subset of network device drivers (tcp, shmem, and a loopback device) is planned for release mid-2004. http://www.open-mpi.org/

Acknowledgments

This work was supported by a grant from the Lilly Endowment, National Science Foundation grants 0116050, EIA-0202048, EIA-9972889, and ANI-0330620, and Department of Energy Contract DE-FG02-02ER25536. Los Alamos National Laboratory is operated by the University of California for the National Nuclear

Security Administration of the United States Department of Energy under contract W-7405-ENG-36. Project support was provided through ASCI/PSE and the Los Alamos Computer Science Institute, and the Center for Information Technology Research (CITR) of the University of Tennessee.

References

1. G. Bosilca, A. Bouteiller, F. Cappello, S. Djilali, G. Fedak, C. Germain, T. Herault, P. Lemarinier, O. Lodygensky, F. Magniette, V. Neri, and A. Selikhov. MPICH-V: Toward a scalable fault tolerant MPI for volatile nodes. In *SC'2002 Conference CD*, Baltimore, MD, 2002. IEEE/ACM SIGARCH. pap298,LRI.
2. D. E. Bernholdt et. all. A component architecture for high-performance scientific computing. *Intl. J. High-Performance Computing Applications*, 2004.
3. G. E. Fagg, E. Gabriel, Z. Chen, T. Angskun, G. Bosilca, A. Bukovski, and J. J. Dongarra. Fault tolerant communication library and applications for high perofrmance. In *Los Alamos Computer Science Institute Symposium*, Santa Fee, NM, October 27-29 2003.
4. R. L. Graham, S.-E. Choi, D. J. Daniel, N. N. Desai, R. G. Minnich, C. E. Rasmussen, L. D. Risinger, and M. W. Sukalksi. A network-failure-tolerant message-passing system for terascale clusters. *International Journal of Parallel Programming*, 31(4):285–303, August 2003.
5. T. Kielmann, R. F. H. Hofman, H. E. Bal, A. Plaat, and R. A. F. Bhoedjang. MagPIe: MPI's collective communication operations for clustered wide area systems. *ACM SIGPLAN Symposium on Principles and Practice of Parallel Programming (PPoPP'99)*, 34(8):131–140, May 1999.
6. Message Passing Interface Forum. *MPI: A Message Passing Interface Standard*, June 1995. http://www.mpi-forum.org.
7. Message Passing Interface Forum. *MPI-2: Extensions to the Message Passing Interface*, July 1997. http://www.mpi-forum.org.
8. Sriram Sankaran, Jeffrey M. Squyres, Brian Barrett, Andrew Lumsdaine, Jason Duell, Paul Hargrove, and Eric Roman. The LAM/MPI checkpoint/restart framework: System-initiated checkpointing. *International Journal of High Performance Computing Applications*, To appear, 2004.
9. Jeffrey M. Squyres and Andrew Lumsdaine. A Component Architecture for LAM/MPI. In *Proceedings, 10th European PVM/MPI Users' Group Meeting*, number 2840 in Lecture Notes in Computer Science, Venice, Italy, Sept. 2003. Springer.
10. Rajeev Thakur, William Gropp, and Ewing Lusk. Data sieving and collective I/O in ROMIO. In *Proceedings of the 7th Symposium on the Frontiers of Massively Parallel Computation*, pages 182–189. IEEE Computer Society Press, Feb 1999.
11. T.S. Woodall, R.L. Graham, R.H. Castain, D.J. Daniel, M.W. Sukalski, G.E. Fagg, E. Gabriel, G. Bosilca, T. Angskun, J.J. Dongarra, J.M. Squyres, V. Sahay, P. Kambadur, B. Barrett, and A. Lumsdaine. TEG: A high-performance, scalable, multi-network point-to-point communications methodology. In *Proceedings, 11th European PVM/MPI Users' Group Meeting*, Budapest, Hungary, September 2004.

Open MPI's TEG Point-to-Point Communications Methodology: Comparison to Existing Implementations

T.S. Woodall[1], R.L. Graham[1], R.H. Castain[1], D.J. Daniel[1], M.W. Sukalski[2],
G.E. Fagg[3], E. Gabriel[3], G. Bosilca[3], T. Angskun[3], J.J. Dongarra[3], J.M. Squyres[4],
V. Sahay[4], P. Kambadur[4], B. Barrett[4], and A. Lumsdaine[4]

[1] Los Alamos National Lab
{twoodall,rlgraham,rhc,ddd}@lanl.gov
[2] Sandia National Laboratories
mwsukal@ca.sandia.gov
[3] University of Tennessee
{fagg,egabriel,bosilca,anskun,dongarra}@cs.utk.edu
[4] Indiana University
{jsquyres,vsahay,pkambadu,brbarret,lums}@osl.iu.edu

Abstract. TEG is a new methodology for point-to-point messaging developed as a part of the Open MPI project. Initial performance measurements are presented, showing comparable ping-pong latencies in a single NIC configuration, but with bandwidths up to 30% higher than that achieved by other leading MPI implementations. Homogeneous dual-NIC configurations further improved performance, but the heterogeneous case requires continued investigation.

1 Introduction

Petascale computing is of increasing importance to the scientific community. Concurrently, the availability of small size clusters (on the order of tens to hundreds of CPUs) also continues to increase. Developing a message-passing system that efficiently deals with the challenges of performance and fault tolerance across this broad range represents a considerable challenge facing MPI developers.

The Open MPI project [3] – an ongoing collaboration between the Resilient Technologies Team at Los Alamos National Lab, the Open Systems Laboratory at Indiana University, and the Innovative Computing Laboratory at the University of Tennessee – is a new open-source implementation of the Message Passing Interface (MPI) standard for parallel programming on large-scale distributed systems [4, 6] focused on addressing these problems. This paper presents initial results from Open MPI's new point-to-point communication methodology (code-named "TEG"[9]) that provides high-performance, fault tolerant message passing. A full description of the design is given in the accompanying paper [9].

Open MPI/TEG's provides an enhanced feature set with support for dropped packets, corrupt packets, and NIC failures; concurrent network types (e.g. Myrinet, InfiniBand, etc.), in a single application run; single message fragmentation and delivery utilizing multiple NIC, including different NIC types, such as Myrinet and InfiniBand; and heterogeneous platform support within a single job, including different OS types,

D. Kranzlmüller et al. (Eds.): EuroPVM/MPI 2004, LNCS 3241, pp. 105–111, 2004.

different addressing modes (32 vs 64 bit mode), and different endianess. All of these features have been adapted into a new modular component architecture that allows for both compile-time and runtime selection of options.

2 Results

The performance of Open MPI's TEG methodology was compared against that of several well-known existing MPI implementations: FT-MPI v1.0.2[2], LA-MPI v1.5.1[5], LAM/MPI v7.0.4[1], and MPICH2 v0.96p2[7]. Experiments were performed using a two processor system based on 2.0GHz Xeon processors sharing 512kB of cache and 2GB of RAM. The system utilized a 64-bit, 100MHz, PCI-X bus with two Intel Pro/1000 NICs (based on the Super P4Dp6, Intel E7500 chipset), and one Myricom PCI64C NIC running LANai 9.2 on a 66MHz PCI interface. A second PCI bus (64-bit, 133MHz PCI-X) hosted a second, identical Myricom NIC. The processors were running the Red Hat 9.0 Linux operating system based on the 2.4.20-6smp kernel. All measurements were made using TCP/IP running over both Gigabit Ethernet and Myrinet.

Data was collected for two critical parameters: latency, using a ping-pong test code (for both MPI blocking and non-blocking semantics) and measuring half round-trip time of zero byte messages; and single NIC bandwidth results, using NetPIPE v3.6 [8].

2.1 Latency

Latency was measured using a ping-pong test code for both blocking and non-blocking MPI semantics. Blocking semantics allow for special optimizations, so more mature implementations will often create specially optimized versions to take advantage of these capabilities. This optimization has not yet been performed for TEG.

Table 1 compares TEG's latencies with those of LAM/MPI 7, LA-MPI, FT-MPI, and MPICH2 for the non-blocking scenario. The latencies for all MPI implementations with TCP/IP over Myrinet are similar, at about 51 μs. Even when TEG makes asynchronous progress with a progress thread, thus greatly reducing the CPU cycles used, latency is basically unchanged.

When the tests are run using Gigabit Ethernet, however, the measured latencies are generally lower and show a greater range of values. TEG's polling mode and LAM/MPI 7 have the lowest latencies at just below 40 μs, or about 25% lower than was obtained over Myrinet. MPICH2 has slightly higher latency, followed by LA-MPI and FT-MPI. The asynchronous progress mode of Open MPI's TEG is much higher than the other entries, with values slightly lower than those obtained with TCP/IP over Myrinet.

Table 2 lists the latency data for the ping-pong results using blocking MPI semantics. In this case, TEG's latencies are not as good as was obtained when using non-blocking semantics, with LAM/MPI 7 and MPICH2 is outperforming TEG better, while LA-MPI and FT-MPI didn't perform as well as TEG. Again, TEG's asynchronous progress mode gives much higher latencies than the polling based approach.

Overall, the initial results obtained with Open MPI/TEG show good latency performance characteristics when compared with more mature implementations. This performance is obtained despite the overhead required to: (a) stripe a single message across

Table 1. Open MPI/TEG latency measurements compared to other MPI implementations (non-blocking MPI semantics – i.e., MPI_ISEND / MPI_IRECV).

Implementation	Myrinet Latency (μs)	GigE Latency (μs)
Open MPI/TEG (Polling)	51.5	39.7
Open MPI/TEG (Async)	51.2	49.9
LAM7	51.5	39.9
LA-MPI	51.6	42.9
FT-MPI	51.4	46.4
MPICH2	51.5	40.3

Table 2. Open MPI/TEG latency measurements compared to other MPI implementations (blocking MPI semantics – i.e., MPI_SEND, MPI_RECV).

Implementation	GigE Latency (μs)
Open MPI/TEG (Polling)	41.3
Open MPI/TEG (Async)	52.6
LAM7	36.0
MPICH2	39.0
LA-MPI	42.8
FT-MPI	46.8

any number of different network types; (b) provide thread safety (MPI_THREAD_-MULTIPLE); and (c) allow proper handling of various failure scenarios. The latency of the blocking calls is not as good as some of the other current implementations. This reflects the relative immaturity of the TEG module and the current lack of optimization.

2.2   Bandwidth

Open MPI/TEG's ability to run in a heterogeneous networking environment makes the study of bandwidth performance very interesting. To establish this condition, a single PTL implementation (TCP/IP) was used for this study and operated on networks with different physical characteristics for the underlying physical transport layer (Myrinet and Gigabit Ethernet). Bandwidths were measured using NetPIPE v3.6 [8] and compared to that obtained from the same set of MPI implementations used in the latency measurements.

Figure 1(a) shows the performance of TEG compared to that from other MPI implementations over Gigabit Ethernet. As the figure shows, TEG, MPICH2, FT-MPI, LA-MPI, and LAM/MPI all have similar performance characteristics, peaking out close to 900 Mb/sec, similar to that obtained from raw TCP. MPICH2, FT-MPI, and LAM/MPI exhibit some irregular behavior with message sizes around 100 Kbyte, but this smoothes out at higher bandwidths and is most likely an artifact of the rendezvous protocol.

However, as Figure 1(b) shows, when running the same tests on a higher bandwidth interconnect (Myrinet), TEG displays much better bandwidths than MPICH2, FT-MPI, LA-MPI, and LAM/MPI. TEG's performance closely tracks the raw TCP/IP benchmark results, except at just above a message size of 100 Kbyte, where the rendezvous

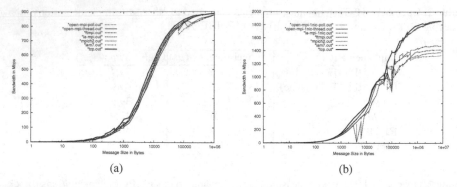

(a)   (b)

Fig. 1. Open MPI/TEG point-to-point ping-pong bandwidth compared to with the MPICH2, FT-MPI, LA-MPI, and LAM/MPI implementations, and raw TCP/IP. (a) Gigabit Ethernet (b) Myrinet.

protocol causes a temporary drop in measured bandwidth. Both TEG and raw TCP/IP peak out at a little above 1800 Kb/sec. In contrast, MPICH2, FT-MPI, LA-MPI, and LAM/MPI all peak out about 30% lower around 1400 Kb/sec. This is due to TEG's reduced overheads associated with multiple packet messages, including generation of the rendezvous protocol's ACK as soon as a match has been made on the receive side, and embedding a pointer to the receive object in the ACK so that subsequent packets can include this in their header for fast delivery on the receive side.

Figure 2 shows the results of running the same bandwidth tests over two Myrinet NICS, with TEG fragmenting and reassembling Open MPI's messages. The single NIC data is included for reference. As one would expect, the advantages of using multiple NIC's are not apparent in this particular test, until the message is large enough to be fragmented. At about an 8Mbyte message size, the dual NIC overall bandwidth is about 30% higher than the single NIC bandwidth. The dual NIC data does not appear to have peaked at the 8Mbyte maximum message size used in the experiment. However, we do not expect the dual NIC configuration to achieve double the rate of the single NIC data, since a single process is handling the TCP/IP stack. Parallelism, therefore, is only obtained over the network and by overlapping the send with the receive in the multiple fragment case.

The dual NIC implementation in Open MPI/TEG is about 10% more efficient than that in LA-MPI. Similar to the single NIC case, TEG benefits from the early generation of the rendezvous protocol ACK and embedding the receive object pointer information in the ACK message.

Figures 3 and 4 show the results of striping a single message over both Gigabit Ethernet and Myrinet using TEG. The single NIC Gigabit Ethernet and Myrinet are also included for reference. The results alway fall between the single NIC Gigabit Ethernet at the low end, and the single NIC Myrinet data at the upper end. This appears to indicate that adding a Gigabit Ethernet NIC to a system with a single Myrinet NIC can result in degraded performance. In contrast, adding a Myrinet NIC to a system with a single Gigabit Ethernet NIC appears to improve performance. Future work will investigate possible explanations for the observed behavior.

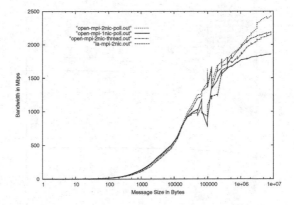

Fig. 2. Open MPI/TEG single and dual NIC point-to-point ping-pong bandwidth over Myrinet. Comparison with LA-MPI.

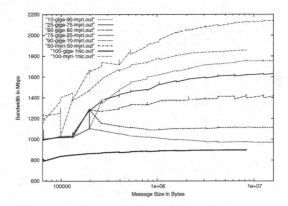

Fig. 3. Message striping across GigE and Myrinet with round-robin scheduling of first fragments across both interfaces and weighted scheduling of remaining data.

Finally, Figure 5 shows the effect of varying the size of the first message fragment on the bandwidth profile of Open MPI/TEG. As one would expect, the asymptotic behavior is independent of the size of the first message fragment. First fragments smaller than 128 KBytes don't require enough processing at the destination to hide the latency of the rendezvous protocol. The early transmission of the rendezvous ACK allows for this overlap to take place.

3   Summary

We have presented the results of running latency and bandwidth test with Open MPI with TCP/IP over Gigabit Ethernet and Myrinet. These results are compared with those obtained running LA-MPI, LAM/MPI, FT-MPI, and MPICH2, and show leading latency results when using non-blocking MPI semantics, and middle of the pack results

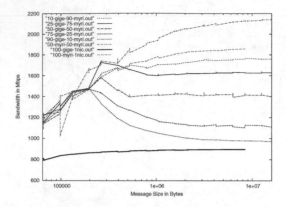

Fig. 4. Message striping across GigE and Myrinet with first fragments delivered via Myrinet and weighted scheduling of remaining data.

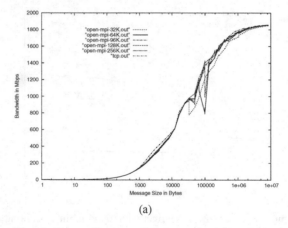

(a)

Fig. 5. Comparison of bandwidth as a function of first message fragment size for Open MPI/TEG.

with still to be optimized blocking MPI semantics. Open MPI's single bandwidths are better than those obtained with these same MPI's.

In addition multi-NIC bandwidth data is obtained with Open MPI. These bandwidths are about 30% better than the single NIC data over Myrinet, and about 10% better than those obtained by LA-MPI. Using Open MPI to send part of a message over networks with different characteristics (Gigabit Ethernet and Myrinet, in this case) needs further study at this early stage of Open MPI development.

Acknowledgments

This work was supported by a grant from the Lilly Endowment, National Science Foundation grants 0116050, EIA-0202048, EIA-9972889, and ANI-0330620, and Department of Energy Contract DE-FG02-02ER25536. Los Alamos National Laboratory is operated by the University of California for the National Nuclear Security Administration of the United States Department of Energy under contract W-7405-ENG-36. Project support was provided through ASCI/PSE and the Los Alamos Computer Science Institute, and the Center for Information Technology Research (CITR) of the University of Tennessee.

References

1. G. Burns, R. Daoud, and J. Vaigl. LAM: An Open Cluster Environment for MPI. In *Proceedings of Supercomputing Symposium*, pages 379–386, 1994.
2. Graham E. Fagg, Edgar Gabriel, Zizhong Chen, Thara Angskun, George Bosilca, Antonin Bukovski, and Jack J. Dongarra. Fault tolerant communication library and applications for high perofrmance. In *Los Alamos Computer Science Institute Symposium*, Santa Fee, NM, October 27-29 2003.
3. E. Garbriel, G.E. Fagg, G. Bosilica, T. Angskun, J. J. Dongarra J.M. Squyres, V. Sahay, P. Kambadur, B. Barrett, A. Lumsdaine, R.H. Castain, D.J. Daniel, R.L. Graham, and T.S. Woodall. Open mpi: Goals, concept, and design of a next generation mpi implementation. In *Proceedings, 11th European PVM/MPI Users' Group Meeting*, 2004.
4. A. Geist, W. Gropp, S. Huss-Lederman, A. Lumsdaine, E. Lusk, W. Saphir, T. Skjellum, and M. Snir. MPI-2: Extending the Message-Passing Interface. In *Euro-Par '96 Parallel Processing*, pages 128–135. Springer Verlag, 1996.
5. R. L. Graham, S.-E. Choi, D. J. Daniel, N. N. Desai, R. G. Minnich, C. E. Rasmussen, L. D. Risinger, and M. W. Sukalksi. A network-failure-tolerant message-passing system for terascale clusters. *International Journal of Parallel Programming*, 31(4), August 2003.
6. Message Passing Interface Forum. MPI: A Message Passing Interface. In *Proc. of Supercomputing '93*, pages 878–883. IEEE Computer Society Press, November 1993.
7. Mpich2, argonne. http://www-unix.mcs.anl.gov/mpi/mpich2/.
8. Q.O. Snell, A.R. Mikler, and J.L. Gustafson. NetPIPE: A Network Protocol Independent Performace Evaluator. In *IASTED International Conference on Intelligent Information Management and Systems*, June 1996.
9. T.S. Woodall, R.L. Graham, R.H. Castain, D.J. Daniel, M.W. Sukalsi, G.E. Fagg, E. Garbriel, G. Bosilica, T. Angskun, J. J. Dongarra, J.M. Squyres, V. Sahay, P. Kambadur, B. Barrett, and A. Lumsdaine. Teg: A high-performance, scalable, multi-network point-to-point communications methodology. In *Proceedings, 11th European PVM/MPI Users' Group Meeting*, 2004.

The Architecture and Performance of WMPI II

Anders Lyhne Christensen[1], João Brito[1], and João Gabriel Silva[2]

[1] Critical Software SA
{alc,jbrito}@criticalsoftware.com
[2] DEI/CISUC University of Coimbra, Portugal
jgabriel@dei.uc.pt

Abstract. WMPI II is the only commercial implementation of MPI 2.0 that runs on both Windows and Linux clusters. It evolved from the first ever Windows version of MPI, then a port of MPICH, but is now fully built from its own code base. It supports both 32 and 64 bit versions and mixed clusters of Windows and Linux nodes. This paper describes the main design decisions and the multithreaded, non-polling architecture of WMPI II. Experimental results show that, although WMPI II has figures comparable to MPICH and LAM for latency and bandwidth, most application benchmarks perform significantly better when running on top of WMPI II.

1 Introduction

The Message Passing Interface (MPI) is the de-facto standard for writing high performance applications based on the message-passing paradigm for parallel computers such as clusters and grids, [MPIR1]. The first version of MPI was published in 1994 by the MPI Forum, [MPIS1], a joint effort involving over 80 people from universities, hardware and software vendors and government laboratories. This initial version specifies functions for point-to-point and collective communication as well as functions to work with data types and structures in heterogeneous environments. The standard has a number of different functions with only slightly different semantics to allow for implementers to take advantage of the special features of a particular platform, like massively parallel computers, clusters of PCs, etc. Naturally, each of these different platforms calls for different communication library designs depending on the underlying hardware architecture. However, even for similar architectures such as the increasingly more popular clusters of PCs, designs can differ considerably, often resulting in radically different performances of real applications. In this paper we take a closer look at the design choices behind WMPI II and how they differ from some other popular implementations, and we present and compare application and micro-benchmark results, for WMPI II, MPICH and LAM [HPI,LAM].

In 1997 the MPI Forum published an ambitious version 2 of the Message Passing Interface Standard (MPI-2), [MPIS2], extending the functionality to include process creation and management, one-sided communication and parallel I/O. These extensions meant that the standard grew from 128 to 287 functions.

D. Kranzlmüller et al. (Eds.): EuroPVM/MPI 2004, LNCS 3241, pp. 112–121, 2004.

Whereas a relatively large number of complete MPI-1 implementations exist, only few complete MPI-2 implementations are available, [LAMW]. WMPI II is one of the few general-purpose MPI-2 implementations for clusters of PCs, and it has by no means been a trivial piece of software to develop. Critical Software has invested more than three years of research and development in the product. We believe that the reason for the relatively limited number of complete MPI-2 implementation is likely due to the complexity of the MPI-2 standard, namely the issues related to extending an existing MPI-1 implementation to support the new MPI-2 features. In MPI-1 each participating process can uniquely be identified in a straightforward manner by its rank in MPI_COMM_WORLD, a static, global communicator valid from startup until shutdown of an application. In this way all processes can have a shared, global view of the MPI environment during the entire run of an application. With the introduction of process creation and management features in MPI-2, this global view assumption is no longer valid since the ability to add and remove processes at runtime has the consequence that each participating process only has a local and partial view of the MPI environment at any given moment.

This article is structured in the following way: In Sect. 2 we give a brief overview of the history of WMPI, in Sect. 3 we explain the main design choices behind WMPI II, and in Sect. 4 we describe the architecture. Sect. 5 discusses our experience with porting from Windows to Linux. Experimental results are presented in Sect. 6 and discussed in Sect. 7. Finally, in Sect. 8 we lay out some lines for the future roadmap for WMPI II.

2   History of WMPI

WMPI started as a research project at the University of Coimbra in 1995. Versions 1.0 to 1.3 of WMPI were based on MPICH and reengineered to run on the Win32 platform. It was the first ever Windows implementation of version 1 of MPI, [MAR].

When the MPI-2 standard was released it became evident that a total rewrite was necessary due to the new functionality related to process creation and management. Version 1.5 consisted of a complete rewrite of WMPI in order to allow for dynamic processes. Virtually all the existing code had to be abandoned, since data structures and associated functions had to be redone to allow for processes to operate with only a partial, local view of the entire application. Although WMPI 1.5 does not have any of the functionality from the MPI 2.0 extensions, its architecture was designed with MPI-2 features in mind.

The first complete implementation of the MPI-2 standard for Windows clusters, WMPI II, was released at the beginning of 2003 and one year later the Linux version of WMPI II followed. WMPI II was based on WMPI 1.5, but extended with all of the functionality of MPI-2, including process creation and management, one-sided communication and parallel I/O. Presently a number of versions exist for various releases of Windows and Linux distributions for 32- and 64 bit systems.

3 Design Choices

With the introduction of MPI-2 it became clear that an entirely new software architecture for WMPI was necessary in order to support the new functionality. This required a significant investment and proved to be a challenging task. It allowed us to rethink a number of fundamental design choices and redesign the core based on our previous experience with WMPI up to version 1.3. We wanted WMPI II to perform well in real applications and be flexible in terms of portability, customization, and extensions. The main goals and consequences of the redesign are listed below:

WMPI II Should Support Multiple, Concurrent Communication Devices: For a cluster of PCs, inter-node communication is done through some type of interconnector, such as Gigabit Ethernet, InfiniBand or Myrinet, whereas processes running on SMP nodes often can benefit from the lower latency and higher bandwidth of shared memory communication. Furthermore, different segments of a cluster can be connected by different types of interconnectors especially as new technologies emerge. Therefore, WMPI II was designed to allow for multiple (two or more) communication devices to be used concurrently. Communication devices are dynamic link libraries (or shared objects in Linux), which are loaded at run-time based on the content of a configuration file. This also enables third parties to implement their own devices thus making WMPI II extendable. In order to allow for multiple communication devices to operate concurrently the Channel Interface from MPICH, [CHA], had to be discarded in favor of a completely new approach with thread-aware callbacks, direct remote memory access and process creation capabilities.

WMPI II Should Not Use Polling and Be Internally Multithreaded: Many implementations, such as MPICH, are not internally multithreaded and rely on a singlethreaded and/or polling approach to ensure message progress. This approach requires that the user application frequently calls the MPI implementation in order to ensure timely progress. In micro-benchmarks, e.g. a ping-pong test, where the MPI library is called constantly, this approach works fine as little else is done besides communicating. Keeping computation and communication in the same thread also avoids one or more context switches, which are necessary if these tasks are performed in separate threads, each time a message is sent or received. However, for real applications frequent calls to the message-passing layer are not guaranteed and polling can steal numerous cycles from the application in cases where it attempts to overlap communication and computation. In a multithreaded approach an application can continue doing calculations while an internal thread in the message-passing library takes care of the communication. This is the only approach consistent with the deep design philosophy of MPI of promoting overlapped communication and computation.

A polling/singlethreaded approach is particularly ill-suited for one-sided communication (OSC) applications, where only one party participates explicitly. In these cases the target process does not issue a matching operation and therefore timely progress is not guaranteed.

WMPI II Should Be Thread-Compliant: Thread compliance is a feature that is often not supported by open-source implementations. A thread-compliant MPI implementation allows an application to call MPI functions concurrently from multiple threads without restriction, [MPIS2]. Since WMPI II was already required to be internally multithreaded, adding thread compliance was relatively easy. It also gives the user the freedom to develop mixed OpenMP/MPI programs, [OMP], where OpenMP is used for fine-grained parallelism on SMP nodes and MPI is used for coarse-grained parallelism. Furthermore, application-level multithreading is becoming increasingly more important in order to take full advantage of modern dual-/multi-core[1] CPUs capable of running two or more threads concurrently.

WMPI II Should Be as Portable as Possible Without Affecting Performance: The initial version of WMPI was exclusively for Windows. Other operating systems, especially Linux, are very popular within the high performance community. Therefore, it was a central design goal to leave the option of porting WMPI II to other operating systems and architectures open, but only to the degree where performance would not be affected. Hence, we did not want to settle for the lowest common denominator at the price of performance, but make it *"as portable as possible"*.

WMPI II Should Be Flexible: Naturally portability between operating systems as different as Linux and Microsoft Windows requires flexibility. However, given that WMPI II is a commercial implementation, additional flexibility is needed to allow for custom fitting. For example, some clients have asked for specific functional extensions while others have requested more advanced features such as well-defined fault semantics. The architecture was therefore designed with such extensions in mind.

4 Architecture

The architecture of WMPI II is shown in Fig. 1. At the top sits the language bindings for C, C++ and Fortran, as well as for the profiling interface (PMPI functions). The WML contains the management logic that takes care of implementing complex communication primitives on top of simpler ones, managing datatypes, expected and unexpected messages and requests. Moreover, the WML emulates functionality not directly supported by some devices. For instance the one-sided communication primitives are implemented on top of point-to-point primitives when communicating with another process through TCP/IP or Globus I/O, since this device does not support remote memory access. The WML represents by far the largest part of the code altogether.

The WML operates with two different types of devices: Boot devices, sitting on top of the Boot Engine as shown in the figure, and communication devices located below the Communication Engine. Communication devices implement

---

[1] Multi-core CPUs: Also called *hyperthreading, chip multithreading* and *symmetric multithreading* depending on the hardware vendor.

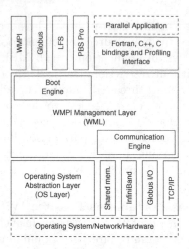

Fig. 1. WMPI II Architecture.

a number of primitives such as sends, receives and functions to open and close connections. These devices are used for communication after startup.

Boot devices, on the other hand, are used during startup, when the MPI environment is initialized. In standalone-mode WMPI II takes care of starting processes. This is done through a custom daemon running on each processing node. Information on the environment, e.g. which nodes are participating, who is who, how to contact the different processes etc. is passed through the daemon to new processes. When a new process starts it reads this information and knows what the world looks like. However, WMPI II does not control the starting of processes if applications are run under a scheduler like Platform LSF, PBS Pro or Globus and therefore cannot pass the necessary information to the participating processes. In each of these other modes processes are started outside the control of WMPI II and newly started processes need to rely on the means provided by the scheduler to obtain the necessary information.

In this way the WML was designed to be a component, which is independent of both the underlying communication hardware and software, as well as the surrounding startup environment. Such a design offers a significant level of flexibility since custom startup mechanisms can be added with minimal effort and new boot devices require only few or no changes to the existing code. For commercial software vendors, this allows for WMPI II to be embedded in libraries where the startup is either completely or partly controlled by the application software and does not require the application to be started through mpiexec, mpirun or alike.

If a WMPI II application is started through a scheduler, the scheduler controls the environment, such as where processes are started and security issues. If applications are started in native mode, e.g. started by WMPI II, the cluster configuration and security issues are handled by WMPI II. We operate with two different types of configuration: A cluster configuration file and a process group file. A cluster configuration file lists the nodes available, which communication devices should be used for inter- and intra-node communication, and under which security context processes should run on each node. Whereas the cluster configuration file describes the nodes available, a process group file lists what nodes should run which executables. The two concepts have been separated due to the introduction of process creation and management in MPI-2. With dynamic processes an application might be started as a singleton MPI process running on a single node and then at runtime spawn additional processes on other nodes. In this way, the potential cluster can differ from the active sub-cluster.

Security can be an issue in some cluster environments; moreover in heterogeneous environments security is handled differently on different operating systems. For UNIX-like operating systems a security context consists of a user name and a password, whereas on Windows the security context also comprises a domain. Since the supported operating systems lack common features for starting processes remotely, e.g. rsh, ssh, DCOM, we rely on a custom daemon. In multi-user environments it is often important to enforce permissions in order to avoid unauthorized access to sensitive data and to protect certain parts of a file system against tampering, voluntary or accidentally. Still the WMPI II daemon, responsible for creating new processes, needs to be running as root/administrator, which gives unrestricted access to a node. Therefore, users are required to specify the security context under which they wish to run MPI processes on the cluster nodes. For the same application different contexts can be used on different nodes. When a process is to be started on a node, the daemon on that node switches to the specified security context, provided that it is valid, and starts the process.

5   Porting from Windows to Linux and Interoperability

The first version of WMPI II was exclusively for the Windows platform; however, as mentioned above, it was built with portability in mind. Besides from the obvious benefit of having to maintain and extend only one source code tree it also greatly simplifies heterogeneous cluster support.

There are a number of ways to develop portable code, such as settling for the lowest common denominator in terms of features or relying on an operating system abstraction layer, where features present in a certain OS are implemented in the abstraction layer. We chose the latter approach for the WML, while for the devices we largely use OS specific code. An event-model similar to the native Windows event-model was implemented on Linux in the OS abstraction layer, since it proved quite convenient given our multithreaded design. Whereas the OS abstraction layer approach works well for WML, the majority of code for the devices has to be OS-specific in order not to compromise performance. The features used in the devices on Windows and on Linux are quite different, e.g. for the TCP device on Windows a combination of overlapped I/O and completion ports are used, whereas on Linux asynchronous sockets and real-time signals are used. The same is true for the shared memory device: On Windows one process can read the memory of another running process, whereas on Linux this is not

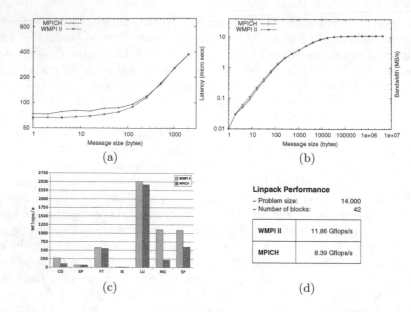

Fig. 2. Windows results for WMPI II and MPICH. (a) shows the latency results. (b) shows the bandwidth results. (c) shows the NAS benchmark results and the table in (d) lists Linpack results.

the case. Moreover, cross-process events are used by the shared memory device on Windows whereas real-time signals are used on Linux. Thus, the code for the WML compiles on both operating systems, whereas the devices have been specially optimized for each operating system. The greatest effort in porting from Windows to Linux was implementing and optimizing the devices for Linux and a large number of tiny compiler and OS differences. All in all it did not require as much effort as anticipated and we have managed to achieve a code sharing of 88%.

6   Experimental Results

In this section we present a number of benchmark results for WMPI II 2.3.0 and compare them with MPICH 1.2.5 on Windows and with MPICH 1.2.5 and LAM 7.0.5 on Linux.

We provide two types of benchmarks: Micro-benchmarks and application benchmarks. Micro-benchmarks attempt to measure the isolated performance of a message-passing library, that is latency and bandwidth, for a number of different MPI functions when nodes do nothing besides communicating. Application benchmarks attempt, as the name suggests, to measure the performance of a message-passing library (and the parallel computer as a whole) under a real application model, e.g. solving a common problem like a dense system of linear equations. For micro-benchmark results we have chosen the PALLAS bench-

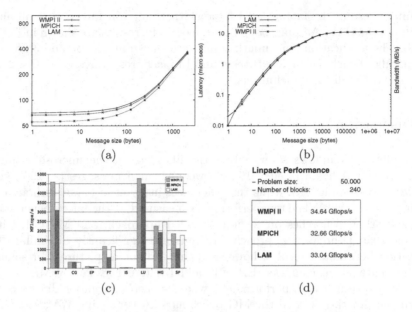

Fig. 3. Linux results for WMPI II, MPICH and LAM. (a) shows the latency results. (b) shows the bandwidth results. (c) shows the NAS benchmark results and the table in (d) lists Linpack results.

mark, [PAL]. The NAS parallel benchmark version 2.4, [NAS], and HP Linpack, [LIN], have been chosen for application benchmarks given their availability for a large number of operating systems and their popularity in terms of evaluating the performance of parallel computers. Notice that we have used default installations of operating systems and MPI implementations for all benchmarks, which means we did not do any tweaking that could potentially increase and/or decrease the performance of any particular MPI implementation.

Windows Results. The Windows benchmarks were run on a heterogeneous cluster of 16 Pentium 4 CPUs running at 1.7 – 2.8 GHz and with 512 MB to 1 GB of RAM. The nodes were running different versions of Windows: 2000, XP, and 2003 Advanced Server and they were connected in a 100 Mb/s, switched network. The latency and bandwidth results obtained by running the PALLAS ping-pong benchmark are shown in Fig. 2a and 2b, respectively. The results of running the NAS parallel benchmarks (class B tests) are shown in Fig. 2c, and the Linpack scores are shown in Fig. 2d. The NAS benchmark was compiled with Absoft Pro FORTRAN 7.0 and the Linpack benchmarks were compiled with Microsoft's C/C++ compiler version 13.10.3077.

Linux Results. The Linux benchmarks were run on a cluster of 16 Pentium 4 2.8 GHz nodes, each with 2 GB of memory running Redhat Linux 9.0. The nodes where connected in a 100 Mb/s switched network. As with the Windows benchmark results the latency and bandwidth results on Linux were obtained by running the PALLAS ping-pong benchmark. The results are shown in Fig. 3a and 3b. The application benchmark results are shown in Fig. 3c and 3d for the NAS parallel benchmark class B tests and Linpack, respectively. GCC 3.2 was used to compile all the benchmarks.

7   Discussion

The results for Windows show only little difference in the micro-benchmark results. WMPI II has a latency 10 μs lower for small messages ($<$ 200 bytes) than MPICH (see Fig. 2a), while the differences in terms of bandwidth are only marginal (see Fig. 2b). The raw performance measured by the micro-benchmarks for WMPI II, which relies on a multithreaded approach, and MPICH, which relies on a singlethreaded approach, are thus not significantly dissimilar. The application benchmark results, however, paint a different picture: For some of the NAS parallel benchmarks, EP, FT, and LU (see Fig. 2c), WMPI II and MPICH have comparable performance, whereas there are major differences in performance for the CG and the MG benchmarks, where using WMPI II yields a performance of 2.5 times and 5.0 times higher than MPICH, respectively. Similarly, the Linpack results show that WMPI II has a performance 41% higher than that of MPICH. This indicates that the ability to overlap computation and communication can be quite advantageous in some applications. Moreover, given the multithreaded, non-polling approach, a WMPI II process will be able to coexist efficiently with other application threads and/or other processes on the same node. We must of course take into consideration that MPICH is relatively old compared to WMPI II; better results for MPICH2 should be expected when stable versions for Windows become available.

The Linux results show that WMPI II has a latency that is slightly worse than MPICH and LAM in particular, while the bandwidth performances for all three implementations are comparable, see Fig. 3a and 3b. For application benchmarks the performance for the three implementations does not vary as much when running on Linux as it is the case for WMPI II and MPICH on Windows. For most of the NAS parallel benchmarks WMPI II and LAM have similar performance while MPICH falls behind on the BT, FT, and SP benchmarks, see Fig. 3c. WMPI II performs 1%-6% better than LAM on the BG, CG, FT and SP benchmarks, while LAM performs 9% better than WMPI II on the MG benchmark. No results are shown in the LU benchmark for LAM, because LAM kept exiting with an error message when we tried to run this benchmark. The results of the Linpack benchmark on Linux in Table 2 shows that WMPI II is slightly faster than both MPICH and LAM, achieving 34.64 Gflops/s compared to 32.66 Gflops/s and 33.04 Gflops/s, respectively.

It is important to stress that the benchmarks for Linux and Windows should not be directly compared, particularly because the latter were run on a cluster with heterogeneous hardware, where the slower machine tends to set the pace for the whole computation for some problems, such as those solved by Linpack.

Based on our results we can conclude that micro-benchmark results, such as raw latency and bandwidth, are not accurate indicators of the performance of an MPI implementation in real applications. Furthermore, our results show that some applications can benefit significantly from a multithreaded approach in the message-passing layer. Moreover, we believe that it is inherent in the MPI standard that implementations should not rely on polling or on a singlethreaded approach. An added benefit of WMPI II's multithreaded architecture is that it can be used for desktop GRIDs/idle time computing since it can coexist with other applications.

8 Future Work

WMPI II is stable, commercially supported middleware that has a number of new features and improvements on its roadmap. For example we want to extend the native support for new communication devices, like Myrinet. We are also currently working on tighter integration with the GRID infrastructure and fault-tolerance, as well as extended support for more POSIX-like operating systems.

References

[MAR]   José Marinho, João Gabriel Silva. "WMPI-Message Passing Interface for Win32 Clusters". Proc. EuroPVM/MPI98, Sept. 1998, Liverpool, Springer Verlag

[MPIR1] Marc Snir et. al., "MPI - The Complete Reference", vol. 1, MIT Press, 1998

[MPIR2] W. Gropp et. al., "MPI - The Complete Reference", vol. 2, MIT Press, 1998

[MPIS1] MPI Forum, "MPI: A Message-Passing Interface Standard", 1994

[MPIS2] MPI Forum, "MPI-2: Extensions to the Message-Passing Interface", 1997

[LAMW]   The LAM group maintains a list of available MPI implementations and their features on http://www.lam-mpi.org.

[HPI]   W. Group et al. "High-performance, portable implementation of the Message Passing Interface Standard", Journal of Parallel Computing, vol. 22, No 6, 1996

[LAM]   Greg Burns et al., "LAM: An Open Cluster Environment for MPI", Proceedings of Supercomputing Symposium, 1994

[CHA]   William Gropp and Ewing Lusk, "MPICH working note: Creating a new MPICH device using the channel interface", Technical Report ANL/MCS-TM-213, Argonne National Laboratory, 1995.

[OMP]   "OpenMP: An Industry-Standard API for Shared-Memory Programming", IEEE Computational Science & Engineering, Vol. 5, No. 1, Jan/Mar 1998

[PAL]   "Pallas MPI Benchmarks - PMB, Part MPI-1", Pallas GmbH, 2000

[NAS]   D. H. Bailey et al. "The NAS Parallel Benchmarks", International Journal of Supercomputer Applications, 1991

[LIN]   Jack J. Dongarra, "Performance of Various Computers Using Standard Linear Equations Software", University of Tennesse, 1995

A New MPI Implementation for Cray SHMEM

Ron Brightwell

Scalable Computing Systems, Sandia National Laboratories*
P.O. Box 5800, Albuquerque, NM 87185-1110
rbbrigh@sandia.gov

Abstract. Previous implementations of MPICH using the Cray SHMEM interface existed for the Cray T3 series of machines, but these implementations were abandoned after the T3 series was discontinued. However, support for the Cray SHMEM programming interface has continued on other platforms, including commodity clusters built using the Quadrics QsNet network. In this paper, we describe a design for MPI that overcomes some of the limitations of the previous implementations. We compare the performance of the SHMEM MPI implementation with the native implementation for Quadrics QsNet. Results show that our implementation is faster for certain message sizes for some micro-benchmarks.

1 Introduction

The Cray SHMEM [1] network programming interface provides very efficient remote memory read and write semantics that can be used to implement MPI. Previously, the SHMEM interface was only available on the Cray T3D and T3E machines and implementations of MPICH using SHMEM were developed specifically for those two platforms [2, 3]. Recently, SHMEM has been supported on other platforms as well, including machines from SGI, Inc., Cray, Inc., and clusters interconnected with the Quadrics network [4].

This paper describes our motivation for this work and presents a design that overcomes some of the limitations of these previous implementations. We compare the performance of the SHMEM MPI implementation with the native implementation for Quadrics QsNet. Micro-benchmark results show that the latency performance of the SHMEM implementation is faster for a range of small messages, while the bandwidth performance is comparable for a range of large messages.

The rest of this paper is organized as follows. The next section describes how this work relates to other published research. Section 3 discusses the motivations for this work. The design of our MPI implementation is presented in Section 4, which is followed by performance results and analysis in Section 5. Section 6 outlines possible future work.

---

* Sandia is a multiprogram laboratory operated by Sandia Corporation, a Lockheed Martin Company, for the United States Department of Energy's National Nuclear Security Administration under contract DE-AC04-94AL85000.

D. Kranzlmüller et al. (Eds.): EuroPVM/MPI 2004, LNCS 3241, pp. 122–130, 2004.
© Springer-Verlag Berlin Heidelberg 2004

2   Related Work

The design and implementation of MPICH for SHMEM described in this paper
is a continuation of previous work for the Cray T3D [2], which was subsequently
ported to the Cray T3E [3]. This new implementation is less complex than the
previous implementations, due to some extended features that are provided in
the Quadrics implementation of SHMEM. In particular, the T3 machines did not
support arbitrarily aligned transfers and required explicit cache-management
routines to be used. The SHMEM implementation for Quadrics provides the
ability to transfer data without any alignment or length restriction, and the
PCI-based network eliminates the need to explicitly manage the data cache.

Recently, some of the techniques that were employed in these earlier MPI/
SHMEM implementations have been used to support MPI implementations for
InfiniBand. For example, the method of using a fixed set of buffers at known
locations for handling incoming messages, which has been referred to as *persistent buffer association*, has been used in the MVAPICH implementation [5].
The similarities in the SHMEM interface and the remote DMA (RDMA) operations provided by the VAPI interface are largely what motivated this updated
MPI/SHMEM implementation. We discuss more details and further motivations
in the next section.

3   Motivation

The native Quadrics implementation of MPI uses the Tports interface to offload
all MPI matching and queue traversal functionality to the network interface
processor. In most cases, this is an ideal approach. However, there are some
cases where matching and queue traversal can be more efficiently handled by
the host processor, as is done in the SHMEM implementation. We have used
this implementation along with the Tports implementation for extensive comparisons that quantify the benefits of independent progress, overlap, and offload
for applications [6–8], using an identical hardware environment.

The MPI/SHMEM implementation also has some features that are not provided by the MPI/Tports implementation. For example, data transfers are explicitly acknowledged in the MPI/SHMEM implementation, while they are not
for the MPI/Tports implementation. This approach may be beneficial for applications where load imbalance causes messages to be produced faster than they
can be consumed. The Tports interface handles buffering of unexpected messages
implicitly, so unexpected messages that arrive from the network are deposited
into buffers that are allocated and managed within the Tports library. This space
never really becomes exhausted, as the Tports library will keep allocating more
memory, relying on virtual memory support to provide more. In some cases, a
protocol that throttles the sender is more appropriate. The MPI/SHMEM implementation provides this throttling since the buffers used for MPI messages
are explicitly managed by both the sender and the receiver.

Finally, the SHMEM interface has capabilities that are very similar to those
provided by the current generation of networking technology that supports

RDMA operations, such as InfiniBand. The distributed shared memory model of SHMEM avoids the need for the initiator and the target of a put or get operation to explicitly exchange information in order to begin a transfer, but the semantics of the transfer and mechanisms used to recognize the completion of transfers are very similar. This is especially true for some of the MPI implementations for InfiniBand that use RDMA operations [5, 9, 10]. We intend to use the SHMEM interface to analyze characteristics, such as strategies for efficiently polling for incoming messages, that may be beneficial to RDMA-based implementations of MPI.

4   Design and Implementation

4.1   Basic Data Transfer Mechanism

Here we describe our basic scheme for message passing using the SHMEM remote memory write (put) and remote memory read (get) operations. A point-to-point transfer between two processes can be thought of as a channel. The sender fills in the MPI envelope information and data in a packet and uses the remote write operation to transfer this packet to the receiver. On the receive side, the receiver recognizes the appearance of a packet and handles it appropriately.

Figure 1 illustrates the contents of a packet. The largest area of a packet is for user data. For our Quadrics implementation, the size of the data field in a packet was 16 KB. Following the data is the MPI envelope information: the context id of the sending MPI communicator, MPI tag, and the length of the message. We also include the local source rank within the communicator as an optimization to avoid a table lookup at the receiver. In addition to this information, a packet also includes two fields, **Send Start** and **Send Complete** that are used for the long message and synchronous message protocols. We will describe their use below in Section 4.2. Finally, the last field in the packet header is the **status** field, which is used to signal the arrival and validity of a packet.

Data
Context Id
Tag
Length
Source Rank
Send Complete
Send Start
Status

Fig. 1. Packet Structure.

There are a few important distinctions about the way in which a packet is constructed. First, the status field must be the last field in the packet, since it signifies the arrival and validity of a packet. Fixing the location of the status within the structure avoids having to use more complex techniques, such as those described in [5], to decipher when a complete packet has arrived. The receiver must only poll on a single memory location to determine packet arrival.

There are two possible strategies for sending a packet. First, two individual put operations could be used. The first put would be used to transfer the user data portion of a packet, and the second put would transfer the MPI envelope information. However, this essentially doubles the network latency performance of a single MPI send operation. For implementations of SHMEM where there is no ordering guarantee for successive put operations, an additional call to a synchronization function, such as shmem_fence may be needed to insure that the second put operation completes after the first. The second strategy copies the user data into a contiguous packet and uses a single put operation to transfer both the data and the MPI envelope. This is the strategy that we have used in our implementation. Since the length of the data portion of a packet is variable, the data is copied into a packet at an offset from the end of the buffer rather than from the beginning. This means that the start of a packet varies with the size of the data, but the end of the packet is always fixed. This way, a packet can be transmitted using a single put operation.

Since the target of a put operation must be known in advance, it is easiest to allocate send packets and receive packets that are *symmetric*, or mapped to the same virtual address location in each process. Our current implementation does this by using arrays that are statically declared such that there are N packets for sending messages to each rank and N corresponding receive packets for each rank. Packets for each rank are managed in a ring, and a counter is maintained for each send and receive packet for each rank to indicate the location of the next free packet.

In order for rank 0 to send a packet to rank 1, rank 0 checks the status field of the current send packet. If the status is set, this indicates that the corresponding receive packet at the destination has not yet been processed. At this point, the sender looks for incoming messages to process while waiting for the status to be cleared. When the status is cleared, the sender fills in the MPI envelope information, copies the data into the packet, sets the status field, and uses a put operation to write the packet to the corresponding receive packet at the destination. The sender then increments the counter to the next send packet for rank 1.

To receive a packet, rank 0 checks the status flag of the current receive packets for all ranks by looping through the receive packets array. Eventually, it recognizes that rank 0 has written a new packet into its current receive packet slot. It examines the contents of the MPI envelope and determines if this message is expected or unexpected and whether it is a short or long message. Once processing of this packet is complete, it clears the local status flag of the receive packet that was just processed, uses a put operation to clear the status flag of the send packet at the sender, and increments the current location of the receive packet for rank 0.

4.2   Protocols

Our implementation employs a traditional two-level protocol to optimize latency for short messages and bandwidth for long messages. Short messages consist of a single packet. Once the packet is sent, a short message is complete. For long messages, the data portion of a packet is not used. The MPI envelope information is filled in, the `Send Start` portion of the packet is set to the location of the buffer to be sent, and the `Send Complete` field is set to the location of the completion flag inside the MPI send request handle. When the packet is received and a matching receive has been posted, the receiver uses a remote read operation to read the send buffer. Once the get operation is complete, it uses a remote write operation to set the value of the completion flag in the send handle to notify the sender that the transfer has finished. The `Send Complete` field is also used to implement an acknowledgment for short synchronous send mode transfers.

4.3   Unexpected Messages

When a packet is received, the posted receive queue is checked for a match. If no match is found, a receive handle is allocated and the contents of the packet are copied to the handle. If the message is short, a temporary buffer is allocated and the data portion of the packet is copied into it. Once a matching receive is posted, the contents of this buffer are copied into the user buffer and the temporary buffer is freed.

   This implementation is very similar to the implementation for the T3E, with a few optimizations. First, the T3E implementation used only one send packet and one receive packet for each destination. We've enhanced this using a ring of packets to allow for several outstanding transfers between a pair of nodes. The T3D implementation used two put operations for each send – one for the MPI envelope information and one for the data. Because of the low latency of the put operation, copying the user data into a packet incurred more overhead than sending the data using a separate remote write operation. However, since the T3E supported adaptive routing, successive remote writes were not guaranteed to arrive in the order they were initiated. As such, the T3E implementation used a memory copy and a single remote write operation, as does our implementation for Quadrics.

4.4   Limitations

While this implementation demonstrates good performance for micro-benchmarks, it does have some drawbacks that may impact scalability, performance, and usability. Since send and receive packets are allocated using host memory, the amount of memory required scales linearly with the number of processes in the job. If we have 8 buffers each of size 16 KB, each rank requires 128 KB of memory. For a 1024 process job, this amounts to 128 MB of memory just for the packets alone. The current implementation does not allocate this memory dynamically, mostly because the Quadrics implementation does not support the Cray shmalloc function for obtaining symmetric memory from a heap[1]. The current library is compiled to support a maximum of 128 processes, so a significant amount of memory is wasted for jobs with fewer processes.

In addition to memory usage, the time required to look for incoming packets increases with the number of processes as well. Polling memory locations is not a very efficient way to recognize incoming transfers. We plan to explore several different strategies for polling and analyze their impact on performance. This limitation is of particular interest because it is also an issue for implementations of MPI for InfiniBand.

The rendezvous strategy that we employ for long messages does not support independent progress, so the opportunity for significantly overlapping computation and communication for large transfers is lost. Our implementation could be enhanced by using a user-level thread to insure that outstanding communication operations make progress independent of the application making MPI library calls.

Our implementation also does not use the non-blocking versions of the put and get operations. While these calls are listed the Quadrics documentation, they are not supported on any of the platforms to which we have access.

Finally, our implementation only supports the SPMD model of parallel programming. This is a limitation imposed by the SHMEM model that does not allow using different executables files in the same MPI job.

5   Performance

5.1   Platform

The machine used for our experiments is a 32-node cluster at Los Alamos National Laboratory. Each node in the cluster contains two 1 GHz Intel Itanium-2 processors, 2 GB of main memory, and two Quadrics QsNet (ELAN-3) network interface cards. The nodes were running a patched version of the Linux 2.4.21 kernel. We used version 1.24-27 of the QsNet MPI implementation and version 1.4.12-1 of the QsNet libraries that contained the Cray SHMEM compatibility library. The SHMEM MPI library is a port of MPICH version 1.2.5. All applications were compiled using Version 7.1 Build 20031106 of the Intel compiler suite. All of our experiments were run using only one process per node using only one network interface, and all results were gathered on a dedicated machine.

5.2   Results

Figure 2 shows the latency performance for the MPI/SHMEM implementation and the vendor-supplied MPI/Tports implementation. The zero-length latency is 5.6 μsecfor MPI/Tports and 6.6 μsecfor MPI/SHMEM. At 256 bytes, the SHMEM implementation begins to outperform the Tports implementation. This trend continues until the size of the message is 4 KB.

---

[1] There is an equivalent function call in the lower-level Elan libraries that could be used.

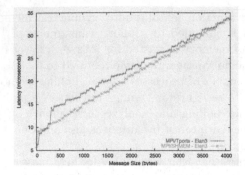

Fig. 2. Message latency for MPI/SHMEM and MPI/Tports.

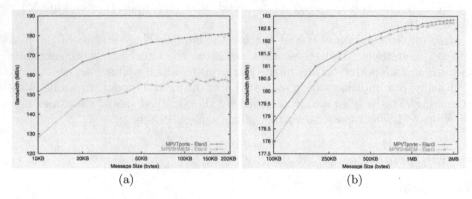

Fig. 3. Medium (a) and Long (b) Message bandwidth for MPI/SHMEM and MPI/Tports.

Figure 3 shows the bandwidth performance of the two implementations. From 10 KB to 200 KB, there is a difference of a little more than 20 MB/s in favor of the Tports implementation. This difference is largely attributable to the ability of the Tports implementation to avoid involving the host processor in large data transfers. In contrast, the SHMEM implementation must rely on the application process to initiate a remote memory read operation. At a message size of 100 KB, the margin between the two implementations has decreased to less than 1 MB/s, and only a minimal difference can be perceived for messages beyond that point.

Figure 4 illustrates how the posted receive queue can affect latency performance. For this measurement, there are 10 requests in the posted receive queue, and the percentage of the queue that must be traversed in order to receive a zero-length message is varied. The latency of the SHMEM implementation is less than the Tports implementation at 8 pre-posted receives. At this point, using the host processor rather than the network interface processor to perform MPI communicator and tag matching operations becomes more efficient.

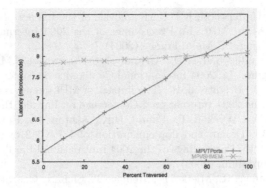

Fig. 4. Pre-posted message latency for MPI/SHMEM and MPI/Tports.

6   Future Work

One of the important areas we intend to pursue with this work is the effect
of memory polling strategies on application performance. For applications that
have relatively few sources of incoming messages, we expect to be able to develop
strategies that allow incoming messages to be discovered more quickly than simply polling all possible incoming message locations. We also intend to explore
these strategies for other RDMA-based implementations of MPI, including InfiniBand.

There is much work that could be done to improve the performance of our
MPI/SHMEM implementation. Collective operations are currently layered on
top of MPI point-to-point functions, so work could be done to leverage the
SHMEM collective routines that are available. SHMEM also has efficient support
for non-contiguous transfers, so there might be some benefit for using these
functions to handle non-contiguous MPI data types. Additionally, there may
be some benefit for using the SHMEM interface for implementing the MPI-2
one-sided operations, since both two-sided and one-sided operations could be
handled by the same transport interface.

References

1. Cray Research, Inc.: SHMEM Technical Note for C, SG-2516 2.3. (1994)
2. Brightwell, R., Skjellum, A.: MPICH on the T3D: A case study of high performance message passing. In: Proceedings of the Second MPI Developers' and Users'
   Conference. (1996)
3. Hebert, L.S., Seefeld, W.G., Skjellum, A.: MPICH on the Cray T3E. In: Proceedings of the Third MPI Developers' and Users' Conference. (1999)
4. Petrini, F., chun Feng, W., Hoisie, A., Coll, S., Frachtenberg, E.: The Quadrics
   network: High-performance clustering technology. IEEE Micro **22** (2002) 46–57
5. Liu, J., Wu, J., Kini, S.P., Wyckoff, P., Panda, D.K.: High performance RDMA-
   based MPI implementation over InfiniBand. In: Proceedings of the 2003 International Conference on Supercomputing (ICS-03), New York, ACM Press (2003)
   295–304

6. Brightwell, R., Underwood, K.D.: An analysis of the impact of overlap and independent progress for MPI. In: Proceedings of the 2004 International Conference on Supercomputing, St. Malo, France (2004)
7. Brightwell, R., Underwood, K.D.: An analysis of the impact of MPI overlap and independent progress. In: 2004 International Conference on Supercomputing. (2004)
8. Underwood, K.D., Brightwell, R.: The impact of MPI queue usage on latency. In: Proceedings of the 2004 International Conference on Parallel Processing. (2004)
9. Liu, J., Jiang, W., Wyckoff, P., Panda, D.K., Ashton, D., Buntinas, D., Gropp, W., Toonen, B.: Design and implementation of MPICH2 over InfiniBand with RDMA support. In: Proceedings of the 2004 International Parallel and Distributed Processing Symposium. (2004)
10. Rehm, W., Grabner, R., Mietke, F., Mehlan, T., Siebert, C.: An MPICH2 channel device implementation over VAPI on InfiniBand. In: Proceedings of the 2004 Workshop on Communication Architecture for Clusters. (2004)

A Message Ordering Problem
in Parallel Programs*

Bora Uçar and Cevdet Aykanat

Department of Computer Engineering, Bilkent University, 06800, Ankara, Turkey
{ubora,aykanat}@cs.bilkent.edu.tr

Abstract. We consider a certain class of parallel program segments in
which the order of messages sent affects the completion time. We give
characterization of these parallel program segments and propose a solution to minimize the completion time. With a sample parallel program,
we experimentally evaluate the effect of the solution on a PC cluster.

1 Introduction

We consider a certain class of parallel program segments with the following characteristics. First, there is a small-to-medium grain computation between two
communication phases which are referred to as pre- and post-communication
phases. Second, local computations cannot start before the pre-communication
phase ends, and the post-communication phase cannot start before the computation ends. Third, the communication in both phases is irregular and sparse.
That is, the communications are performed using point-to-point send and receive operations, where the sparsity refers to small number of messages having
small sizes. These traits appear, for example, in the sparse-matrix vector multiply $y = Ax$, where matrix A is partitioned on the nonzero basis and also in the
sparse matrix-chain-vector multiply $y = ABx$, where matrix A is partitioned
along columns and matrix B is partitioned conformably along rows. In both examples, the x-vector entries are communicated just before the computation and
the y-vector entries are communicated just after the computation.

There has been a vast amount of research in partitioning sparse matrices
to effectively parallelize computations by achieving computational load balance
and by minimizing the communication overhead [2–4, 7, 8]. As noted in [7], most
of the existing methods consider minimization of the total message volume. Depending on the machine architecture and problem characteristics, communication overhead due to message latency may be a bottleneck as well [5]. Furthermore, the maximum message volume and latency handled by a single processor
may also have crucial impact on the parallel performance [10, 11]. However, optimizing these metrics is not sufficient to minimize the total completion time
of the subject class of parallel programs. Since the phases do not overlap, the
receiving time of a processor, and hence the issuing time of the corresponding
send operation play an important role in the total completion time.

---

* This work is partially supported by the Scientific and Technical Research Council of
Turkey (TUBITAK) under grant 103E028.

D. Kranzlmüller et al. (Eds.): EuroPVM/MPI 2004, LNCS 3241, pp. 131–138, 2004.

There may be different solutions to the above problem. One may consider balancing the number of messages per processor both in terms of sends and receives. This strategy would then has to partition the computations with the objectives of achieving computational load balance, minimizing total volume of messages, minimizing total number of messages, and also balancing the number of messages sent/received on the per processor basis. However, combining these objectives into a single function to be minimized would challenge the current state of the art. For this reason, we take these problems apart from each other and decompose the overall problem into stages, each of which involving a certain objective. We first use standard models to minimize the total volume of messages and maintain the computational load balance among processors using effective methods, such as graph and hypergraph partitioning. Then, we minimize the total number of messages and maintain a loose balance on the communication volume loads of processors, and in the meantime we address the minimization of the maximum number of messages sent by a single processor. After this stage, the communication pattern is determined. In this paper, we suggest to append one more stage in which the send operations of processors are ordered to address the minimization of the total completion time.

2   Message Ordering Problem and a Solution

We make the following assumptions. The computational load imbalance is negligible. All processors begin the pre-communication phase at the same time because of the possible global synchronization points and balanced computations that exist in the other parts of the parallel program. The parallel system has a high latency overhead so that the message transfer time is dominated by the start-up cost due to small message volumes. By the same reasoning, the receive operation is assumed to incur negligible cost to the receiving processor. For the sake of simplicity, the send operations are assumed to take unit time. Under these assumptions, once a send is initiated by a processor at time t_i, the sending processor can continue with some other operation at time t_{i+1}, and the receiving processor receives the message at time t_{i+1}. This assumption extends to concurrent messages destined for the same processor. The rationale behind these assumptions is that, the start-up costs for all messages destined for a certain processor truly overlap with each other.

Let *send-lists* $S_1(p)$ and $S_2(p)$ denote the set of messages, distinguished by the ranks of the receiving processors, to be sent by processor P_p in pre- and post-communication phases, respectively. For example, $\ell \in S_1(p)$ denotes the fact that processor P_ℓ will receive a message from P_p in the pre-communication phase. For $\ell \in S_1(p)$, we use $s_1(p, \ell)$ to denote the completion time of the message from P_p to P_ℓ, i.e., P_p issued the send at time $s_1(p, \ell) - 1$, and P_ℓ received the message at time $s_1(p, \ell)$. We use $s_2(p, \ell)$ for the same purpose for the post-communication phase. Let W be the amount of computation performed by each processor. Let

$$r_1(p) = \max_{j:p \in S_1(j)} \{s_1(j, p)\} \tag{1}$$

denote the point in time at which processor P_p receives its latest message in the pre-communication phase. Then, P_p will enter the computation phase at time

$$c_1(p) = \max\{|S_1(p)|, r_1(p)\}, \tag{2}$$

i.e, after sending all of its messages and receiving all messages destined for it in the pre-communication phase. Let

$$r_2(p) = \max_{j:p\in S_2(j)} \{s_2(j,p)\} \tag{3}$$

denote the point in time at which processor P_p receives its latest message in the post-communication phase. Then, processor P_p will reach completion at time

$$c_p = \max\{c_1(p) + W + |S_2(p)|, r_2(p)\}, \tag{4}$$

i.e., after completing its computational task as well as all send operations in the post-communication phase and after receiving all post-communication messages destined for it. Using the above notation, our objective is

$$minimize\{\max_p\{c_p\}\}, \tag{5}$$

i.e, to minimize the maximum completion time. The maximum completion time induced by a message order is called the bottleneck value, and the processor that defines it is called the bottleneck processor. Note that the objective function depends on the time points at which the messages are delivered.

In order to clarify the notations and assumptions, consider a six-processor system as shown in Fig. 1(a). In the figure, the processors are synchronized at time t_0. The computational load of each processor is of length five-units and shown as a gray rectangle. The send operation from processor P_k to P_ℓ is labeled with $s_{k\ell}$ on the right-hand side of the time-line for processor P_k. The corresponding receive operation is shown on the left-hand side of the time-line for processor P_ℓ. For example, processor P_1 issues a send to P_3 at time t_0 and completes the send at time t_1 which also denotes the delivery time to P_3. Also note that P_3 receives a message from P_5 at the same time. In the figure, $r_1(1) = c_1(1) = t_5$, $r_2(1) = t_{10}$ and $c_1 = t_{15}$. The bottleneck processor is P_1 with the bottleneck value $t_b = t_{15}$.

Reconsider the same system where the messages are sent according to the order as shown in Fig. 1(b). In this setting, P_1 is also a bottleneck processor with value $t_b = t_{11}$.

Note that if a processor P_p never stays idle then it will reach completion at time $|S_1(p)| + W + |S_2(p)|$. The optimum bottleneck value cannot be less than the maximum of these values. Therefore, the order given in Fig. 1(b) is the best possible. Let P_q and P_r be the maximally loaded processors in the pre- and post-communication phases respectively, i.e., $|S_1(q)| \geq |S_1(p)|$ and $|S_2(r)| \geq |S_2(p)|$ for all p. Then, the bottleneck value cannot be larger than $|S_1(q)| + W + |S_2(r)|$. The setting in Fig. 1(a) attains this worst possible bottleneck value.

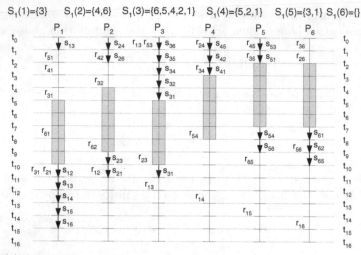

(a) A sample message order which produces worst completion time

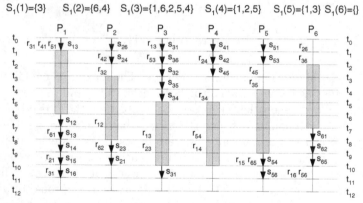

(b) A sample message order which produces best completion time

Fig. 1. Worst and best order of the messages.

Observe that in a given message order, the bottleneck occurs at a processor with an outgoing message. Meaning that, for any bottleneck processor that receives a message at time t_b, there is a processor which finishes a send operation at time t_b. Therefore, for a processor P_p to be a bottleneck processor we require

$$c'_p = c_1(p) + W + |S_2(p)| \qquad (6)$$

as a bottleneck value. Hence, our objective reduces to

$$minimize\{\max_p\{c'_p\}\}. \qquad (7)$$

Also observe that the bottleneck processor and value remains as is, for any order of the post-communication messages. Therefore, our problem reduces to ordering the messages in the pre-communication phase. From these observations we reach the intuitive idea of assigning the maximally loaded processor in the post-communication phase to the first position in each pre-communication send-list. This will make the processor with maximum $|S_2(\cdot)|$ enter the computation phase as soon as possible. Extending this to the remaining processors we develop the following algorithm. First, each processor P_p determines its key-value $key(p) = |S_2(p)|$. Second, each processor obtains the key-values of all other processors with an all-to-all communication on the key-values. Third, each processor P_p sorts its send-list $S_1(p)$ in descending order of the key-values of the receiving processors. These sorted send-lists determine the message order in the pre-communication phase, where the order in the post-communication phase is arbitrary.

Theorem 1. *The above algorithm obtains the optimal solution that minimizes the maximum completion time.*

Proof. We take an optimal solution and then modify it to have each send-list sorted in descending order of key-values.

Consider an optimal solution. Let processor P_b be the bottleneck processor finishing its sends at time t_b. For each send-list in the pre-communication phase, we perform the following operations.

For any P_ℓ with $key_b \leq key_\ell$ where P_b and P_ℓ are in the same send-list $S_1(p)$, if $s_1(p, \ell) \leq s_1(p, b)$, then we are done, if not swap $s_1(p, \ell)$ and $s_1(p, b)$. Let $t_s = s_1(p, \ell)$ before the swap operation. Then, we have $t_s + W + key_\ell \leq t_b$ before the swap. After the swap we will have $t_s + W + key_b$ and $t_h + W + key_\ell$ for some $t_h < t_s$, for processors P_b and P_ℓ. These two values are less than t_b.

For any P_j with $key_j \leq key_b$ where P_j and P_b are in the same send-list $S_1(q)$, if $s_1(q, b) \leq s_1(q, j)$, then we are done, if not swap $s_1(q, b)$ and $s_1(q, j)$. Let $t_s = s_1(q, b)$ before the swap operation. Then, we have $t_s + W + key_b \leq t_b$. After the swap operation we will have $t_s + W + key_j$ and $t_h + W + key_b$ for some $t_h < t_s$ for processors P_j and P_b, respectively. Clearly, these two values are less than or equal to t_b.

For any P_u and P_v that are different from P_b with $key_u \leq key_v$ in a send-list $S_1(r)$, if $s_1(r, v) \leq s_1(r, u)$, then we are done, if not swap $s_1(r, u)$ and $s_1(r, v)$. Let $t_s = s_1(r, v)$ before the swap operation. Then, we have $t_s + W + key_v \leq t_b$. After the swap operation we will have $t_s + W + key_u$ and $t_h + W + key_v$ for some $t_h < t_s$, for P_u and P_v respectively. These two values are less than or equal to t_b. Therefore, for each optimal solution we have an equivalent solution in which all send-lists in the pre-communication phase are sorted in decreasing order of the key values. Since the sorted order is unique with respect to the key values, the above algorithm is correct.

3   Experiments

In order to see whether the findings in this work help in practice we have implemented a simple parallel program which is shown in Fig 2. In this figure, each processor first posts its non-blocking receives and then sends its messages

```
MPI_Barrier(MPI_COMM_WORLD);
startTime = MPI_Wtime();
for(iter = 0; iter < MAXITER; iter++){
  communication(preSendList, preSendCount, preRecvList, preRecvCount,
               sendBuf, recvBuf, iter);
  computation(sendBuf, recvBuf);
  communication(postSendList, postSendCount,postRecvList,postRecvCount,
               sendBuf, recvBuf, iter + 1);
  MPI_Barrier(MPI_COMM_WORLD);
}
totTime = 1000.0*MPI_Wtime() - 1000.0*startTime;
```

(a) Parallel program segment

```
void computation(MSSGTYPE *sendBuf, MSSGTYPE *recvBuf){
int i;
for(i = 0; i < numProcs; i++){
  int j, indi = mssgSizes * i;
  for(j = 0; j < mssgSizes; j++)
    sendBuf[indi+j]=(sendBuf[indi+j]+recvBuf[indi+j])/(MSSGTYPE)2;
}
}
```

(b) Local computation performed at each processor

```
void communication(int *sList, int sCnt, int *rList, int rCnt,
                   MSSGTYPE *sBuf, MSSGTYPE *rBuf, int tag){
int i;
MPI_Request reqs[rCnt]; MPI_Status stats[rCnt];
for(i = 0 ; i < rCnt; i++){
  int p = rList[i], ind = p*mssgSizes;
  MPI_Irecv(&rBuf[ind], mssgSizes, bMPITYPESTR, p,
            tag, MPI_COMM_WORLD,&reqs[i]);
}
for(i = 0; i < sCnt; i++){
  int p = sList[i], ind = myId * mssgSizes;
  MPI_Send(&sBuf[ind], mssgSizes,bMPITYPESTR, p, tag,MPI_COMM_WORLD);
}
if(rCnt > 0) MPI_Waitall(rCnt, reqs, stats);
}
```

(c) Implementation of pre- and post-communication phases

Fig. 2. A simple parallel program.

in the order as they appear in the send-lists. In order to simplify the effects
of the message volume on the message transfer time, we set the same volume
for each message. We have used LAM [1] implementation of MPI and mpirun
command without -lamd option. The parallel program were run on a Beowulf
class [9] PC cluster with 24 nodes. Each node has a 400MHz Pentium-II proces-

Table 1. Communication patterns and parallel running times on 24 processors.

Data	Communication pattern			Mssg order	unit Max $\{c'_p\}$	Completion time milliseconds			
						Message length (bytes)			
	min	max	tot			8	64	512	1024
1-PRE	5	21	290	best	38	4.3	4.4	5.5	7.2
1-POST	6	22	358	worst	42	4.8	5.0	6.2	7.8
2-PRE	3	23	313	best	39	4.9	5.0	6.0	7.3
2-POST	11	22	370	worst	45	5.3	5.4	6.7	7.8
3-PRE	10	23	490	best	45	6.3	6.4	7.8	9.7
3-POST	15	23	504	worst	46	6.6	6.6	8.2	10.1
4-PRE	6	22	312	best	41	4.5	4.6	5.9	7.3
4-POST	10	20	356	worst	42	5.3	5.6	6.8	8.2
5-PRE	5	23	228	best	36	4.0	4.1	4.9	5.9
5-POST	7	13	228	worst	36	4.4	4.6	5.6	6.6
6-PRE	1	23	212	best	35	4.1	4.1	5.1	6.0
6-POST	4	17	236	worst	40	4.5	4.6	5.8	6.7
7-PRE	3	20	226	best	29	3.7	3.7	4.5	5.3
7-POST	7	17	253	worst	37	3.9	3.9	5.0	5.9
8-PRE	2	23	267	best	43	4.7	4.7	6.1	7.6
8-POST	4	22	278	worst	45	5.7	5.9	7.0	8.1
9-PRE	3	16	167	best	35	3.7	4.0	4.8	5.6
9-POST	4	20	273	worst	36	4.3	4.3	5.3	6.0
10-PRE	2	23	300	best	46	4.7	4.7	6.3	8.0
10-POST	10	23	316	worst	46	5.6	5.7	7.1	8.3
W (Computation time):						0.00	0.01	0.06	0.11

sor and 128MB memory. The interconnection network is comprised of a 3COM SuperStack II 3900 managed switch connected to Intel Ethernet Pro 100 Fast Ethernet network interface cards at each node. The system runs Linux kernel 2.4.14 and Debian GNU/Linux 3.0 distribution.

We extracted the communication patterns of some row-column-parallel sparse matrix-vector multiply operations on 24 processors. Table 1 lists minimum and maximum number of send operations per processor under columns *min* and *max*. Total number of messages is given under the column *tot*.

For each test case, we have run the parallel program of Fig. 2 with small message lengths of 8, 64, 512, and 1024-bytes to justify the practicality of the assumptions made in this work. We have experimented with the best and worst orders. The best message orders are generated according to the algorithm proposed in § 2. The worst message orders are obtained by sorting the pre-communication send-lists in increasing order of the key-values of the receiving processors. In all cases, we used the same message order in the post-communication phase. The running are presented in milliseconds in Table 1. We give the best among 20 runs (see [6] for choosing best in order to obtain reproducible results). In the table, we also give $\max_p\{c'_p\}$ for worst and best orders with $W = 0$. In all cases, the best order always gives better completion time than the worst order. In theory, however, we did not expect improvements for the 5th and 10th cases, in which the two orders give the same bottleneck value. This unexpected outcome may be resulting from the internals of the process that handles the communication requests. We are going to investigate this issue.

4   Conclusion

In this work, we addressed the problem of minimizing maximum completion time of a certain class of parallel program segments in which there is a small-to-medium grain computation between two communication phases. We showed that the order in which the messages are sent affects the completion time and showed how to order the messages optimally in theory. Experimental results on a PC cluster verified the existence of the specified problem and the validity of the proposed solution. As a future work, we are trying to set up experiments to observe the findings of this work in parallel sparse matrix-vector multiplies. A generalization of the given problem addresses parallel programs that have multiple computation phases interleaved with communications. This problem is in our research plans.

References

1. G. Burns, R. Daoud, and J. Vaigl. LAM: an open cluster environment for MPI. In John W. Ross, editor, *Proceedings of Supercomputing Symposium '94*, pages 379–386. University of Toronto, 1994.
2. Ü. V. Çatalyürek and C. Aykanat. Hypergraph-partitioning based decomposition for parallel sparse-matrix vector multiplication. *IEEE Transactions on Parallel and Distributed Systems*, 10(7):673–693, 1999.
3. Ü. V. Çatalyürek and C. Aykanat. A fine-grain hypergraph model for 2d decomposition of sparse matrices. In *Proceedings of International Parallel and Distributed Processing Symposium (IPDPS)*, April 2001.
4. Ü. V. Çatalyürek and C. Aykanat. A hypergraph-partitioning approach for coarse-grain decomposition. In *Proceedings of Scientific Computing 2001 (SC2001)*, pages 10–16, Denver, Colorado, November 2001.
5. J. J. Dongarra and T. H. Dunigan. Message-passing performance of various computers. *Concurrency—Practice and Experience*, 9(10):915–926, 1997.
6. W. Gropp and E. Lusk. Reproducible measurements of mpi performance characteristics. Tech. Rept. ANL/MCS-P755-0699, Argonne National Lab., June 1999.
7. B. Hendrickson and T. G. Kolda. Graph partitioning models for parallel computing. *Parallel Computing*, 26:1519–1534, 2000.
8. B. Hendrickson and T. G. Kolda. Partitioning rectangular and structurally unsymmetric sparse matrices for parallel processing. *SIAM J. Sci. Comput.*, 21(6):2048–2072, 2000.
9. T. Sterling, D. Savarese, D. J. Becker, J. E. Dorband, U. A. Ranaweke, and C. V. Packer. BEOWULF: A parallel workstation for scientific computation. In *Proceedings of the 24th International Conference on Parallel Processing*, 1995.
10. B. Uçar and C. Aykanat. Minimizing communication cost in fine-grain partitioning of sparse matrices. In A. Yazıcı and C. Şener, editors, *in Proc. ISCISXVIII-18th Int. Symp. on Computer and Information Sciences*, Antalya, Turkey, Nov. 2003.
11. B. Uçar and C. Aykanat. Encapsulating multiple communication-cost metrics in partitioning sparse rectangular matrices for parallel matrix-vector multiplies. *SIAM J. Sci. Comput.*, 25(6):1837–1859, 2004.

BSP/CGM Algorithms for Maximum Subsequence and Maximum Subarray*

C.E.R. Alves[1], E.N. Cáceres[2], and S.W. Song[3]

[1] Universidade São Judas Tadeu, São Paulo, SP, Brazil
prof.carlos_r_alves@usjt.br
[2] Universidade Federal de Mato Grosso do Sul, Campo Grande, MS, Brazil
edson@dct.ufms.br
[3] Universidade de São Paulo, São Paulo, SP, Brazil
song@ime.usp.br

Abstract. The maximum subsequence problem finds the contiguous subsequence of n real numbers with the highest sum. This problem appears in the analysis of DNA or protein sequences. It can be solved sequentially in $O(n)$ time. In the 2-D version, given an $n \times n$ array A, the maximum subarray of A is the contiguous subarray that has the maximum sum. The sequential algorithm for the maximum subarray problem takes $O(n^3)$ time. We present efficient BSP/CGM parallel algorithms that require a constant number of communication rounds for both problems. In the first algorithm, the sequence stored on each processor is reduced to only five numbers, so that the resulting values can be concentrated on a single processor which runs an adaptation of the sequential algorithm to obtain the result. The parallel algorithm requires $O(n/p)$ computing time. In the second algorithm, the input array is partitioned equally among the processors and we first reduce each subarray to a sequence, and then apply the first algorithm to solve it. The parallel algorithm takes $O(n^3/p)$ computing time. The good performance of the parallel algorithms is confirmed by experimental results run on a 64-node Beowulf parallel computer.

1 Introduction

Given a sequence of real numbers, the problem of identifying the (contiguous) subsequence with the highest sum is called the *maximum subsequence problem* [3]. If the numbers are all positive, the answer is obviously the entire sequence. It becomes interesting when there are also negative numbers in the sequence.

The maximum subsequence problem arises in several contexts in Computational Biology in the analysis of DNA or protein sequences. Many such applications are presented in [8], for example, to identify transmembrane domains

* Partially supported by FINEP-PRONEX-SAI Proc. No. 76.97.1022.00, FAPESP Proc. No. 1997/10982-0, CNPq Proc. No. 55.2028/02-9, 30.5218/03-4, 47.0163/03-8, FUNDECT-MS.

D. Kranzlmüller et al. (Eds.): EuroPVM/MPI 2004, LNCS 3241, pp. 139–146, 2004.

in proteins expressed as a sequence of amino acids. Karlin and Brendel [5] define scores ranging from -5 to 3 to each of the 20 amino acids. For the human β_2-adrenergic receptor sequence, disjoint subsequences with the highest scores are obtained and these subsequences correspond to the known transmembrane domains of the receptor. Efficient $O(n)$ time sequential algorithms are known to solve this problem [2, 3].

A variation is the 2-D maximum-sum subarray problem, where we wish to obtain the maximum sum over all rectangular contiguous subregions of a given $n \times n$ array. The 2-D version of this problem is solved by an algorithm of Smith [9] in $O(n^3)$ time. Parallel algorithms for the 1-D and 2-D versions are presented by Wen [12, 6] for the PRAM model, in $O(\log n)$ time using, respectively, $O(n/\log n)$ and $O(n^3/\log n)$ processors (EREW PRAM). On the other hand, Qiu and Akl [7] developed parallel algorithms for the 1-D and 2-D versions of the problem on several interconnection networks such as the hypercube, star and pancake interconnection networks of size p. The 1-D algorithm takes $O(n/p + \log p)$ time with p processors and the 2-D algorithm takes $O(\log n)$ time with $O(n^3/\log n)$ processors.

In this paper we propose efficient parallel algorithms on the BSP/CGM computing model for both problems (1-D and 2-D versions). The first algorithm takes $O(n/p)$ parallel time with p processors and a constant number of communication rounds in which $O(p)$ numbers are transmitted. The second algorithm takes $O(n^3/p)$ parallel time with p processors and a constant number of communication rounds. Experimental results are obtained by running the algorithm on a 64-node Beowulf parallel machine. Very promising results are presented at the end of this paper. To our knowledge, there are no BSP/CGM algorithms for these problems in the literature. A preliminary version of the 1-D algorithm has appeared in [1].

2   The Maximum Subsequence Problem

Consider a sequence of n real numbers or scores $(x_1, x_2, \ldots, x_n)$. A *contiguous subsequence* is any interval $(x_i, \ldots, x_j)$ of the given sequence, with $1 \le i \le j \le n$. For simplicity, we use the term *subsequence* to mean *contiguous subsequence* throughout this paper. In the maximum subsequence problem we wish to determine the subsequence $M = (x_i, \ldots, x_j)$ that has the greatest total score $T_M = \sum_{k=i}^{j} x_k$. Without loss of generality, we assume at least one of the x_i to be positive.

Obviously if all the numbers in the sequence are positive, then the maximum subsequence is the entire original sequence. We allow the scores to be negative numbers. For instance, given the sequence $(3, 5, 10, -5, -30, 5, 7, 2, -3, 10, -7, 5)$, the maximum sequence is $M = (5, 7, 2, -3, 10)$ with total score $T_M = 21$.

There is a simple and elegant sequential algorithm of $O(n)$ for the maximum subsequence problem [2, 3]. It is based on the idea that if we have already determined the maximum subsequence M of total score T_M of the sequence $(x_1, x_2, \ldots, x_k)$, then we can easily extend this result to determine the maximum subsequence of the sequence $(x_1, x_2, \ldots, x_k, x_{k+1})$.

3   The Parallel Maximum Subsequence Algorithm

We propose a parallel algorithm for the maximum subsequence problem for a given sequence of n scores. We use the BSP/CGM (coarse-grained multicomputer) model [4, 11], with p processors, where each processor has $O(n/p)$ local memory. This algorithm requires a constant number of communication rounds.

Consider a given sequence of n scores $(x_1, x_2, \ldots, x_n)$. Without loss of generality, we assume that n is divisible by p. They are partitioned equally into p intervals, such that each of the p processors stores one interval. Thus the interval $(x_1, \ldots, x_{n/p})$ is stored in processor 1, the interval $(x_{n/p+1}, \ldots, x_{2n/p})$ is stored in processor 2, and so on. We now show that each interval of n/p numbers can be reduced to only five numbers.

Without loss of generality, denote the interval of n/p numbers stored in each processor by $I = (y_1, y_2, \ldots, y_{n/p})$.

We show that it is possible to partition I into five subsequences, denoted by P, N_1, M, N_2, S where

1. $M = (y_a, \ldots, y_b)$ is the maximum subsequence of I, with score $T_M \geq 0$.
2. $P = (y_1, \ldots, y_r)$ is the maximum prefix of I, with score $T_P \geq 0$.
3. $S = (y_s, \ldots, y_{n/p})$ is the maximum suffix of I, with score $T_S \geq 0$.
4. N_1 is the interval between P and M, with score $T_{N_1} \leq 0$.
5. N_2 is the interval between M and S, with score $T_{N_2} \leq 0$.

Each processor finds the maximum subsequence M of I, the maximum prefix P of I and the maximum suffix S of I. We have several cases to consider.

If all the y_i are negative numbers, then we assume M, P and S empty with $T_M = T_P = T_S = 0$, N_1 is the entire I and N_2 empty with $T_{N_2} = 0$.

We now show that

Lemma 1. *If M is not empty, then one of the following cases must hold.*

1. *P is to the left of M, with $r < a$, and with N_1 in between.*
2. *M is equal to P, with $a = 1$ and $b = r$. We have no N_1.*
3. *M is a proper subsequence of P, with $a > 1$ and $b = r$. We have no N_1.*

Proof. If $r < a$, case 1 holds. Let us suppose that $r \geq a$. We have to prove that $r = b$, showing that 2 or 3 holds.

With $r \geq a$, if $r < b$ then the score of $(y_a, \ldots, y_r)$ is smaller than T_M, so the score of $(y_{r+1}, \ldots, y_b)$ is positive. Then the prefix $(y_1, \ldots, y_b)$ would have a score greater than T_P, a contradiction.

Similarly, with $r \geq a$ and $b < r$, $(y_{b+1}, \ldots, y_r)$ would have a positive score and $(y_a, \ldots, y_r)$ would have a score greater than T_M, again a contradiction. So $r \geq a$ leads to $r = b$.

We have also the following lemma regarding the maximum suffix S with a similar proof.

Lemma 2. *If M is not empty, then one of the following cases must hold.*

1. *S is to the right of M, with $s > b$, and with N_2 in between.*
2. *M is equal to S, with $a = s$ and $b = n/p$. We have no N_2.*
3. *M is a proper subset of S, with $a = s$ and $b < n/p$. We have no N_2.*

The five values T_P, T_{N_1}, T_M, T_{N_2} and T_S for each interval are used in the parallel algorithm. When M and P are not disjoint, that is, M is a subsequence of P, whether proper or not, we redefine T_P to be 0 and T_{N_1} to be the non-positive score of the prefix that immediately precedes M. A similar adaptation in done with S and T_{N_2} when M and S are not disjoint. It is easy to see that after this redefinition,

$$T_P + T_{N_1} + T_M + T_{N_2} + T_S = \sum_{i=1}^{n/p} y_i,$$
$$\text{score of } P = \max\{T_P, \ T_P + T_{N_1} + T_M\}, \text{ and}$$
$$\text{score of } S = \max\{T_M + T_{N_2} + T_S, \ T_S\}.$$

Thus, in this way, we build a sequence of five numbers with the same scores as in the original interval, regarding the total score (entire sequence), maximum subsequence, maximum prefix and maximum suffix. The seemingly useless zeros are kept to simplify the bookkeeping in the last step of the parallel algorithm.

Having computed the five numbers above, each processor sends them to processor 1. Processor 1 solves the maximum subsequence problem of the $5p$ numbers sequentially, in $O(p)$ time and reports the solution. See Algorithm 1.

Algorithm 1 Parallel Maximum Subsequence.

Input: The input sequence of n numbers $(x_1, x_2, \ldots, x_n)$ equally partitioned among the p processors.

Output: The maximum subsequence of the input sequence.

1: Let the sequence stored in each processor be $I = (y_1, y_2, \ldots, y_{n/p})$. Each processor obtains the maximum subsequence M of I with score T_M.
2: Each processor obtains the maximum prefix P with score T_P, and obtains the maximum suffix S with score T_S. The interval between P and M is N_1 with score T_{N_1}; the interval between M and S is N_2 with score T_{N_2}.
3: Consider Lemma 2 and redefine the appropriate values of $T_P, T_{N_1}, T_M, T_{N_2}, T_S$ if necessary.
4: Each processor sends the five values $T_P, T_{N_1}, T_M, T_{N_2}, T_S$ to Processor 1.
5: Processor 1 receives the $5p$ values and computes the maximum subsequence of the received values.
6: Let the maximum subsequence obtained be $m_1, \ldots, m_k$. The processor that stores m_1 can easily compute the start index of the maximum subsequence corresponding to the original input, while the processor that stores m_k can compute the end index of the answer.

We have the main result.

Theorem 1. *Algorithm 1 correctly computes the maximum subsequence of $(x_1, x_2, \ldots, x_n)$ in a constant number of communication rounds involving the transmission of $O(p)$ numbers and $O(n/p)$ local computation time.*

Proof. The correctness of the parallel algorithm is based on Lemma 1 and Lemma 2. It is easy to see that the maximum subsequence considering the $5p$ values corresponds to the maximum subsequence of the original sequence. If the latter is entirely contained in one of the p intervals, the correspondence is direct. Otherwise, it starts within an interval (being its maximum suffix), spans zero or more entire intervals, and ends within another interval (being its maximum prefix). The $5p$ values contain all the necessary information to find this subsequence. The local computation time is $O(n/p)$.

4   The Parallel Maximum Subarray Algorithm

Let a_{ij} be the elements of an $n \times n$ array A, $1 \leq i, j \leq n$. Denote by $A[i_1..i_2, j_1..j_2]$ the subarray of A composed by rows i_1 to i_2 and columns j_1 to j_2. Given integers g, h from 1 to n, denote by $R^{g,h}$ the set all the subarrays $A[i_1..i_2, g..h]$, $1 \leq i_1 \leq i_2 \leq n$. The maximum of the sums of subarrays belonging to $R^{g,h}$ can be found as follows. For given $1 \leq g \leq h \leq n$, consider the subarray $A[1..n][g, h]$. Define the column sequence $C^{g,h}$ of size n as the sequence formed by the sums of each row of $A[1..n][g, h]$, i.e., $C_i^{g,h} = \sum_{j=g}^{h} a_{ij}$.

It can easily be shown [6] that the maximum sum of all the subarrays of $R^{g,h}$ is equal to the maximum subsequence sum of $C^{g,h}$. Let us illustrate this by a simple example.

Let $A = \begin{pmatrix} -10 & -10 & 20 & -20 & -30 \\ 5 & -20 & 10 & 40 & 10 \\ 30 & -40 & 20 & -10 & -15 \\ -20 & 4 & -5 & 50 & 10 \\ 10 & -20 & 10 & -40 & 10 \end{pmatrix}$. We have $A[1, 5][g, h] = \begin{pmatrix} 20 & -20 & -30 \\ 10 & 40 & 10 \\ 20 & -10 & -15 \\ -5 & 50 & 10 \\ 10 & -40 & 10 \end{pmatrix}$

for $g = 3$ and $h = 5$ and $C^{3,5} = (-30, 60, -5, 55, -20)$.

The maximum subsequence of $C^{3,5}$ is $(60, -5, 55)$ and its sum is 110. This is also the sum of the maximum subarray among all subarrays belonging to $R^{3,5}$, that is, the subarray

$\begin{pmatrix} 10 & 40 & 10 \\ 20 & -10 & -15 \\ -5 & 50 & 10 \end{pmatrix}$

The problem of computing the maximum subarray sum of A is equivalent to obtaining the the maximum sum of all the subarrays of $R^{1,n}$. By considering all possible pairs of g, h, with $1 \leq g \leq h \leq n$, the maximum subarray sum of A is the maximum of the corresponding maximum subsequence sums of such $C^{g,h}$. In our example, the maximum subarray is obtained for $g = 3$ and $h = 5$.

The sequences $C^{g,h}$ can be computed efficiently by a preprocessing the rows of A. Each row is replaced by its prefix sum. With this, the sum of elements of a row between any two columns g and h can be computed in constant time.

We summarize the parallel subarray algorithm in the following.

Algorithm 2 Parallel Maximum Subarray.

Input: Each processor i, $1 \leq i \leq p$, receives the subarray $A[(i-1)n/p..i\ n/p][1..n]$.
Output: The maximum subarray with the largest sum.

1: Each processor computes the prefix sums of the rows stored in it.
2: Each processor computes all the $C^{g,h}$'s (of the subarray).
3: For each $C^{g,h}$, compute the five values (as in Algorithm 1) in order to obtain the maximum subsequence. (Each processor computes $n^2/2$ of these values.)
4: Each processor i sends to each processor j the j-th part of the array computed in step 3 and receives the i-th part of the array from each processor j.
5: Each processor i computes the subarray of maximum sum with the data computed/received (a total of $(n^2/2)/p$ columns).
6: Processor 1 obtains the subarray with the largest sum among the subarrays.

In step 4, all the processors send/receive a vector of size n^2/p (h-relation of size n/p). Step 3 is executed for each one of the O(n^2) $C^{g,h}$. Since there are n/p elements in the column, we have the time complexity of O(n^3/p).

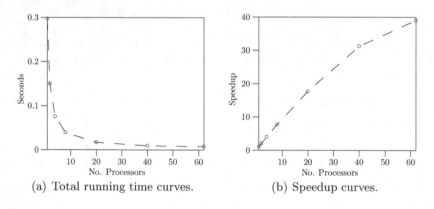

(a) Total running time curves.   (b) Speedup curves.

Fig. 1. Maximum subsequence algorithm - input size $n = 1,830,000$.

5   Experimental Results

We have run the two parallel algorithms on a 64-node Beowulf machine consisting of low cost microcomputers with 256MB RAM, 256MB swap memory, CPU Intel Pentium III 448.956 MHz, 512KB cache. The cluster is divided into two blocks of 32 nodes each. The nodes of each block are connected through a 100 Mb fast-Ethernet switch. Our code is written in standard ANSI C using the LAM-MPI library. We assumed an input size of $n = 1,830,000$. (This size corresponds to the number of nucleotide pairs of the bacterium *Haemophilus influenzae*, the first free-living organism to have its entire genome sequenced [10].) We have used randomly generated data. Results of the 1-D algorithm are shown in Figure 1. Results of the 2-D algorithm are shown in Figures 2 and 3. In the 2-D case, for smaller arrays (up to 512×512), the gain in computing time (O(n^3/p)) is

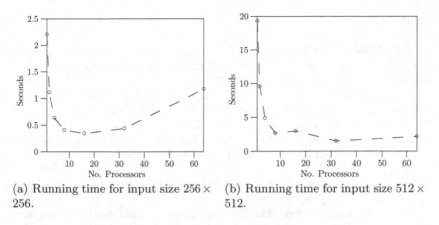

(a) Running time for input size $256 \times$ 256.

(b) Running time for input size $512 \times$ 512.

Fig. 2. Maximum subarray algorithm - input sizes 256×256 and 512×512.

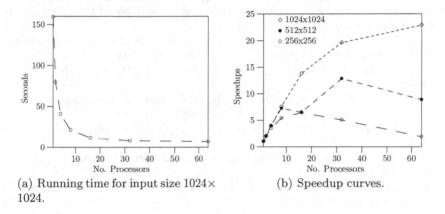

(a) Running time for input size $1024 \times$ 1024.

(b) Speedup curves.

Fig. 3. Maximum subarray algorithm - input sizes 1024×1024 and speedups.

not sufficient to compensate the volume of messages transmitted ($O(n^2/p)$), and the performance obtained is not so good. The results are improved when we use larger arrays such as 1024×1024.

6   Conclusion

We propose efficient parallel solutions to the maximum subsequence and the maximum subarray problems. In the proposed algorithms, the input is partitioned equally among the processors. Both algorithms require a constant number of communication rounds. The proposed algorithms not only find the maximum score of the subsequence or subarray, but also the subsequence or the subarray proper. The good performance of the parallel algorithms is confirmed by experimental results run on a 64-node Beowulf parallel computer. Finally we notice that since the sequential algorithms are very efficient (respectively $O(n)$ and $O(n^3)$ time for the 1-D and 2-D problems), the parallel versions are justified for large sequences or large arrays.

References

1. C. E. R. Alves, E. N. Cáceres, and S. W. Song. Computing maximum subsequence in parallel. In *Proceedings II Brazilian Workshop on Bioinformatics - WOB 2003*, pages 80–87, December 2003.
2. J. L. Bates and R. L. Constable. Proofs as programs. *ACM Transactions on Programming Languages and Systems*, 7(1):113–136, January 1985.
3. J. Bentley. *Programming Pearls*. Addison-Wesley, 1986.
4. F. Dehne, A. Fabri, and A. Rau-Chaplin. Scalable parallel geometric algorithms for coarse grained multicomputers. In *Proc. ACM 9th Annual Computational Geometry*, pages 298–307, 1993.
5. S. Karlin and V. Brendel. Chance and significance in protein and dna sequence analysis. *Science*, 257:39–49, 1992.
6. K. Perumalla and N. Deo. Parallel algorithms for maximum subsequence and maximum subarray. *Parallel Processing Letters*, 5(3):367–373, 1995.
7. K. Qiu and S. G. Akl. Parallel maximum sum algorithms on interconnection networks. Technical report, Queen's Unversity, Department of Computer and Information Science, 1999. No. 99-431.
8. W. L. Ruzzo and M. Tompa. A linear time algorithm for finding all maximal scoring subsequences. In *Proceedings of the Seventh International Conference on Intelligent Systems for Molecular Biology*, pages 234–241. AAAI Press, August 1999.
9. D. R. Smith. Applications of a strategy for designing divide-and-conquer algorithms. *Sci. Comput. Programs*, 8:213–229, 1987.
10. D. P. Snustad and M. J. Simmons. *Principles of Genetics*. John Wiley and Sons, 2000.
11. L. Valiant. A bridging model for parallel computation. *Communication of the ACM*, 33(8):103–111, 1990.
12. Zhaofang Wen. Fast parallel algorithm for the maximum sum problem. *Parallel Computing*, 21:461–466, 1995.

A Parallel Approach
for a Non-rigid Image Registration Algorithm*

Graciela Román-Alonso,
Norma Pilar Castellanos-Abrego, and Luz Zamora-Venegas

Departamento de Ing. Eléctrica, Universidad Autónoma Metropolitana, Izt.
Ap. Postal 55-534, D.F. 09340, México
{grac,npca}@xanum.uam.mx, luz@ixil.izt.uam.mx

Abstract. Currently, the non-rigid registration of medical images has become an important issue in medical image processing. This work deals with the MPI parallel implementation of a non-rigid algorithm to speed-up its performance in processing image sub-regions with an Evolutionary Algorithm (EA). We are parallelizing the EA and implementing a distributed version of a Divide and Conquer algorithm, using groups of processes. Results show the influence of the number of groups and the number of processors in the execution time. Our implementation offers a speed-up of 15.6 on a cluster of 15 PC's decreasing the execution time from 39min to 2.5min.

1 Introduction

Medical image registration is a technique for the spatial alignment of anatomical structures. The non-rigid registration employs non-linear transformations to correct deformations caused by image scanners [8], by the soft tissue movement when a tumor is growing [7] or by lung volume changes [1], etc. This techniques usually incorporate several Degrees of Freedom (DOFs), increasing the dimensionality of the search space and the complexity of finding the optimal global parameters of the spatial transformation. The current trend of this techniques is towards hierarchical algorithms to deal with the high complexity of multidimensional spaces. In this work we present a parallel approach using the MPI library for a non-rigid image registration algorithm on a cluster of PCs. Hierarchical non-rigid registration algorithms demand powerful hardware or long execution time. Unlike other solutions using parallel processing [6][12], the algorithm used here is based on the Divide and Conquer method (DC) [9], allowing image processing for independent regions and an easier parallel implementation, without decreasing the outcome quality. Results show the influence of the number of groups of processes and the number of processors in the execution time. We also show the marked improvement with this parallel implementation comparing it with the sequential algorithm, which is described in the next sections, and its applicability to medical images is showcased [4].

* This work is supported by the CONACyT Project 34230-A: Infraestructura para la construcción de aplicaciones fuertemente distribuidas.

D. Kranzlmüller et al. (Eds.): EuroPVM/MPI 2004, LNCS 3241, pp. 147–154, 2004.

2 Non-rigid Image Registration

2.1 Background

In the registration of a pair of images, the generation of a spatial functional relationship to map the positions of similar objects or patterns is required. Let $f_s(x_s)$ and $f_t(x_t)$, be the intensity distribution, where $x_s \in \Omega_s$ and $x_t \in \Omega_t$, are the positions of the source and target domains, respectively. A spatial transformation generates this map between the domain positions of the source and target image $G(x_s) = x_t$, in a domain defined by their overlapping region. In this work we use a hierarchical algorithm previously reported [4], that estimates a global deformation $G : \Re^2 \to \Re^2$, by maximizing a similarity measure based on the intensity of the objects in the source and target domain, $f_s(G(x_s)) \approx f_t(x_t)$. The global deformation function G is estimated by the composition of L local spatial transformations,

$$G = g_L \circ g_{L-1} \circ \ldots \circ g_2 \circ g_1 \tag{1}$$

where $g_{i+1} \circ g_i$, denotes the composition of local transformations in sub-regions of the image going from global to local. Once the corresponding spatial domains or sub-regions have been localized in the source and target images, searching for the optimal vector of parameters $\mathbf{p}$ that defines g, can be seen as a nonlinear optimization problem. Thus, it is possible to get the spatial correspondence between similar patterns in both images by maximizing the similarity function $I(f_t, f_s(g))$. The searching strategy imposes constraints on $\mathbf{p}$ to guarantee continuity, differentiability and one-to-one correspondence.

2.2 Deformation Model on Sub-regions

The local spatial transformation g_i used here, has the advantage over other techniques of using 5 DOFs to produce an independent deformation on a sub-region,

$$g(x_s) = [(1 - \lambda^\alpha)\mathbf{A} + \lambda^\alpha \mathbf{I}]x_s \tag{2}$$

where I represents the identity map, is called the smoothness parameter, A is an affine map,

$$\mathbf{A} = \begin{bmatrix} 1+a & b \\ c & 1+d \end{bmatrix}, \tag{3}$$

where $a, b, c, d \in \Re$. At last, $\lambda : \Re^2 \to \Re$ is defined by a circle with a unitary diameter, $\lambda(x_s) = 2[(x_s - 1/2)^2 + (y_s - 1/2)^2]^{1/2}$ that defines the region to be transformed.

2.3 Divide and Conquer

The DC method allows image processing for independent regions. The algorithm starts with the global transformation of the source image (f_s, which can be again transformed if a better global similarity measure is required between the transformed source image (f_{st}) and the target image (f_t). Then, new transformations must be applied on sub-regions by equally dividing the image by 4. Thus, this dividing procedure can go on, generating smaller sub-regions until a local similarity measure is obtained (which means that no more processing is needed for this sub-region) or a minimum size has been reached. All the transformations generated in each sub-region compose the global transformation function G described in equation (1).

2.4   Optimization

The non-rigid registration task is solved with an Evolutionary Algorithm (EA) that generates an initial random population and uses a random mutation strategy. The reproduction is carried out by a function which is guided by the best current fitness value to search for the five global optimal parameters of g_i (equation (2)). The fitness function represents the similarity between the transformed source image and the target image , described by the Normalized Mutual Information (NMI)[11]. This hybrid evolutionary algorithm is an adaptation of that reported by Pham et al. [10], for a variable mutation rate and validation of constraints. Each individual of the population is implemented by an array of 6 float numbers, where the first entry holds the fitness value (similarity between images) and the next entries hold the 5 parameters of the transformation to optimize, corresponding to the A matrix in equation (3) and the smoothness parameter (α). This EA runs in parallel and its implementation is described below.

fitness	*a*	*b*	*c*	*d*	α

3   Parallel Implementation

In this work we apply the parallelism at two levels: a)In the optimization. For each image (or sub-region) we use a Parallel Evolutionary Algorithm (PEA) to search for the optimal vector of parameters that defines a specific local transformation. The PEA is executed by one group of processes (PEA group). b)In the DC method. The task of the DC algorithm is parallelized by employing 2, 3 and 4 groups of processes to transform the 4 sub-regions obtained in the first partition of the image.

3.1   The Parallel Evolutionary Algorithm

Several parallel implementations of evolutionary algorithms have been reported. They usually divide the total population into the processors and employ different termination conditions [5] [3],[2]. In this work, as a first attempt of parallelization we use a basic master-slave model (Fig. 1) where a central process controls the number of repetitions of the best individual of the total population to stop

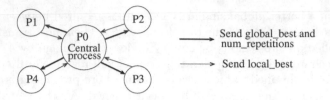

Fig. 1. Centralized model of the Parallel Evolutionary Algorithm (PEA group).

the execution. Based on the parameters that obtained the best fitness in the sequential version, we used 15 and 16 population sizes on 5,8 and 15 processors (3,2 and 1 individual per processor respectively).We define a PEA group composed of n processes eachone running in different processors. Each process runs the following algorithm in parallel:

```
Generate the initial local_population
num_repetitions = 0
while num_repetitions < MAX_REPETITIONS
{   evaluate local_population
    local_best = get the best individual of local_population
    Gather(local_best,all_local_bests, central_processor)
    if ( my_id == central_processor)
    {   global_best = get global_best of all_ local_bests
        update num_repetitions  }
    Bcast(global_best, central_processor)
    Bcast(num_repetitions,central_processor)
    if (mutation_rate)  Mutate_local_population
    Replace_Reproduction(global_best) }
```

Each process generates in parallel an initial local population of equal size corresponding to the division of the total population size into the number of processors. Then, the population is evaluated and the best fitness value of the local population is identified. For each new generation, a best global individual is obtained from all the best local individuals that were sent to a central process. Thus, the central process of the PEA sends to all processors the best global individual (using Broadcast instructions) for the evaluation of the reproduction and mutation operators. The repetitions counter value of the central processor is also sent to all processes which is the indicator to stop the algorithm when it equals the MAX_REPETITIONS value. The final result is given from the central process.

3.2   Image Processing for Independent Regions

We propose two parallel implementations of the algorithm: a centralized and a distributed version of the DC method. In the centralized implementation a unique centralized process generates all the divisions of subregions and it also

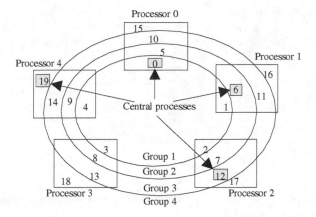

Fig. 2. Four overlapped groups of processes on a cluster of five processors.

works as the centralized processor in a PEA group executing the parallel evolutionary algorithm to obtain the best individual of each generated subregion. The distributed implementation that we are proposing here, considers the use of several PEA groups to carry out in parallel the image registration of the first 4 sub-regions obtained after the first transformation of the source image , as it was described in section 2.3. In order to exploit all the processors on the cluster, the groups of processes are overlapped such that each central process is located on a different processor. In Fig.2 , an example is presented of the configuration with 4 groups of processes on 5 processors (5 processes per group), being processes 0, 6, 12 and 19 the central processes of each PEA group. We tested our approach using 2, 3 and 4 groups of processes on a cluster of 5, 8 and 15 processors. After the Group 1 generates the first 4 sub-regions, they are sent to the central processes of each group (as is shown in Fig. 2 by the arrows), to execute in parallel the DC algorithm on their sub-regions. When we use 4 groups, each group works on a quarter of the first transformed image . For 3 groups, the Group 1 works with 2 quarters and Groups 2 and 3, each one works with only one quarter. In the case of 2 groups, Groups 1 and 2, each one works with 2 quarters of the image.

4   Results and Discussion

Results of the parallel implementation of the non-rigid registration algorithm were obtained with the following calibrated parameters of the EA: 15 repetitions of the best global individual and a Population size of 15 or 16. We tested this parallel implementation with magnetic resonance images of 256x256 pixels and 256 gray levels. The minimun size of sub-regions was 16x16 pixels and its miminum similarity measure of 0.71. The NMI was implemented by aproximating the probability distribution with the image histogram of 10 gray levels. We chose three configurations of 5, 8 and 15 processors in order to asign 3, 2 and

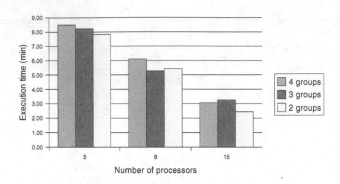

Fig. 3. Comparison of the execution time using 2, 3 and 4 groups of processes on 5, 8 and 15 processors, respectively.

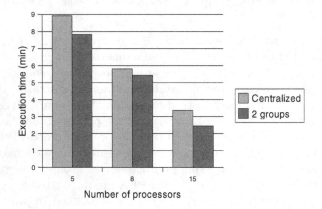

Fig. 4. Comparison of the obtained execution time by the centralized DC and the 2 groups parallel version on 5, 8 and 15 processors.

1 individuals per processor, respectively. The processors on the cluster are homogenous (Pentium 4, 512 MB RAM and 2.8 GHz, connected with a super stack 3 switch). We obtained the execution time (10 executions per configuration) by using 2, 3 and 4 groups of processes on the configuration of 5, 8 and 15 processors (Fig. 3), respectively. An improvement of the execution time can be observed when we increase the number of processors. The best results were obtained using 15 processors, in this configuration the PEA performance was better because it considers only one individual per process. The implementation with 4 and 3 groups increased the parallelism but introduce more communication cost. On the contrary,by using only 2 groups we got the best results. As can be seen in Fig. 4, using 2 groups gave better results than the centralized DC algorithm (only one group of processes) for any number of processors. Comparing the execution time with 2 groups on a cluster of 15 PC's (2.5 min) and the sequential DC version (39 min), an speed-up of a factor of 15.6 was achieved. An image example is

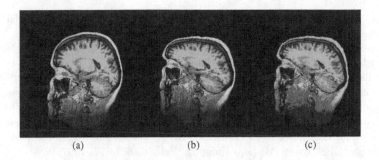

(a)   (b)   (c)

Fig. 5. Results were obtained applying the non-rigid registration method to Magnetic Resonance images of a human head. (a) The source image, (b) the target image and (c) the final transformed source image, are shown.

shown in Fig. 5, achieving a fitness value of 0.7520, which means that we got a similarity of 75.20transform source image and the target image.

5   Conclusions and Future Work

In this paper we present a parallel approach for a non-rigid registration algorithm of medical images that employs a Divide and Conquer method to process an image by independent sub-regions. An evolutionary algorithm finds the optimal vector of parameters to process each image or subregion. Our approach was based on the definition of two parallel implementations of the algorithm: a centralized and a distributed version of the DC method. In the centralized version we defined one group of processes with a central process that generates all the subregions based on the DC method. This implementation also controls the processes of the group executing a parallel evolutionary algorithm for each generated subregion. The distributed version considers 2, 3 and 4 groups of processes which are overlapped on a homogenous cluster of 5, 8 and 15 processors, each group has a process per processor. The groups are executed in parallel, the centralized DC algorithm on the 4 subregions are generated in the first division of the initial transformation of the source image. The 2 groups version got the best results using a cluster of 15 PC's, in contrast with the 4 and 3 groups that increased the parallelism, but on the other hand introduced more communication and synchronization cost. Future work is related to some implementation aspects that will allow the improvement of the algorithm's performance. One of them is the incorporation of load balancing techniques to distribute the generated subregions between the central processes, taking into account the workload of processors.

References

1. Boldea V., Sarrut D., Clippe S. *Lung Deformation Estimation with Non-Rigid Registration for Radiotherapy Treatment. MICCAI'03.* Lecture Notes in Computer Science. Vol. 2878 , 770-777, 2003.

2. Cantú-Paz E. *Markov Chain Models of Parallel Genetic Algorithms*. IEEE Transactions on evolutionary computation, Vol. 4, Num.3, September 2000.
3. Cantú-Paz E., Goldberg D.E. *On the Scalability of Parallel Genetic Algorithms*. Evolutionary Computation, Vol. 7, Num.4.,1999.
4. Castellanos, N.P., Del Angel P.L., Medina, V. *Nonrigid Image Registration Technique as a Composition of Local Warpings*. Pattern Recognition, In Press, 2004.
5. Chipperfield, A., Fleming, P. *Parallel Genetic Algorithms*. Parallel and Distributed Computing Handbook, A. Y. H. Zomaya, editor, McGraw-Hill, New York, 1996.
6. Christensen, G. E., Joshi, S. C., Miller, M. I. *Volumetric transformation of brain anatomy*. IEEE Trans. Medical Imaging. Vol. 16. 864-877, 1997.
7. van Engeland, S., Snoeren, P., Hendriks, J., Karssemeijer, N. *Comparison of Methods for Mammogram Registration*. EEE Trans. Medical Imaging. Vol. 22 , 1436-1444, 2003.
8. Kybic, J., Thévenaz, P., Nirkko, A., Unser, M. *Unwarping of Unidirectionally Distorted EPI Images*. IEEE Trans. Medical Imaging. Vol. 19, 80-93, 2000.
9. Michael, A., Barry, W. *Parallel Programming, Techniques and Applications Using Networked Workstations and Parallel Computers*. Prentice Hall, 1999.
10. Pham, D.T., Karaboga, D. *Intelligent Optimisation Techniques*. Springer-Verlag, Londres, 51-61, 2000.
11. Pluim, J.P.W., Maintz, J.B.A., Niessen, W.J. *Mutual-Information-Based Registration of Medical Images*. IEEE Trans. Medical Imaging. Vol. 22, 986-1004, 2003.
12. Rohde, G. K., Aldroubi, A., Dawant, B.M. *The Adaptive Bases Algorithm for Intensity-Based Nonrigid Image Registration*. IEEE Trans. Medical Imaging. Vol. 22, 1470-1479, 2003.
13. Stefanescu, R., Pennec X., Ayache, N., *Parallel Non-rigid Registration on a Cluster of Workstations*. In *HealthGrid'03*. In *Sofie Norager*, 2003.

Neighborhood Composition:
A Parallelization of Local Search Algorithms

Yuichi Handa, Hirotaka Ono, Kunihiko Sadakane, and Masafumi Yamashita

Dept. of Electrical Engineering and Computer Science, Kyushu University
{u1,ono,sada,mak}@tcslab.csce.kyushu-u.ac.jp

Abstract. To practically solve NP-hard combinatorial optimization problems, local search algorithms and their parallel implementations on PVM or MPI have been frequently discussed. Since a huge number of neighbors may be examined to discover a locally optimal neighbor in each of local search calls, many of parallelization schemes, excluding so-called the multi-start parallel scheme, try to extract parallelism from a local search by distributing the examinations of neighbors to processors. However, in straightforward implementations, when the next local search starts, all the processors will be assigned to the neighbors of the latest solution, and the results of all (but one) examinations in the previous local search are thus discarded in vain, despite that they would contain useful information on further search.

This paper explores the possibility of extracting information even from unsuccessful neighbor examinations in a systematic way to boost parallel local search algorithms. Our key concept is *neighborhood composition*. We demonstrate how this idea improves parallel implementations on PVM, by taking as examples well-known local search algorithms for the Traveling Salesman Problem.

1   Introduction

Most of combinatorial problems which frequently arise in various real-world situations, such as machine scheduling, vehicle routing and so on, are known to be NP-hard [3], and are believed that there would not exist polynomial time algorithms to find optimal solutions. Many researchers are hence interested in approximation algorithms [8] that can find near-optimal solutions in reasonable time.

Among them are metaheuristics algorithms based on local search very popular [4, 1, 10] because of their simplicity and robustness. A generic outline of a local search algorithm starts with an initial feasible solution and repeats replacing it with a better solution in its neighborhood until no better solution is found in the neighborhood. Although local search algorithms are much faster than exact algorithms, they may still require an exponential time for some instances. Furthermore we have started considering that just a polynomial time algorithm is no longer practical for huge instances in real applications.

Parallel implementations of local search algorithms are promising to satisfy the above requirement and hence many parallel algorithms have been proposed

D. Kranzlmüller et al. (Eds.): EuroPVM/MPI 2004, LNCS 3241, pp. 155–163, 2004.

mainly 1) to reach a better solution and 2) to re.duce the processing time, although they are apparently related with each other. A well-known paradigm called multi-start has every processor independently execute the same algorithm from randomly selected initial solution and returns the best solution among those obtained by the processors, mainly to increase the quality of solution (see e.g., [2]). A parallelized GRASP is an application of multi-start paradigm; while original GRASP randomly generates initial solutions in greedy manner then applies local search for each initial solution, in parallelized GRASP it is considered that several processors do GRASP for different seeds of randomness (e.g., [7]).

As mentioned, a local search algorithm repeats a local search starting with the current solution. Since the area of neighborhood can be huge as the size of instance becomes large, a local search consists of many independent searches (inside neighborhood) for a locally optimal solution. Many parallel implementations thus try to extract parallelism from each of the local searches by distributing the searches (in each of local searches) to processors, mainly to reduce the processing time. However, in straightforward implementations, all the processors will be assigned to the search of neighborhood of the current solution when the next local search starts, and the results of all (but one) searches in the previous local search are thus discarded in vain, despite that they would contain useful information on further search.

This paper explores the possibility of extracting useful information even from unsuccessful neighborhood searches to boost parallel local search algorithms. Our key concept is *neighborhood composition*. In a local search algorithm, a local search starts with a solution x and tries to obtain a better solution y. Suppose that a search i suggests the replacement of a subsolution u_i of x with a v_i to obtain a better solution x_i. Since the size of an instance is huge and search i can explore only very limited neighborhood of x, for many i and j, u_i and u_j do not overlap each other. Obviously, in many problems, x_i can be improved further by replacing u_j in x_i by v_j. This trivial fact is the essence. Suppose that searches i and j are executed in different processors p_i and p_j in a parallel implementation and that x_i achieves the locally optimal solution. Why don't we use u_j, v_j pair to improve the current solution, instead of discarding it? This is our claim.

The neighborhood composition gives us a concrete idea how to realize this idea on PVM or MPI. It works on the master/slave model. First a master processor divides the solution space and distributes them to slave processors. Then each slave searches the solution space independently. The master keeps the current best solution found in the whole search, and the slaves keep their own current solutions. When a slave finds an improved solution, it sends the difference between its own current solution structure and the improved solution structure, as *improvement information*, to the master. The master then tries to apply the improvement information to its own current solution. If the trial is successful, the master gets a new improved solution and sends it to the slaves. Otherwise if it is not, the section of solution space assigned to the slave is updated (because the slave's current solution may be completely different from the current solution). Since our method does not need synchronize the slaves, each slave can search its own neighborhood while other slaves communicate with the master, which hide the communication overheads.

To confirm the availability of our method, we choose the Traveling Salesman Problem [6] (TSP, for short) and 2OPT (or Or-OPT) and Lin-Kernighan [5] neighborhood local search algorithms as a model problem and its algorithms. We implemented these algorithms on PVM, and then conducted computational experiments. The reasons why we adopt TSP and these neighborhood set are as follows: 1) TSP is one of the most well-known problems, and we can easily obtain many benchmark problems from TSPLIB[1]. Also many algorithms based on local search are proposed. 2) 2OPT, Or-OPT and Lin-Kernighan neighborhoods in TSP are suitable to explain our idea visually. Through the computational experiments, we see that our parallelized local searches achieve good performance in several cases; that is, the processing time of parallelized algorithm is much smaller than the original one.

2   TSP and Local Search

In this section, we introduce some basic ideas and notations used throughout this paper. Although we should actually give more general definitions and notations for combinatorial optimization and local search algorithms, we restrict our explanations to TSP and its local search problems due to the space limitation. One can easily apply them to many other problems and many other local search algorithms.

TSP is described as follows: Given a set of n cities and an $n \times n$ distance matrix D, where d_{ij} denotes the distance from city i to city j, with $i, j = 1, ..., n$, find a tour that visits each city exactly once, and is of minimum total length.

In order to solve TSP, many local search based algorithms are proposed. A local search starts from an initial solution σ and repeats replacing σ with a better solution in its *neighborhood* $N(\sigma)$ until no better solution is found in $N(\sigma)$, where $N(\sigma)$ is a set of solutions obtainable by slight perturbations. The local search from an initial solution σ_0, in which the neighborhood N is used, is formally described as follow.

Algorithm Local Search(N, σ_0)

step1. Set $\sigma := \sigma_0$.
step2. Search a feasible solution $\sigma' \in N(\sigma)$ such that $cost(\sigma') < cost(\sigma)$ (SEARCH). If such σ' exists, set $\sigma := \sigma'$ (IMPROVE) and return to step2. Otherwise go to step3.
step3. Output σ and stop.

Obviously, the performance of local search algorithm depends on which neighborhood we use. If we adopt wider neighborhood, the local search may find the better solutions, however, it probably requires more computational time. In this paper, we consider local search algorithms for TSP whose neighborhood types are 2OPT, 3OPT, and Or-OPT. A 2OPT move deletes two edges, thus breaking

---

[1] http://www.crpc.rice.edu/softlib/tsplib/

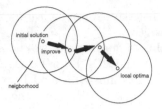

Fig. 1. Local Search.

the tour into two paths, and then reconnects those paths in the other possible way (Fig.2). Thus, A 2OPT neighborhood of a solution (tour) σ is defined as a set of tours that can be obtained by 2OPT moves from tour σ. Similarly, 3OPT move and neighborhood are defined by the exchange replaces up to three edges of the current tour. An Or-OPT neighborhood is a subset of 3OPT (Fig.3). The Lin-Kernighan heuristic allows the replacement of an arbitrary number of edges in moving from a tour to neighboring tour, where again a complex greedy criterion is used in order to permit the search to go to an unbounded depth without an exponential blowup. The Lin-Kernighan heuristic is generally considered to be one of the most effective methods for TSP.

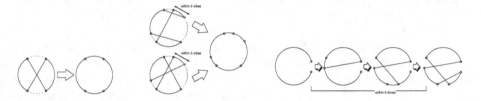

Fig. 2. 2OPT move.   **Fig. 3.** Or-OPT move.   **Fig. 4.** k LK-OPT move.

All of these are the most basic types of neighborhoods for TSP, and indeed quite a many number of local search and metaheuristics algorithms are proposed and studied[1]. Also, these neighborhood structures can be seen in many other (NP-hard) combinatorial problems. Indeed, GAP(general assignment problem)'s basic neighborhoods, swap-neighborhood, ejection-neighborhood have quite similar structures to 2OPT and LK-OPT of TSP, respectively.

3   Parallelization of Local Search

As described in Section 2, local search has two important phases, SEARCH and IMPROVE. In our parallelization, these two phases are imposed by different processors: IMPROVE is done by a master processor, and SEARCH is by slave processors; a rough idea of acceleration is that the master processor maintains the current best solution (do IMPROVE) and slave processors share SEARCH operations.

For this purpose, we consider to divide the neighborhood in SEARCH, and to assign them to slave processors. Such a division is easy to design if all the slave processors are synchronized. One simple way is as follows: The master divides search space, and assigns them to slaves. Each slave executes SEARCH operation for the assigned space. Once a slave finds a better solution, it sends the result to the master. Then it performs IMPROVE by replacing the current solution with the received solutions, and divides the new search space for the improved solution again. In this method, search space is divided into all n slaves, so that the time spent by SEARCH will be ideally reduced to $1/n$. In fact, this method however needs much time in communication between the master and slaves. Moreover, only a result of one slave is used for updating the master's current solution; most of SEARCHs by slave processors are in vain.

To overcome this, we propose the following method:

Algorithm Master (IMPROVE)

Step1. Set $\sigma_{master} := \sigma_0$ and divide $N(\sigma_{master})$ into N_1, ..., N_n, then send σ_{master} and N_i to each slave $slave[i]$.
Step2. Wait. When receiving a data from slave, go to step3.
Step3. Improve σ_{master} with the improvement information received from $slave$ $[i]$, if possible(CHECK).
Step4. Divide $N(\sigma_{master})$ and send the σ_{master} and new N_i to $slave[i]$ from which the improvement information were received just before. Return to step2.

Algorithm Slave (SEARCH)

Step1. Receive the initial solution σ_0 and N_i. $\sigma_{slave[i]} := \sigma_0$.
Step2. If there is a better solution σ' in N_i, send to the master the improvement information of σ'.
Step3. Wait. When receiving σ_{master} and N_i from the master, replace N_i and set $\sigma_{slave[i]} := \sigma_{master}$, and go to step2.

Improvement information in the above description is defined as the difference between the original solution and its improved solution. For example, in case of 2OPT, improvement information is represented by two edges in the improved solution but not in the original solution.

As an example, suppose that there are one master and three slaves available. First, the master and slaves have the same initial solution. Slaves begin to search the same neighborhood of the initial solution from different neighbor (Fig.5).

If one slave finds a better solution, the slave sends to the master not the better solution but the improvement information. Receiving the improvement information, the master checks if it is consistent for the current solution (i.e., the improvement keeps the solution feasible). If CHECK is yes, the master improves the current solution based on it. The master then returns the new current solution to only the slave that generated the improvement information; The other slaves (Slave A and C at Fig. 6) do not receive the new solution. That is, these slaves continue to their own neighborhood.

Here, it should be noted that the solution kept by the master may be different from ones of the slaves. If the neighborhood structure of the slave's solution is

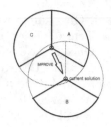
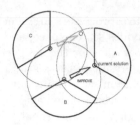

Fig. 5. Each slave begins a search in a different space.

Fig. 6. Slave B finds a better solution.

Fig. 7. Slave A finds a better solution (CHECK is needed).

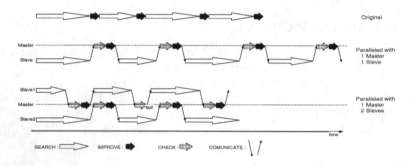

Fig. 8. Comparison of Non-Parallelized, Parallelized by 1 master and 1 slave and Parallelized by 1 master and 2 slave.

quite similar to the one of master's solution, its improvement information can be applied and the improvement is successful (Fig. 9). Otherwise, the improvement information cannot be applied and it fails (Fig. 10). Actually, since our algorithm is based on local-search manner, we can expect that the neighborhood structures are not so different. We call this devise of improving solutions *neighborhood composition*.

Note that our parallelized algorithm does not need synchronization, which causes the following properties: As the defect, the master needs to do the extra task CHECK for every improvement of slave's solution. However, the overheads may become negligible because while one slave communicates with the master and the master does CHECK, other slaves can continue SEARCH (Fig. 8).

4   Experimental Results

To evaluate the performance of this parallelization method, we implement parallel algorithms based on a local search algorithm with basic neighborhood structures [2] [3]. The algorithms which we have parallelized are the following 3 types:

---

[2] http://www-or.amp.i.kyoto-u.ac.jp/members/ibaraki/today/tsp1.c/   Società Italiana di Fisica 2001

[3] http://tcslab.cscc.kyushu-u.ac.jp/%7Eū1/program/lkh3_1.c

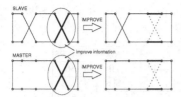

Fig. 9. The master's solution has the same two edges that the slave cut for IMPROVE.

Fig. 10. The master's solution does not have the same two edges that the slave cut for IMPROVE.

Table 1. Average run time (s) and its ratio to the original of parallelized "2OPT", "2,Or-OPT" and "k-LK-OPT".

instance	neighborhood	original	2 slaves	4 slaves	8 slaves
att532 (532cities)	2OPT	5.26(1.00)	3.61(0.68)	2.73(0.51)	2.43(0.46)
	2,Or-OPT	62.83(1.00)	35.98(0.57)	25.16(0.40)	20.36(0.32)
	3-LK-OPT	4.65(1.00)	4.24(0.91)	3.18(0.68)	2.53(0.54)
	5-LK-OPT	4.70(1.00)	4.30(0.91)	2.85(0.60)	2.54(0.54)
pr1002 (1002cities)	2OPT	23.36(1.00)	14.23(0.60)	10.06(0.43)	8.25(0.35)
	2,Or-OPT	349.42(1.00)	199.03(0.56)	118.58(0.34)	83.19(0.24)
	3-LK-OPT	21.11(1.00)	16.77(0.79)	12.56(0.59)	9.75(0.46)
	5-LK-OPT	21.68(1.00)	17.79(0.82)	12.66(0.58)	9.57(0.44)
nrw1379 (1379cities)	2OPT	38.61(1.00)	25.49(0.66)	18.64(0.48)	16.11(0.41)
	2,Or-OPT	881.49(1.00)	494.68(0.56)	309.04(0.35)	203.28(0.23)
	3-LK-OPT	34.04(1.00)	28.10(0.82)	21.79(0.64)	16.69(0.49)
	5-LK-OPT	35.42(1.00)	30.23(0.85)	24.41(0.68)	17.78(0.50)

A. searching 2OPT neighborhood
B. searching 2OPT and Or-OPT neighborhood
C. searching Lin-Kernighan neighborhood

As a problem instance, we used a standard TSP instance, att532, pr1002, nrw1379 in TSPLIB. Our parallel programs run on a cluster of 2.26GHz Pentium4 CPUs.

Table 1 shows the run times of the original and the parallelized algorithms "A", "B" and "C". In all cases, as the number of slaves increases, the run times become much shorter. However, 1) the speed-up effects get small in the parallelization with many slaves. Also 2) the effects get worse in order of "B", "A" and "C". It is considered that both of them are related to the failure of IMPROVE. Table 2 shows the relationship between the failure ratio of IMPROVE and the number of slaves or the depth of the neighborhood. From this, it is observed that the failure ratio becomes higher as the number of slaves increases. The reason may be that the solution structures of slaves easily vary if the number of slaves is large; the similarity is lost. This high ratio of the failure in the parallelization with many slaves explains the phenomenon 1). On the other hand, the failure ratio of "C" is larger than that of "A" or "B". We consider the reason is that

Table 2. Comparison of ratio (%) of failing to improve with the improve solution at parallelized 5-LinKernighan (Instance = att532).

neighborhood	improve information	2 slaves	4 slaves	8 slaves
2,OrOPT	2 edges	2.44	4.79	8.86
	3 edges	13.52	33.18	42.88
	total	2.59	5.35	9.57
5-LK-OPT	2 edges	10.93	25.99	41.48
	3 edges	12.74	28.28	44.59
	4 edges	13.62	29.68	47.29
	5 edges	16.92	37.25	55.40
	total	12.80	29.32	45.98

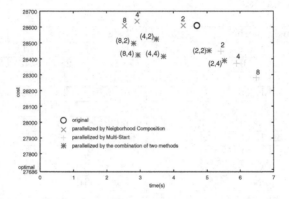

Fig. 11. Comparison of the parallelized by Neighborhood Composition and by Multi-Start and by both two method (5-LKH, Instance = att532).

the IMPROVE of "C" can change the solution structure drastically. (One IMPROVE of 5-LK-OPT cuts at most 5 edges, while one IMPROVE of 2,Or-OPT cuts only 2 or 3 edges). This may be the reason of a part of 2): parallelization of "C" is less effective than "A" and "B".

We also examined the comparison and the combination with another parallelization method, say multi-start parallelization, where several processors execute an identical algorithm with randomness independently, and the best solution among them is adopted. Figure 11 plots the result by the two method. The horizontal axis represents the run time and the vertical axis gives the costs obtained by the algorithms. The symbol "o" represents the result of the original Lin-Kernighan algorithm. The symbol "×" and "+" depict the results by Lin-Kernighan algorithms of our parallelization and of multi-start parallelization, and the numbers above the symbols mean the numbers of slave processors. The symbol "*" shows the results by the combination of two methods, where "(x, y)" means that y teams of 1 master and x slaves execute our parallelized Lin-Kernighan algorithm. While our parallelization makes the local search faster as shown before, multi-start parallelization makes its solution better. By the combination of the two methods, we can get highly better solution in short time.

This implies that our parallelization and multi-start parallelization have different characteristics and then do not conflict.

5   Conclusion

In this paper, we proposed a parallelization method for local search algorithms, which is applicable to many combinatorial optimization algorithms. The purpose of our parallelization is to reduce the run time. We then implemented typical parallelized local search algorithms for TSP, and conducted computational experiments. The result of experiment shows that our parallelization method greatly reduces the run time: our method or the idea of our method are potentially useful for many combinatorial optimization problems and many local search algorithms. As a future work, we need to confirm that our parallelization method is useful for more sophisticated algorithms, and compare or combine other parallelization methods. Another direction is the construction of more systematic devices to reduce overheads, such as the work of [9].

Acknowledgements

The authors would like to thank anonymous referees for their helpful comments. This work was partially supported by the Scientific Grant-in-Aid by the Ministry of Education, Science, Sports and Culture of Japan.

References

1. E. Aarts and J. K. Lenstra, *Local Search in Combinatorial Optimization*, John Wiley & Son, 1997.
2. Y. Asahiro, M. Ishibashi, and M. Yamashita, Independent and Cooperative Parallel Search Methods for the Generalized Assignment Problem, *Optimization Methods and Software*, 18 (2), 129-141, 2003.
3. M. R. Garey and D. S. Johnson, *Computers and Intractability: A Guide to the Theory of NP-Completeness*, Freeman, 1979.
4. P.E. Gill, W. Murray and M.H. Wright, *Practical Optimization*, Academic Press. 1981.
5. K.Helsgaun, An Effective Implementation of the Lin-Kernighan Traveling Salesman Heuristic, *European Journal of Operational Research* 126 (1), 106-130, 2000.
6. E. L. Lawler, J. K. Lenstra, A. H. G Rinnooy Kan and D. B. Shmoys, The Traveling Salesman Problem, *A Guided Tour of Combinatorial Optimization*, John Wiley and Sons 1985.
7. S. Vandewalle, R. V. Driesschie, and R. Piessens, The parallel performance of standard parabolic marching schemes, *International Journals of Super Scomputing*, 3 (1), 1-29, 1991.
8. V. V. Vazirani, *Approximation Algorithms*, SpringerVerlag, 200.
9. A. S. Wagner, H. V. Sreekantaswamy, S. T. Chanson, Performance Models for the Processor Farm Paradigm, *IEEE Transactions on Parallel and Distributed Systems*, 8 (5), 475-489, 1997.
10. M. Yagiura and T. Ibaraki, On Metaheuristic Algorithms for Combinatorial Optimization Problems, *Systems and Computers in Japan*, 32 (3), 33-55, 2001.
11. M. Yagiura, T. Ibaraki and F. Glover, An Ejection Chain Approach for the Generalized Assignment Problem, *INFORMS Journal on Computing*, 16, 133-151, 2004.

Asynchronous Distributed Broadcasting in Cluster Environment

Sándor Juhász and Ferenc Kovács

Department of Automation and Applied Informatics
Budapest University of Technology and Economics
1111 Budapest, Goldmann György tér 3. IV. em., Hungary
{juhasz.sandor,kovacs.ferenc}@aut.bme.hu

Abstract. Improving communication performance is an important issue in cluster systems. This paper investigates the possibility of accelerating group communication at the level of message passing libraries. A new algorithm for implementing the broadcast communication primitive will be introduced. It enhances the performance of fully-switched cluster systems by using message decomposition and asynchronous communication. The new algorithm shows the dynamism and the portability of the software solutions, while it has a constant asymptotic time complexity achieved only with hardware support before. Test measurements show that the algorithm really has a constant time complexity, and in certain cases it can outperform the widely used binary tree approach by 100 percent. The presented algorithm can be used to increase the performance of broadcasting, and can also indirectly speed up various group communication primitives used in standard message passing libraries.

1 Introduction

Clusters play an increasingly important role in solving problems of high computational challenge, because they scale well, and provide high performance, and good fault tolerance at lower cost than traditional super computers. The speed of internode communication is in focus of many research efforts, because it often hinders efficient implementation of communication intensive algorithms. Thanks to hardware improvements, cluster systems of our days can take benefit of extensions of Ethernet standard (Fast and Gigabit Ethernet), as well as new standards providing high performance (more Gbps) and low latency (< 10 μs) such as Myrinet, SCI, Quadrics, or InfiniBand. The prices of active network elements were also dropped, thus clusters systems usually use a fully-switched network topology, reducing competition for the physical bandwidth and thus providing a collision-free environment for communication.

The peak performance of communications is limited by the physical properties of the underlying network, but previous studies [1,2] concluded that the performance of real-life parallel applications is much more sensitive on different software overheads. Due to the inefficiencies and overheads at the levels of application, message passing

D. Kranzlmüller et al. (Eds.): EuroPVM/MPI 2004, LNCS 3241, pp. 164–172, 2004.

subsystems (such as PVM [3] or MPI [4]), and operating systems, the physical transfer time itself – especially for smaller messages – is only a fraction of the total application-level delay. This paper seeks to speed up the group communication of message
passing libraries. While providing basic elements for sending and receiving messages,
these libraries also offer group communication primitives to ease the creation of
complex communication patterns at the application level. Among these primitives
broadcasting plays an emphasized role, because it is widely used in itself, and also as
a building block of other communication primitives (*allgather, alltoall, allreduce*).
This paper presents a method for enhancing the performance of broadcasting by software means. Our new algorithm uses message decomposition and asynchronous
communication to achieves an execution time complexity of O(1) without hardware
support.

The rest of the paper is organized as follows: Section 2 introduces the commonly
used broadcasting methods – both with and without hardware support – used in cluster environments. Section 3 describes our symmetrical algorithm providing a new
approach of data distribution in fully-switched cluster systems. Section 4 compares
the performance of the widely used tree and the new method, and verifies the correspondence of the measured curves and the performance predicted by the theory. The
paper concludes with summarizing the results and showing their application possibilities.

2   Overview of Broadcasting Methods

Following the recommendations of the MPI standard [4] most communication
subsystems implement all the group communication primitives based on the point-to-
point transfer functions. This technique allows a fast and portable implementation of
the group primitives (only the bottom, point-to-point layer must be rewritten for other
platforms). The efficiency of the different implementations is strongly influenced by
the topology of the underlying connection network. Because of its wide practical use,
this paper focuses on the virtual crossbar (fully switched) topology. All execution
time estimations use the widely accepted [5,6] linear model, where the communication time t_c equals to

$$t_c(n) = t_0 + nt_d ,\tag{1}$$

where n is message size, t_0 is the initial latency, and t_d is the time needed to transfer
one data unit (reciprocal of the effective bandwidth).

The simplest way of broadcasting is the linear method, where the data transfer is
controlled from a single source. This is the most straightforward method, following
exactly the philosophy of one to all data distribution: the source node sends the data
to be distributed to each of the partner nodes one by one. This technique is simple and
easy to implement, but not very efficient: it has a linear increase of execution time as
the number p of destination nodes grows:

$$t_c(n, p) = p(t_0 + nt_d) \Rightarrow O(p) ,\tag{2}$$

Because the source node plays a central role as a single sender, this algorithm has
the advantage of reducing collisions on a shared medium, that is why it was preferred

in the early (middle of the '90-es) implementations of communication libraries (MPICH [7], LAM/MPI [8]). The linear complexity of the broadcasting can be reduced by taking advantage of distributed implementation. The most general way is to parallelize the control of the communication using a binary tree topology. In this case the originator node only sends its data to two other nodes, and all the receiver nodes will act as secondary sources sending the message to two more nodes, doubling each step the number of senders. This achieves an execution time of

$$t_c(n,p) = \lfloor 2 * \log_2(p+2) - 2 \rfloor * (t_0 + nt_d) \Rightarrow O(\log_2 p) . \tag{3}$$

Compared to the linear implementation the main advantages are reduced complexity, and better load balancing. This implementation scales better, and is used in most current MPI implementations (e.g. MagPIe [9]).

Some newer approaches reduce the broadcasting time by using completely different communication patterns. Using message fragmentation combined with pipeline data distribution has a communication time of

$$t_c(n,p) = (p-1)*t_0 + nt_d \Rightarrow O(p) , \tag{4}$$

which becomes $O(1)$ if t_0 is low enough to be neglected (large message sizes, implementations with very low initial latencies). Another promising method is to implement the broadcasting with a Scatter and a consecutive Allgather operation as proposed by Van de Geijn and his colleagues [10]. Thakur in [11] further improves this idea by using recursive doubling in the scattering and gathering phase. With this change the communication time becomes:

$$t_c(n,p) = (\log_2 p + p - 1)t_0 + n\frac{p-1}{p}t_d \Rightarrow O(p) , \tag{5}$$

which also becomes of complexity $O(1)$ for large p and negligible t_0.

With additional implementation efforts, environment specific protocol stacks (such as GAMMA [2]) can be developed to further enhance the network throughput. Many network infrastructures support a form of hardware *multicast* or *broadcast*. In this case the originator has to send the data only once, and the hardware layer takes care of the rest. In theory this results in a complexity of $O(1)$ meaning high performance with perfect scalability. However each receiver node gets exactly the same messages, implying that the problems of reliability, handling large messages, and forming arbitrary groups must be solved by the developers at the software level.

The problem of **reliability** is a well-treated topic [12,13,14], and is usually solved by a kind of acknowledgment mechanism. As all the reply messages are sent back to a single originator, on a large scale their processing may be costly (ACK flooding [14]). Using lazy-acknowledgment protocols and multilevel ACK collection [13] can alleviate this problem. **Large messages** must be broken down into smaller pieces that can be handled by the lower network layers. The originator always has to be aware of the amount of empty buffer space in all the destination nodes, and the whole broadcasting can advance only at the pace of the slowest partner. Higher level message passing libraries allow the **formation of any groups**, but in case of using hardware support the location and the available set of the destination tasks might be restricted by the network infrastructure. This situation is often worked around by sending all the broadcast messages to each node, thus everybody has to process every single message and decide whether they were really meant to receive that information. Although the desired $O(1)$ complexity is usually not reached in the practice, the hardware support offers the fastest and most scalable solution, but the price is paid by the higher implementation efforts and the total loss of portability.

3 Asynchronous Symmetric Broadcasting Algorithm

The best software solutions described in (4) and (5) manage to eliminate the dependence of the message transfer time itself $(n*t_d)$ on the number of nodes, but cannot remove the effect of the initial latency t_0. The symmetrical algorithm introduced in this section proposes a solution for problem in crossbar clusters. Our work improves the idea of Van de Geijn [10] by overlapping the two phases of scattering and gathering with the use of asynchronous communication.

The group primitives of most message passing libraries work in a synchronous mode, where the total time of communication is the sum of the consecutive message sending steps as described in (2), (3), (4) and (5). Previous studies [12,15] showed, that software inefficiencies can be hidden by allowing various communication steps to overlap with the computation and as well as with each other. To outperform the traditional methods we suggest combining message decomposition with asynchronous communication. While message decomposition allows transferring the different message fragments in parallel, asynchronous communication eliminates the effect of the additional initial t_0 latencies introduced by the greater number of message fragments. The proposed symmetrical algorithm consists of a complex communication schema divided into two overlapping phases presented separately in Fig. 1.

In the first phase the source node sends a different part of the original message to each destination node, respectively. The receiver nodes all act as secondary sources after receiving their message fragment in the second phase. With a number p of the destination nodes the algorithm takes the following steps:

1. The source node cuts the message of size n to be broadcasted into p pieces. To avoid rounding errors, fragment i is formed as

$$addr(i) = \left\lfloor \frac{(i-1)n}{p} \right\rfloor \quad len(i) = \left\lfloor \frac{i*n}{p} \right\rfloor - \left\lfloor \frac{(i-1)n}{p} \right\rfloor, \tag{6}$$

where $addr$ and len refer to the position of the ith fragment in the original source buffer. The messages are reconstructed in the same way in the destinations.

2. The message fragments are completed with some administrative information, such as the total size of the original message (n), the address of the current fragment, and a transaction identifier (id) marking the parts belonging together. The source node forwards fragment i to destination node i, but does all transfers in parallel thereby eliminating the effect of the additional t_0 latencies:

$$t_c(n, p) = t_0 + p\frac{nt_d}{p} = t_0 + nt_d. \tag{7}$$

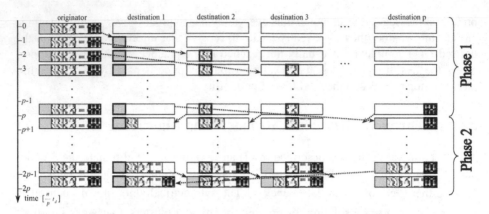

Fig. 1. Symmetrical, asynchronous broadcasting schema

3. If a destination node *j* receives a message part with a new identifier, it allocates a memory space for the whole message of size *n*, and copies the arrived part to its final place (*addr*). If the arriving message possesses an already known identifier, then its content is copied to the relevant address in the already existing buffer.

4. If the incoming message arrived from the original source (i.e. belongs to phase 1), then this destination node *k* is responsible for forwarding this fragment to the remaining *p-1* destination nodes. While distributing this message fragment to the partners, node *k* is still able to receive other message fragments.

The algorithm is finished when all the message fragments arrived at all the destination nodes. The algorithm does not make any assumption on the order of when the message parts will arrive. Destination nodes may receive some of the message parts from the second phase before getting their part to distribute from the source node.

Considering a cluster system where the nodes are connected through a switching hub using full-duplex mode, the ideal execution time of the algorithm is only limited by the physical bandwidth of the links. The critical execution path is the distribution of the last fragment leaving the source node according to (7). The secondary distributor node must forward this part of length *n/p* to the remaining *p-1* nodes, resulting in a total execution time of

$$t_c(n,p) = t_0 + nt_d + t_0 + \frac{(p-1)nt_d}{p} = 2t_0 + nt_d(2 - \frac{1}{p}). \tag{8}$$

The execution time shown in (8) has an asymptotical complexity of $O(1)$ for any t_0 and t_d parameters. In theory, the algorithm is perfectly scalable, and provides significantly better execution times than the widely used tree algorithm. The broadcasting takes the time of two full message transfers, which is double of the time expected when using hardware support of similar complexity.

Despite of its elegance, the algorithm suffers from some drawbacks limiting its usability. The algorithm is not efficient for short messages, because small packets carry relatively more overhead, and the symmetrical algorithm forces the generation of great number (~p^2) of n/p sized fragments. In case of small message sizes it is better to distribute the whole information in a single step during the first phase. A second problem is the validity of the presumption that all the nodes are able to send and receive at the full speed of their link capacity. Current computers can easily cope with the full-duplex bandwidth of the network adapter, but the switching hub can prove to be a bottleneck. For perfect scalability the switching hub must be able provide full speed at all of its ports at the same time. Although the amount of the incoming and outgoing data is evenly distributed in time, the switching hub has to handle the competition of more nodes sending data packets to the same destination. This competition state is resolved with buffers having limited speed and size. As the scalability of the algorithm is limited by the saturation point of the active network equipments, the use of our algorithm is practically limited to small cluster environments (a few tens of nodes).

4 Performance Measurements

This section compares the execution time of the widely spread binary tree method to the new symmetrical asynchronous algorithm. The same hardware and software environment were used in both cases. The testbed was built up out of 15 uniform PCs having an Intel Pentium IV processor of 2.26 GHz, 256 MB RAM, and an Intel 82801DB PRO/100 VE network adapter of 100 Mbits. The nodes were connected through a 3Com SuperStack 4226T switching hub. All the nodes were running Windows XP operating system, and an implementation NT-MPICH v1.3.0 [15] was used to implement and test both broadcast algorithms.

We chose to handle message sizes between 2 bytes and 256 kB with logarithmical steps, as most real-word applications exchange their data in this range. Two kinds of scenarios were measured. In the first case, the built-in broadcast primitive of the MPICH library was tested. This primitive is based on tree topology, and the call on the source node returns only if all the partners receive the message. In the second case the asynchronous algorithm was implemented using the asynchronous message transfer primitives (`MPI_Irecv`, `MPI_Isend`). To allow measuring the communication time on the source node each destination node sends a short reply message when all the broadcasted data has successfully arrived. In both cases, the total communication time spans from the invocation of the broadcast primitive to the arrival of the last acknowledgement. To equalize the variations in execution times 10 separate measurements were made in each data point, and their arithmetical mean is considered to be the result. The time measurements were made using the `RDTSC` instruction of the Pentium processor, which allows measuring the elapsed time with high precision.

The complexity is more apparent on a linear scale, but the range of message sizes span over multiple orders of magnitude, thus Fig. 2 only shows the execution times for messages up to a size of 64 kB. In practice the symmetric algorithm showed a constant complexity for a large range of message sizes. For small messages there is a visible growth of execution time as the number of nodes increases (Fig. 3). This is caused by the overhead of message fragmenting. As expected from (3) and (8), the

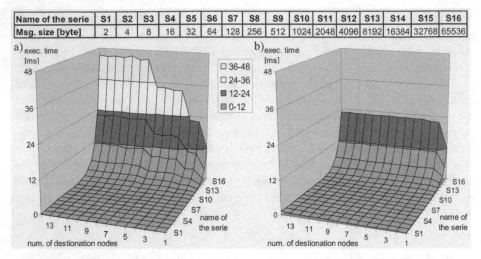

Name of the serie	S1	S2	S3	S4	S5	S6	S7	S8	S9	S10	S11	S12	S13	S14	S15	S16
Msg. size [byte]	2	4	8	16	32	64	128	256	512	1024	2048	4096	8192	16384	32768	65536

Fig. 2. Execution times of the tree (a) and of the symmetrical (b) algorithms on linear scale

tree algorithm generally shows a logarithmic increase and the symmetric one a hyperbolic increase respectively. For small messages (≤ 2 kB) the tree algorithm is slightly better, but for the remaining part of domain the performance of the symmetric algorithm is always superior to that of the tree approach. When broadcasting to more than 9 nodes the symmetric algorithm proves to be even twice as fast as the traditional tree method.

To verify the efficiency of the implementation the measured results were also compared to the limits in theory. Equation (8) shows the expected peak performance of the algorithm. The values t_0 and t_d are characteristic to the message passing layer, and are obtained by executing a ping-pong benchmark and using linear regression [6]. For MPI implementation used here [16] the values $t_0 = 263$ µs and $t_d = 0{,}087$ µs were measured. The performance in theory and the measured curves are compared in Fig. 3. The implemented algorithm is reasonably close to the peak performance, significant difference can only be detected on the curve of messages with a length of 256 kB. The anomaly is likely to be caused by some inefficiencies of the MPI library, as this irregularity disappears for higher number of nodes.

5   Conclusion

This paper showed a method for improving communication performance in fully-switched cluster systems without changing the network infrastructure and the used protocols. We focused our attention on improving group communication, and proposed a new broadcasting algorithm. The new method is easy to implement, and benefits from a complexity of $O(1)$ having been achieved only with hardware support so far. The presented algorithm builds on asynchronous, reliable point-to-point communication and uses message decomposition. The symmetrical pattern allows good

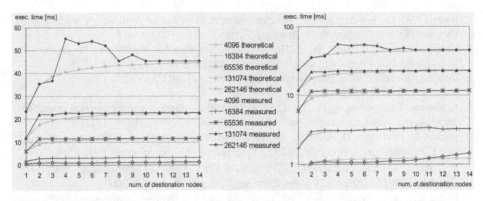

Fig. 3. Theoretical and measured performance using linear (a) and logarithmic (b) scales

scalability and automatic load balancing. The algorithm is intended to work in the very common, fully-switched cluster environment. As their ideal performance is achieved by continuous communication between each pair of nodes, the performance strongly relies on the capacity of the network switch. Its saturation point puts a practical limit on the scalability perfect in theory, thus the algorithm can be used efficiently in a small cluster environment (a few tens of nodes) only. The number of generated messages grows with the square of the number of the destination nodes. More messages generate more overhead, causing the algorithm to perform worse for very small message sizes. The usability of the new method was demonstrated by comparing its performance to the traditional binary tree implementation. During the tests the tree version was outperformed significantly by the new symmetric method for message sizes greater than 2 kB. In some cases the new algorithm had a double performance compared to the built-in communication primitive.

As broadcasting is an important building block of other message passing primitives, the results presented in this paper can be used directly for improving the performance of group communication in message passing libraries for cluster environments.

References

1. Martin et al.: Effects of Communication Latency, Overhead and Bandwidth in a Cluster Architecture. 24th Annual Symposium on Computer Architecture, Denver, (1997) 85-97
2. Chiola, G., Ciaccio G.: Efficient Parallel Processing on Low-Cost Clusters with GAMMA Active Ports. Parallel Computing 26, Elsevier Science, (2000) 333-354
3. Geist, A., Beguelin, A., Dongarra J., Jiang, W., Sunderam V.: Parallel Virtual Machine – A User's Guide and Tutorial for Networked Parallel Computing. MTI Press, London (1994)
4. Snir, M., Otto, S., Huss-Lederman, S., Walker, D., Dongarra J.: MPI–The Complete Reference. Volume 1 - The MPI-1 Core, 2nd edition, MIT Press (1998)
5. Meyer, U. et al.: Algorithms for memory hierarchies. LNCS 2625, Springer-Verlag, Berlin (2003) 320-354

6. Baugh J.W. Jr., Konduri R.K.S.: Discrete element modeling on a cluster of workstations. Engineering with Computers 17, Springer-Verlag, London (2001) 1-15

7. Gropp, W., Lusk, E., Doss, N.: A High-Performance, Portable Implementation of the MPI Message Passing Interface Standard. Parallel Computing, 22(6), (1996) 789–828

8. Indiana University, Indiana University's Open Systems Lab: LAM/MPI. http://www.lam-mpi.org/

9. Kielmann, T., Hofman, F. H.,. et al.: MagPIe: MPI's collective communication operations for clustered wide area systems. ACM SIGPLAN Notices, 34(8), (1999) 131-140

10. Barnett M., S. Gupta, D. Payne, L. Shuler, R. van de Geijn, J. Watts: Interprocessor Collective Communication Library (InterCom), Proceedings of Supercomputing'94, 1994.

11. R. Thakur, W. Gropp: Improving the Performance of Collective Operations in MPICH, 10th EuroPVM/MPI Conference, Venice, Italy, September 29-October 2, 2003.

12. Floyd S.. et al: A Reliable Multicast Framework for Light-Weight Sessions and Application Level Framing. IEEE/ACM Transactions on Networking, 5(6), (1997) 784–803

13. Pingali, S., Towsley, D., Kurose, J. F.: A Comparison of Sender-Initiated and Receiver-Initiated Reliable Multicast Protocols. Sigmetrics Conference on Measurement and Computer Systems, ACM Press, New York, USA, (1994) 221–230

14. Buntinas D., Panda, D. K., Brightwell R.: Application-Bypass Broadcast in MPICH over GM. International Symposium on Cluster Computing and the Grid, (2003)

15. Juhász S., Charaf, H.: Exploiting Fast Ethernet Performance in Multiplatform Cluster Environment. 19th ACM Symposium on Applied Computing, Nicosia, (2004) 1407-1411

16. RWTH Aachen: Multi-Platform MPICH, http://www.lfbs.rwth-aachen.de/mp-mpich/

A Simple Work-Optimal Broadcast Algorithm for Message-Passing Parallel Systems

Jesper Larsson Träff

C&C Research Laboratories, NEC Europe Ltd.
Rathausallee 10, D-53757 Sankt Augustin, Germany
traff@ccrl-nece.de

Abstract. In this note we give a simple bandwidth- and latency optimal algorithm for the problem of broadcasting m units of data from a distinguished root processor to all $p - 1$ other processors in one-ported (hypercubic) message-passing systems. Assuming linear, uniform communication cost, the time for the broadcast to complete is $O(m + \log_2 p)$, more precisely no processor is involved in more than $\lceil \log_2 p \rceil$ communication operations (send, receive, and send-receive), and for any constant *message size threshold* b each processor (except the root) sends at most $m - b' + (\lceil \log_2 p \rceil - \ell)b'$ units of data, where b' is determined by the smallest $\ell \leq \lceil \log_2 p \rceil$ such that $b' = m/2^\ell \leq b$ (the root sends $2m - b' + (\lfloor \log_2 p \rfloor - \ell)b'$ units of data). Non-root processors receive m units of data.

Building on known ideas, the salient features of the algorithm presented here is its *simplicity of implementation*, and *smooth transition* from latency to bandwidth dominated performance as data size m increases. The implementation performs very well in practice.

1 Introduction

We consider the problem of broadcasting m units of data from a distinguished *root* processor to all other processors in p processor message-passing parallel systems. Unlike other recent studies [2, 8] in the *LogP*-model [4] or related, we assume a simple one-ported (half- or full duplex) communication model with uniform, linear communication cost. Using (non-pipelined) binary or binomial trees, broadcasting m units takes $O(m \log_2 p)$ time (for the last processor to receive the m data). We are interested in algorithms that are *latency- and bandwidth-optimal* in the sense of taking time $O(m + \log_2 p)$ (for the last processor to complete) with small constant factors. Borrowing terminology from parallel algorithmics, we call such algorithms *work-optimal* [5], since copying m units sequentially p times takes $O(pm)$ time. In the algorithm presented in this paper all processors do at most $\lceil \log_2 p \rceil$ communication operations (send, receive, and send-receive), and for a given *message size threshold* b each processor (except the root) sends at most $m - b' + (\lceil \log_2 p \rceil - \ell)b'$ units of data, where b' is determined by the smallest $\ell \leq \lceil \log_2 p \rceil$ such that $b' = m/2^\ell \leq b$ (the root sends $2m - b' + (\lceil \log_2 p \rceil - \ell)b'$ units of data). The algorithm gives a smooth transition from latency (the $\log_2 p$ term)

D. Kranzlmüller et al. (Eds.): EuroPVM/MPI 2004, LNCS 3241, pp. 173–180, 2004.

to bandwidth (the m term) dominated performance as m increases. Non-root processors receive m units of data.

The algorithm is easy to implement and is theoretically attractive compared to the commonly implemented binomial tree and simple, pipelined binary tree algorithms. For medium sized data it may be better than the advanced pipelined approaches in [6–8]. A comparison to other broadcast algorithms developed for the linear communication cost model is given in Section 1.1. We leave it open whether it is possible to modify and analyze the algorithm in the $LogP$-model, and what the the practical impact hereof might be.

The basic idea of the work-optimal algorithm presented in this paper derives from [1], in which the idea of broadcasting large data volumes by a *scatter* followed by a *collect* operation was presumably first introduced, as well as the idea of *hybrids* between latency efficient tree algorithm and bandwidth efficient composite algorithm. The composite algorithm (scatter followed by collect) has recently been implemented for the MPI implementation mpich2 [10], but in a non-hybrid, either-or fashion: a binomial tree algorithm is used for small data, and for large data volumes the composite algorithm is used.

The contribution of this paper is a simple and easy to implement formulation of the *hybrid* idea of [1] with better bounds than the implementation in [10]. The implementation performs very well in practice compared to a binomial tree broadcast implementation. For $p = 2^n$ the algorithm is a natural hypercube algorithm.

In the following we assume that the m units of data to be broadcast are stored contiguously in an array M, which we index from 0. A segment of the array of k units starting at index a is denoted by $M[a, a + k[$. We number the p processes from 0 to $p-1$. Without loss of generality we may assume that processor 0 is the root process (otherwise shift the processor numbering cyclically toward 0), that p is a power of 2 (otherwise, solve the broadcast problem on the largest power of 2 processors $p' < p$, and let the $p - p'$ first processors send the data to the remaining $p - p'$ processors – although we can and do do substantially better), and that both m and the threshold b are powers of 2 (the general case is not substantially more difficult to handle).

1.1   Comparison to Other Broadcast Algorithms

In the linear cost model sending a message of m units takes $\alpha + \beta m$ time, where α is the latency or start-up cost, and β the inverse bandwidth. Under this model, assuming furthermore single-ported, full-duplex communication, we give in Table 1 the time to completion for the slowest processor for some (well-)known, relevant broadcast algorithms.

Under the single-ported assumption, the simple binary tree algorithm is a factor 2 slower than the binomial tree, but in contrast is easy to pipeline. With N being the block size used for pipelining, the time for last leaf processor to complete is $2(\lfloor \log_2 p \rfloor - 1)(\alpha + \beta N) + 2(m/N - 1)(\alpha + \beta N) = 2(\lceil \log_2 p \rceil - 2)(\alpha + \beta m) + 2\alpha m/N + 2\beta m$ since the non-root processors after the pipeline has been filled, receive N units of data in every second communication step. Balancing

Table 1. Performance of some broadcast algorithms in the linear communication cost model. Here $n = \log_2 p$, and ℓ is the smallest ℓ such that $m/2^\ell \leq b$.

Algorithm	Time for last processor to complete
Binomial tree	$\lceil n \rceil (\alpha + \beta m)$
Binary tree (BT)	$2\lfloor n \rfloor (\alpha + \beta m)$
Pipelined BT	$2(\lfloor n \rfloor - 1)(\alpha + \sqrt{\frac{\alpha\beta}{\lfloor n \rfloor - 1}}\sqrt{m}) + 2\sqrt{\alpha\beta(\lfloor n \rfloor - 1)}\sqrt{m} + 2\beta m$
Work-optimal [10]	$\begin{cases} \lceil n \rceil(\alpha + \beta m) & \text{if } m \leq b \\ 2(\lceil n \rceil \alpha + \beta m(1 - 1/p)) & \text{if } m > b \end{cases}$
Work-optimal (this paper)	$\begin{cases} n(\alpha + \beta m) & \text{if } m \leq b \\ (2n - \ell)\alpha + 2\beta m(1 - 1/2^\ell) + (n - \ell)\beta m/2^\ell & \text{if } m > b, m/2^\ell \leq b \\ 2n\alpha - \alpha + 2\beta m(1 - 1/p) & \text{if } m > b \end{cases}$
Fractional tree [7]	$O(\alpha \log p) + \beta m(1 + O(\sqrt[3]{\alpha \log p/m}))$
ESBT [6] (hypercube only)	$n\alpha + 2\sqrt{n\alpha\beta}\sqrt{m} + m\beta$

the two N-dependent terms the minimum completion time is achieved yielding $N = \sqrt{\frac{\alpha}{(\lceil \log_2 p \rceil - 1)\beta}}\sqrt{m}$ with completion time as given in Table 1. The algorithm of this paper has a better latency-term than the pipelined binary tree (and marginally better than the algorithm in [10]), and in contrast to the implementation in [10] per design gives a smooth transition from latency to bandwidth dominated behavior. These algorithms all have a factor 2β in the message size, such that the speed-up over the binomial tree broadcast is $\geq 1/2 \log_2 p$ for large m. Theoretically better algorithms do exist, eg. the (classical) *edge-disjoint spanning binomial tree algorithm* (ESBT) of [6], and the recent *fractional tree algorithm* given in [7], both of which have only a factor β in the message size. The ESBT-algorithm is a hypercube algorithm and (apparently) only trivially generalizes to the case where p is not a power of two. The fractional tree algorithm seems somewhat more difficult to implement.

2   The Algorithm

The work-optimal broadcast algorithm uses a binomial tree. As will become clear, the chosen numbering of the processors is such that all communication is among processors that are neighbors in a hypercube. For $m \leq b$ the algorithm is just an ordinary broadcast over the binomial tree, and thus takes time $O(m \log_2 p)$ which (for constant b) is $O(\log_2 p)$. As m grows beyond the threshold b, the idea is that a processor v which receives $m' > b$ units successively halves the amount of data passed on to each of its children. Each processor thus terminates this *scatter phase* in time $O(m + \log_2 p)$. After the scatter phase the processors that have received only a fraction $m/2^k$ of the data (simultaneously for $k = 1, 2, \ldots$) perform a *collect operation* (also known as allgather) which gathers all m data on each of these processors. This also takes $O(m + \log_2 p)$ time. Figure 1 illustrates the communication pattern for $p = 16$ and $m = 8$ units, and may be helpful to consult.

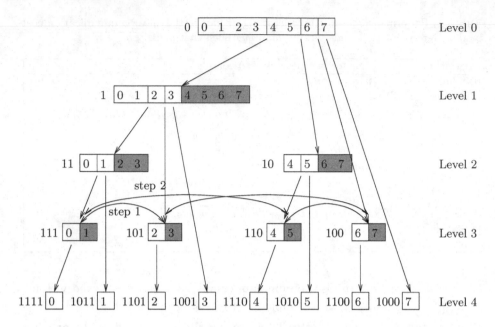

Fig. 1. The broadcast tree and communication pattern for $p = 16$, and $m \gg b$ a multiple of 8. Data received and sent in the scatter phase are shown in the rectangles (with processor numbers in binary to the left); the shaded parts are only received, but not sent further on (this observation is exploited to improve the completion time). As can be seen, all data are available on each level, and can be collected by a butterfly exchange (shown for level 3), during which only hypercube neighbors need to communicate.

To describe the algorithm more formally some notation is needed. We number the bits of processor v from 0 to $\log_2 p - 1$, bit 0 being the least significant bit. We let $\text{msb}(v)$ denote the *most significant bit* of v (with $\text{msb}(0) = 0$). The binomial tree is rooted at 0. Each processor v is assigned a *level* in the tree which shall be

$$\text{level}(v) = \text{msb}(v) + 1$$

The maximum level d is $\text{msb}(p) = \log_2 p$. Processor v has $d - \text{level}(v)$ children at successive levels with

$$\text{child}(v, i) = v + 2^{\text{msb}(v)+i}$$

for $i = 1, \ldots, d - \text{level}(v)$. The parent of processor $v \neq 0$ is

$$\text{parent}(v) = v - 2^{\text{msb}(v)}$$

On each level a consecutive numbering $\text{block}(v)$ of the processors is needed for addressing data segments of M, defined by

$$\text{block}(v) = \text{level}(v) - \text{mirror}_{[1,\text{level}(v)-1]}(v - 2^{\text{msb}(v)})$$

Root 0:

```
// Scatter phase
s ← m
for i = 1, . . . , d do
        Send(M[m − s, m[, child(i, 0))
        if s > b then s ← s/2
```

Non-root v:

```
// Receive from parent: compute size s of vs block
s, k ← m, 0
for j = 1, . . . , level(v) − 1 do
        if s > b then s ← s/2 else k ← k + 1
s′ ← (block(v)/2^k)s
Recv(M[s′, s′ + s[, parent(v))
// Scatter phase
s″ ← s
for i = 1, . . . , d − level(v) do
        if s″ > b then s″ ← s″/2
        Send(M[s′, s″[, child(v, i))
        s′ ← s′ + s″
// Collect phase: butterfly-algorithm
s′ ← (block(v)/2^k)s
for i = level(v) − 1, . . . , k do
        u ← v ⊕ 2^i // hypercube neighbor for this step
        if block(u) > block(v) then s″ ← s′ + s else s″ ← s′ − s
        Send(M[s′, s′ + s[, u) ∥ Recv(M[s″, s″ + s[, u)
        if s″ < s′ then s′ ← s″
        s ← 2s // double collect size
```

Fig. 2. The work-optimal broadcast algorithm.

where $\text{mirror}_{[a,b]}(v)$ denotes v with the bits between a and b reversed (eg. bit a becomes bit b, bit $a + 1$ becomes bit $b − 1$, and so on). As an example consider the processors at level 3 in Figure 1, from left to right:

$$\text{block}(111_2) = 3 − \text{mirror}_{[1,2]}(111_2 − 100_2) = 3 − 3 = 0$$
$$\text{block}(101_2) = 3 − \text{mirror}_{[1,2]}(101_2 − 100_2) = 3 − 2 = 1$$
$$\text{block}(110_2) = 3 − \text{mirror}_{[1,2]}(110_2 − 100_2) = 3 − 1 = 2$$
$$\text{block}(100_2) = 3 − \text{mirror}_{[1,2]}(100_2 − 100_2) = 3 − 0 = 3$$

The work-optimal broadcast algorithm with threshold b is given in Figure 2. Here $⊕$ denotes bitwise exclusive or, so by $v ⊕ 2^i$ the ith bit of v is flipped. Each non-root receives from its parent a block of M of size $s = m/2^{\text{level}(v)−1−k}$, where $k > 0$ only if the block size at some point above level(v) has dropped below the threshold b. The start of the block received by v is $(\text{block}(v)/2^k)s$, from which it can be shown that all s-sized blocks of M are present at level(v), with each block occurring k times. Thus, a collect operation suffices to complete the broadcast at each level. In the butterfly-like collect operation the blocks at neighboring processors in each step are consecutive, so there is no need for moving data around within M, and the indices s' and s'' are easy to compute for the next step. Also note that level(v) and block(v) can be computed in time $O(\text{level}(v))$ (or faster, if hardware support for integer logarithms is available), and since each processor v has to wait level(v) communication rounds before receiving its first data, these computations do not affect the time bound.

The maximum number of communication operations (receive, send, send-receive) per non-root processor is

$$\underbrace{1}_{\text{Recv}} + \underbrace{d - \text{level}(v)}_{\text{Send}} + \underbrace{\text{level}(v) - 1 - k}_{\text{Send}\|\text{Recv}} = d - k \le d = \log_2 p$$

The root performs $\log_2 p$ send operations.

Now assume that $m/2^{d-1} = 2m/p \ge b$, such that halving is done at all levels. Each non-root processor receives $s = \frac{m}{2^{\text{level}(v)-1}}$ units, sends $\sum_{i=1}^{d-\text{level}(v)} s/2^i = s - 2m/p$ units to its children, and in the collect phase sends and receives $m - s$ units respectively, for a total of $m(1 - 2/p)$ units sent and received per non-root processor. The time a non-root processor must wait for data from its parent is proportional to the time to send $\sum_{i=1}^{\text{level}(v)-1} m/2^{i-1} = 2m - s$ units of data. In time proportional to $s - 2m/p$ each non-root sends data to its children, and completes its collect phase in time proportional to the time needed to send and receive $m - s$ units of data. Thus the completion time for a non-root is proportional to $3m - s - 2m/p$. The root sends $2(1 - 1/p)m$ units of data. Now observe that only half of the s units of data received by each non-root actually have to be propagated further for the scatter phase (in Figure 1 the shaded parts of the data blocks are not needed for the scatter phase). This allows a significant improvement of the bound. In the scatter phase (see Figure 2) instead of sending the s'' units to the next child immediately, only half is sent with the remainder postponed till after the scatter phase. Before the collect phase can start these remainders are sent in reverse order to the children. Instead of having to wait for time proportional to $2m - s$, non-roots now only have to wait for time $m - s/2$. The time needed to scatter data further remains proportional to $s - 2m/p$ (but divided over two sub-phases), and the time needed to receive the "missing" remaining half of the s units is proportional to $s/2$. The collect phase remains unchanged for a total time proportional to $2(1 - 1/p)m$. Putting the pieces together gives the $2\alpha \log_2 p - \alpha + 2\beta m(1 - 1/p)$ completion time of Table 1.

2.1   Collect Phase for Large m

It is worth pointing out that also the collect phase at each level can be equipped with a *large message threshold* B after which the butterfly exchange is replaced by a ring-algorithm. More concretely, butterfly-exchange is done until the size s of collected data exceeds B; at that point a number of disjoint sets of processors which together have all m units exist, and within each of these sets, a ring is used for collecting all m units on each processor.

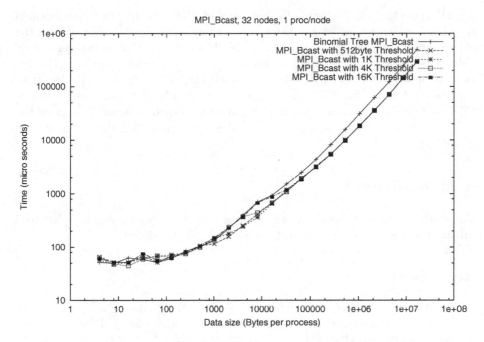

Fig. 3. The work-optimal broadcast (with thresholds 512Bytes, 1KByte, 4KBytes, 16Kbytes, respectively), compared to the binomial tree broadcast. A doubly logarithmic plot has been used.

2.2   The Case When p Is Not a Power of Two

Let p' be the largest power of two smaller than p. The most efficient solution for this case is for the $p - p'$ first processors after the root to send $m/(p - p')$ individual units data to the last $p - p'$ processors prior to the collect phase. Simultaneously with the collect phase of the first p' processors, the remaining $p - p'$ processors perform an allgather operation using for instance the algorithm in [3]. The additional cost of this is one communication round in which $p - p'$ processors send $m/(p - p')$ units of data. Computing buffer placement for the data received by each of the last $p - p'$ processors is somewhat technical.

3   Performance Results

The algorithm as formalized in Figure 2 can immediately be implemented, and used for the MPI_Bcast primitive of the *Message Passing Interface* (MPI) [9]. We have compared the performance of the work-optimal broadcast algorithm to the binomial tree algorithm still found in many MPI implementations. Results from a 32-node AMD-based SMP cluster with Myrinet interconnect (1 active process per node) are shown in Figure 3, where we have used four different thresholds, $b = 512\text{Bytes}, 1\text{KByte}, 4\text{KBytes}, 16\text{KBytes}$. Best performance is achieved with the 1KByte threshold. Already for messages of 4KBytes size, the work-optimal algorithm is about a factor 1.5 faster, and for large messages above 1MBytes the improvement approaches a factor 2. An MPI implementation in full generality, able to handle arbitrary numbers of processors, message sizes, thresholds, as well as non-consecutive data (MPI user defined data types [9, Chapter 3]) is less than 300 lines of C code. For SMP-clusters where each node may host more than one MPI process the work-optimal algorithm should be used to broadcast across the SMP-nodes, with a shared-memory broadcast algorithm for the node-internal broadcast.

Acknowledgment

The author wants to thank an anonymous reviewer whose comments led to a substantial clarification of the contribution of this paper.

References

1. M. Barnett, S. Gupta, D. G. Payne, L. Schuler, R. van de Geijn, and J. Watts. Building a high-performance collective communication library. In *Supercomputing'94*, pages 107–116, 1994.
2. J. Bruck, L. D. Coster, N. Dewulf, C.-T. Ho, and R. Lauwereins. On the design and implementation of broadcast and global combine operations using the postal model. *IEEE Transactions on Parallel and Distributed Systems*, 7(3):256–265, 1996.
3. J. Bruck, C.-T. Ho, S. Kipnis, E. Upfal, and D. Weathersby. Efficient algorithms for all-to-all communications in multiport message-passing systems. *IEEE Transactions on Parallel and Distributed Systems*, 8(11):1143–1156, 1997.
4. D. E. Culler, R. M. Karp, D. Patterson, A. Sahay, E. E. Santos, K. E. Schauser, R. Subramonian, and T. von Eicken. LogP: A practical model of parallel computation. *Communications of the ACM*, 39(11):78–85, 1996.
5. J. JáJá. *An Introduction to Parallel Algorithms*. Addison-Wesley, 1992.
6. S. L. Johnsson and C.-T. Ho. Optimum broadcasting and personalized communication in hypercubes. *IEEE Transactions on Computers*, 38(9):1249–1268, 1989.
7. P. Sanders and J. F. Sibeyn. A bandwidth latency tradeoff for broadcast and reduction. *Information Processing Letters*, 86(1):33–38, 2003.
8. E. E. Santos. Optimal and near-optimal algorithms for k-item broadcast. *Journal of Parallel and Distributed Computing*, 57(2):121–139, 1999.
9. M. Snir, S. Otto, S. Huss-Lederman, D. Walker, and J. Dongarra. *MPI – The Complete Reference*, volume 1, The MPI Core. MIT Press, second edition, 1998.
10. R. Thakur and W. D. Gropp. Improving the performance of collective operations in MPICH. In *Recent Advances in Parallel Virtual Machine and Message Passing Interface. 10th European PVM/MPI Users' Group Meeting*, volume 2840 of *Lecture Notes in Computer Science*, pages 257–267, 2003.

Nesting OpenMP and MPI in the Conjugate Gradient Method for Band Systems

Luis F. Romero[1], Eva M. Ortigosa[2], Sergio Romero[1], and Emilio L. Zapata[1]

[1] Department of Computer Architecture
University of Málaga, 29071, Málaga, Spain
[2] Department of Computer Architecture and Technology
University of Granada, 18071, Granada, Spain
sromero@ac.uma.es

Abstract. An analysis of the data dependencies in the Conjugate Gradient iterative method for the solution of narrow band systems is performed in order to develop and implement a mixed OpenMP-MPI code which takes into account the computer architecture and memory hierarchy at three different levels: processor, shared–memory and network levels. This kind of hybrid parallelization allows code porting and tuning for different multiprocessors and grids.

1 Introduction

Many problems in physics and engineering are described by means of either ordinary or partial differential equations which need to be solved in a spatio–temporal domain. Usually, these equations are discretized in either structured or unstructured grids, and, by using implicit integration techniques and linearization, a system of linear algebraic equations is obtained at each time step. This system can be efficiently solved by means of a variety of methods, such as the preconditioned Conjugate Gradient algorithm. Such a solution demands high performance computing, including parallelism.

A parallel computer is roughly described, using two abstraction layers as a very powerful machine and as a set of powerful processing elements (PE) which are interconnected by some means which allows interprocessor data exchange. It is not very frequent to find such a simple description of a parallel computer, but, commonly, the architectural details are widely explained. Nevertheless, most papers dealing with efficient codes for these machines simplify their approaches and formulations by using this two–layers point a view; at the highest (network) level, achieving effective parallelism requires that work be divided equally amongst the processors in a way which minimizes interprocessor communication. In this domain–decomposition step, the grid points are divided into spatially local regions of approximately equal work. At the lowest (processor) level, achieving an intensive processor exploitation requires a good scheduling of the operations to be performed in the local region of each processor. In ODE derived problems the schedule is driven by the order of the data sets. This ordering should consider

D. Kranzlmüller et al. (Eds.): EuroPVM/MPI 2004, LNCS 3241, pp. 181–190, 2004.

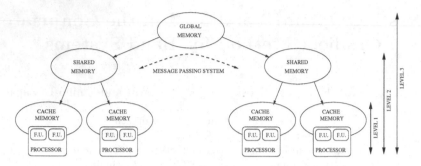

Fig. 1. Three architectural levels. Processor level: functional units and cache memories; shared–memory level: processors and main memories; network level: distributed memories and computational nodes.

both the processor architecture and the memory hierarchy. For a grid–based model, the two–level architectural view of the problem can be summarized in a two–stage program optimization: a partition phase and an ordering phase.

In this work, we analyze the effect of the insertion of a third nested level in the architectural view in order to achieve an efficient solution for narrow band system of equations derived from the numerical integration of differential equations. An appropriate ordering of the equations will increase the cache performance at the processor level, a good partition will minimize the communications at the network level, and, finally, an intermediate shared-memory level will increase the efficiency in many multiprocessor systems and make easy the programming stage of load balanced, distributed codes. OpenMP and MPI have been used in a nested fashion in order to accede to the two highest parallel levels, i.e., the network and shared–memory levels. The hybrid programming technique employed here has been previously used by several authors [1, 2].

The model proposed here has been used for the solution of two banded systems. The first one arises from the finite difference discretization of a system of three–dimensional reaction–diffusion equations in a regular (cubic) domain, whereas the second one has been derived from a finite element discretization of clothes using unstructured grids.

2   A Three-Level Architectural View

An advanced modern computer is usually made up of several computational nodes interconnected by a network. A computational node consists of one or more processors sharing a common space of memory. Finally, each processor includes several functional units in it, which share one or more levels of caches. This three–level architectural view is shown in Figure 1.

For an efficient exploitation of the underlying hardware, a code must take into account this structure and should fulfill the following requirements:

– While optimizing an application, the semantics of the code should not be modified. At the processor level, out of order instruction completion occurs, although the processor architecture will internally eliminate data hazards. At the shared memory level, the programmer inserts synchronization points to eliminate these hazards (e.g., read–after–write and write–after–read). Finally, at the network level, the same problem is solved by correctly inserting a message delivery only when the data are prepared, and receiving them before these data are used, without replacing the still needed ones.

- To exclusively perform the required computations. Most of this work relies on the compiler at the processor level, e.g. loop fusion and loop coalescence. At the highest levels, the programmer should explicitly take care of this.
- To avoid collapsed or lazy processing elements. Again, apart from intrinsic program properties, this is one of the programmer's responsibility at the two highest levels, and well–balanced parallel codes should be written. Note that this is internally avoided inside the processors by eliminating structural hazards and using advanced ILP techniques. The compiler does much of this work with techniques like loop–unrolling and software pipelining.
- To minimize the data access time. In this case, cache optimization is required at the processor level to maximize the spatial and temporal locality. Padding and automatic prefetching can be used in some architectures. In shared–memory models, page migration must be taken into account in order to avoid excessive cache invalidations, and explicit cache prefetching is recommended, if it can be applied. At the network level, the messages should be sent as soon as possible for communication/computation overlapping purposes and should be directly stored (if at all possible) in the lower cache levels. The "owner computes" rule must be applied. In any case, a memory–centered approach for the code programming is essential to take care of this requirement [3].

3   Design of Optimal Algorithms/Codes

In the previous section, some important requirements that need to be taken into account in the development of efficient codes have been presented. It should be pointed out, however, that some of these requirements may compete with each other in a given application. For example, repeating computations is sometimes preferred to searching for the previously computed result in a remote place.

In this section, the three level architectural view presented in the previous section will be considered, and a strategy which allows for the best optimization among different possibilities will be presented. A tiling technique (i.e., decomposition of computations into smaller blocks) and task graphs are used as tools to analyze the computations to be performed and the data dependencies to be considered at each stage.

At the network level, the program or algorithm is described by means of a coarse task graph. This graph should be partitioned into subgraphs by considering the following issues: load balance, minimal temporal dependence, and minimal communication amongst subgraphs. Perhaps one of the most important issues to consider at this level is *simplicity*, because the parallelism and communication between nodes are explicitly managed by the programmer. Moreover, at this level, a coarser task graph will make easier the search for an optimal partition of such graph, but a finer graph will produce an optimal algorithm. It is necessary to find a trade–off between these issues, depending on the desired programming time/program use ratio [4].

This is a *first level partition* in which communication among computational nodes (and, therefore, the program semantics) are managed through a message–passing model. In the examples presented here, the message–passing model has been implemented by using MPI libraries.

At the shared–memory level, each one of the resulting subgraphs is partitioned again. The main issues to consider at this level are: load balance, minimal temporal dependence and communications in the previous partition. At this level, the task graph for each node should be refined, i.e., re–tiled, in order to avoid load imbalance problems. Minimal communication amongst subgraphs is only required if we deal with a NUMA architecture. However, at this level, it is not necessary to deal with simple partitions, because communications are implicitly managed by the cache–coherent system [3]. Nevertheless, if an explicit prefetching technique is used, simplicity would be more desirable.

This is a *second level partition* in which communication amongst the *threads* attached to the processors are managed through a shared–memory map. In the examples shown in this paper, the shared–memory map has been implemented by using OpenMP directives.

In some cases, one or both of the above partition stages can be eliminated, but the design of an architecture–independent code should consider both stages simultaneously. Partition parameters can be adjusted in order to select the best combination of nodes and threads for any computer or problem size.

At the processor level, an scheduling of the computations is performed. This scheduling considers the previous partition stages in order to minimize the communication overhead in size and number of messages, and makes an optimal usage of the cache. In particular, the spatio–temporal locality of the ordering should be compatible with the communication-computation overlapping requirements, i.e., data required in remote processors should be computed first, and data coming from other processors should be used at the end. If these requirements cannot be simultaneously fulfilled, some architectural parameters such as, for example, the size of the cache memories, the network bandwidth, etc., must be considered in order to develop an efficient code. In the latter case, communications corresponding to the distribution in the first level partition are usually very expensive and, therefore, should be given priority. In addition, some tasks should be replicated, if required.

For an optimal cache usage, each data should be used as soon as possible after it has been computed. In a task graph, this means that two graph nodes connected by an arc should be processed together if it is at all possible [5].

In any case, the semantics of the code at this *sorting level* are maintained by using messages and flags for synchronization purposes in order to take into account communications in the first and second level partitions, respectively.

4   The Parallel Conjugate Gradient Method

The CG technique [11] is an iterative method for the solution of linear systems of equations, i.e., $\mathbf{Ax} = \mathbf{b}$, which converges to the exact solution in exact arithmetic through a sequence of at most n vector approximations, where n is the dimension of $\mathbf{x}$ and $\mathbf{b}$. In practice, only a few iterations are employed to obtain a good estimate of the solution. In the PCG method, shown in Figure 2, both the condition number of $\mathbf{A}$ and the number of iterations may be reduced by pre–multiplying the system $\mathbf{Ax} = \mathbf{b}$ by a preconditioner $\mathcal{A}^{-1}$, which is usually an approximation to the inverse of $\mathbf{A}$.

$$
\begin{aligned}
&\mathbf{r}_0 = \mathbf{b} - \mathbf{Ax}_0 \\
&\mathbf{z}_0 = \mathcal{A}^{-1} \cdot \mathbf{r}_0 \\
&\rho_0 = \mathbf{r}_0^{T} \cdot \mathbf{z}_0 \\
&\mathbf{p}_1 = \mathbf{z}_0 \\
&\mathbf{q}_1 = \mathbf{Ap}_1 \\
&\eta_1 = \mathbf{p}_1^{T} \cdot \mathbf{q}_1 \\
&\text{for } i = 2, 3, \ldots, L \\
&\quad \mathbf{r}_{i-1} = \mathbf{r}_{i-2} - \tfrac{\rho_0}{\eta_{i-1}}\mathbf{q}_{i-1} \quad \text{(T1 in threesome 1)} \\
&\quad \mathbf{z}_{i-1} = \mathcal{A}^{-1} \cdot \mathbf{r}_{i-1} \quad\quad\;\; \text{(T2 in threesome 1)} \\
&\quad \rho_{i-1} = \mathbf{r}_{i-1}^{T} \cdot \mathbf{z}_{i-1} \quad\quad\; \text{(T3 in threesome 1)} \\
&\quad \mathbf{x}_{i-1} = \mathbf{x}_{i-2} + \tfrac{\rho_0}{\eta_{i-1}}\mathbf{p}_{i-1} \\
&\quad \mathbf{p}_i = \mathbf{p}_{i-1} + \tfrac{\rho_0}{\rho_{i-1}}\mathbf{z}_{i-1} \quad \text{(T1 in threesome 2)} \\
&\quad \mathbf{q}_i = \mathbf{Ap}_i \quad\quad\quad\quad\quad \text{(T2 in threesome 2)} \\
&\quad \eta_i = \mathbf{p}_i^{T} \cdot \mathbf{q}_i \quad\quad\quad\;\; \text{(T3 in threesome 2)} \\
&\text{end} \\
&\mathbf{x}_L = \mathbf{x}_{L-1} + \tfrac{\rho_0}{\eta_{L-1}}\mathbf{p}_L
\end{aligned}
$$

Fig. 2. Preconditioned Conjugated Gradient method.

In the implementation of the PCG method presented above [6], the computations in the iterative section have been divided into two threesomes with similar computations in each ($\mathbf{p}$, $\mathbf{q}$ and η on the one hand, and $\mathbf{r}$, $\mathbf{z}$ and ρ on the other). Note that the preconditioned stage (T2 in threesome 1) is also a matrix–vector product (as is T2 in threesome 2) for explicit preconditioners. Hereafter, we shall focus our discussion on the operations in threesome 2.

4.1   Sample Problems

Two different examples have been chosen to illustrate the ideas presented in the previous section. The first example corresponds to the Oregonator model [7] of wave propagation in excitable media which is governed by two nonlinearly coupled partial differential equations which have been discretized by means of second–order accurate (in both space and time) finite difference methods in a three-dimensional regular grid.

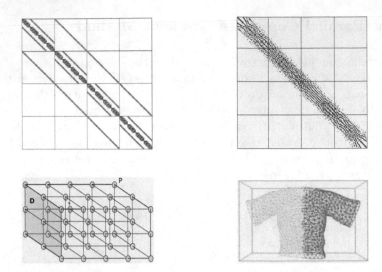

Fig. 3. System matrices for two kinds of problems. On the left, structured three-dimensional grid with natural ordering. On the right, unstructured grid for a piece of cloth with striped ordering.

The second example corresponds to a force–based physical model of cloth dynamics in virtual scenarios [8] which has been discretized by means of an unstructured finite element method in two dimensions.

The equations resulting from the discretization of these two examples yield large systems of algebraic equations at each time step which have been solved by using the PCG method with Jacobi and block-Jacobi preconditioners in the first and second examples respectively, Figure 3. For the block-Jacobi preconditioner, the number of variables per node have been used to form 3×3 blocks. A natural ordering for the structured grid and a striped ordering in the unstructured case result in narrow band matrices when the number of grid points is large enough.

5   Implementation

At the first level partition, the narrowness of the matrix bandwidth of the problems considered in this paper has been taken into account by considering the connectivity between different grid points. For grids in which any point is only connected to neighboring points, there always exists an ordering which keeps the coefficients of the system matrix in a diagonal band. For a large number of grid points, the band can be very narrow. Since at this partition level, simplicity is an important issue, a block partition of the system matrix has been used. For the PCG method, the communication pattern will be clearly simplified in this case, especially if the ratio of the number of variables assigned to a processor to the matrix bandwidth is greater than two. In such a case, the matrix–vector product in threesome T2 will only require communication between neighboring computational nodes.

In structured grids, the PCG method has been tiled by first considering the vector operations in algorithm (Figure 2) and then performing computations in z–planes for each vector operation. Since a seven point stencil has been used in the finite difference discretization of the reaction–diffusion equations, the sparse matrix–vector product for any plane only requires data from neighboring planes. The PCG algorithm in unstructured grids has been tiled by first taking into account the different objects of a scenario, and then using a stripe partition of the objects with wide enough stripes so as to ensure that every computation for a finite element only requires data from, e.g., vertices, edges, etc., located, at most, in two adjacent stripes.

For the tiles employed in this paper, the task graphs for each threesome in the iterations of PCG are very simple as shown in Figure 4. If the total number of stripes or z–planes is much larger than the number of computational nodes (this can be easily adjusted in many DSM architectures), the first level partition should be performed by using vertical cuts in the graph, thus resulting in simple communications (as required by this partition) and a relatively good load balance. In Figure 4, only the task graph for a single computational node and communications with other processors is presented.

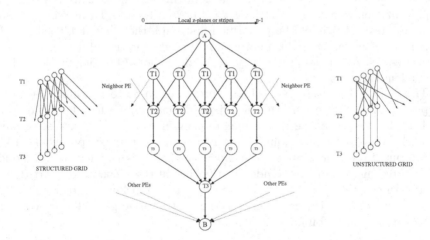

Fig. 4. Task graph for threesome 2 in the iterative section of the PCG method. Each node in the central graph corresponds to operations for a z–plane or a stripe in the structured or unstructured grid, respectively. Details of tasks T1, T2 and T3 for a given plane or stripe, using structured and unstructured grids, are shown on the left and right, respectively. A=ρ, B=η.

The second level partition is slightly different from the first one. For this second partition, the number of remaining unsplit stripes or z–planes is small; therefore, the risk of load imbalance is higher, and an efficient solution should consider a recursive tiling of the nodes of the graph shown in Figure 4. This recursive tiling can be performed by either dividing each node into a number of subnodes corresponding to the computations for each grid point, or considering a small set of grid points (for example, a row in the structured grid). It can be shown that the structured grid maintains a simple pattern for the data dependencies in the re-tiled graph, but this is not the case for the unstructured grid, in which an irregular pattern for the graph node dependencies is present. However, as communications need not be simple in the second level partition, a vertical cut in the refined graph may offer a good load balance, but other specialized graph partitioners can also minimize the amount of data to be exchanged. Note that it is easy to implement a prefetching strategy in the structured grid.

In the sorting stage, an efficient scheduling of the computations is required, and this has been performed in the following steps:

- First, by considering the communication pattern in the first partition stage, perform tasks T1 in the "halo" parts of each node and send a message as soon as possible so as to maximize the communication–computation overlapping of the sparse matrix–vector product [10]. For the same reason, tasks T2 and T3 are carried out in the proximity of the borders and should be performed at the end of this procedure.
- Second, by considering the communications in the second level partition, insert flags after tasks T1 in the halos of this level partition, and, for synchronization purposes, test those flags before T2 and T3 in remote halos. Again, tasks T1 should be performed first, and, then, T2 and T3 should be performed in remote halos in order to minimize the risk of delays in remote processors. At this level, this task scheduling should be compatible with an efficient cache usage.
- Third, sort the tasks assigned to a processor so as to maximize cache exploitation. By observing the tiles dependencies in the first level partition, T1, T2 and T3 can be carried out in combination [11] on successive parts of the vectors corresponding to each z–plane or each stripe, where T1 is computed one plane or stripe in advance (i.e., this is a fusion technique). This technique can be extended to the tiles in the second level partition in structured grids, and corresponds to well-known blocking strategies for cache optimization. For the grid points in the unstructured grid, efficient ordering strategies can be found in the literature [5].

Note that the similarities between the first and second steps facilitate the programming of the mixed model proposed in this paper. Indeed, this model consists of a repetition of the partition phase (using the chunks after the first step as sources in the second step). Shared–memory or message–passing communication models may be chosen, depending of the position of a processor in a node. Reductions for computing T3 have been performed using both models in two steps, i.e., thread reduction and node reduction.

6   Results and Conclusions

The mixed model proposed in this paper has been implemented using MPI libraries for the first level partition, and OpenMP directives in the second level

Table 1. Execution times for one time step of simulation of three–dimensional reaction–diffusion using a 75×75×75 point grid.

	MPI			Mixed				OpenMP		
	NN	Time (s)	Effic.	NN	NT/Node	Time (s)	Effic.	NT	Time (s)	Effic.
sequential	1	5.563	1	1	1	5.658	0.983	1	5.673	0.981
parallel with 8 PEs				1	8	0.953	0.730	8	0.969	0.718
				2	4	0.849	0.819			
				4	2	0.777	0.895			
	8	0.718	0.968	8	1	0.729	0.954			

Table 2. Execution times for one second simulation of eight (599 point grid) flags in a virtual scenario.

	MPI			Mixed				OpenMP		
	NN	Time (s)	Effic.	NN	NT/Node	Time (s)	Effic.	NT	Time (s)	Effic.
sequential	1	75.3	1	1	1	75.3	1	1	75.3	1
parallel with 8 PEs				1	8	7.74	1.22	8	7.74	1.22
				2	4	5.89	1.60			
				4	2	5.06	1.86			
	8	6.10	1.54	8	1	6.10	1.54			

partition (threads). For the structured problem, a Sunfire 6800 parallel computer has been employed. This is a UMA-SMA system based on the 750MHz UltraSparc III processor with 24 CPUs. MPI-2 library has been used in this machine. The unstructured problem has been implemented on a SGI Origin2000 multiprocessor, with 8 nodes, each one includes two 400-MHz R12000 processors sharing a memory module. In this architecture, only the MPI-1 library was available. Since the MPI-1 library is a thread-unsafe one, we have limited the first partition to a distribution of different clothes in a virtual scenario.

Tables 1 and 2 show the comparative results of three kind of parallelization: using only MPI, using only OpenMP and the mixed code. In these table, NN represents the number of computational nodes (one or more processors which can share memory space using OpenMP), NT represents the total number of threads and PE is the number of processors.

Table 1 shows that, for the structured problem, the mixed model does not improve the performance of the MPI implementation. In both cases, the parallel efficiency is penalized by the memory bandwidth, but OpenMP is more sensitive because of the sparsity in the data access. For the unstructured problems, the best results have been obtained with two threads and four nodes in an Origin2000 (table 2), as expected for its nodal architecture. In conclusion, the shared–memory and message–passing versions of previous, grid-based codes have been combined into a single hierarchical approach for the implementation of the PCG algorithm. The algorithm presented here can adjust the distribution parameters in two phases in order to obtain the best performance for a given computer architecture and grid size.

References

1. Smith, L. & Bull, M. Development of mixed mode MPI / OpenMP applications, Scientific Programming, Vol. 9, No 2-3, 2001, pp. 83-98.
2. Cappello, F., Etiemble, D. MPI versus MPI+OpenMP on the IBM SP for the NAS benchmarks. SC2000, Supercomputing 2000, Dallas.
3. Gropp, W.D., Kaushik, D.K., Keyes, D.E. & Smith, B.F. High performance parallel implicit CFD. *Journal of Parallel Computing*, **27**, 2001, pp. 337–362.
4. Romero, L.F., Ortigosa, E.M. & Zapata, E.L. Data-task parallelism for the VMEC program. *Journal of Parallel Computing*, **27**, 2001, pp. 1347–1364.
5. Douglas, C.C., Hu J., Kowarschik, M., Rüde, U. & Weiss, C. Cache optimization for structured and unstructured grid multigrid, *Electronic Transactions on Numerical Analysis*, **10**, 2000, pp. 21–40.
6. Ortigosa, E.M., Romero, L.F., & Ramos, J.I. Parallel Scheduling of the PCG Method for Banded Matrices Arising from FDM/FEM. *Journal of Parallel and Distributed Computing*, vol. 63, iss. 12, December 2003, pp. 1243-1256.
7. Winfree, A.T. *The Geometry of Biological Time*, Springer–Verlag: New York, 2001.
8. Gutierrez, E., Romero, S., Plata, O., & Zapata, E.L. Parallel Irregular Reduction Techniques for Cloth Simulation, *Adaptivity in Parallel Scientific Computing (Seminar no. 03211)*, Dagstuhl, Saarland, Germany, May 18-23, 2003.
9. Demmel, J., Heath, M. & van der Vorst, H. Parallel linear algebra. *Acta Numerica*, **2**, 1993, pp. 111–197.
10. Basserman, A. Parallel sparse matrix computations in iterative solvers on distributed memory machines. *Journal of Parallel and Distributed Computing*, **45**, pp. 46–52, 1997.
11. Dongarra, J., Duff, I., Sorensen, D. & van der Vorst, H. *Numerical Linear Algebra for High-Performance Computers*, SIAM: Philadelphia, 1998.

An Asynchronous Branch and Bound Skeleton
for Heterogeneous Clusters

J.R. González, C. León, and C. Rodríguez*

Dpto. de Estadística, I.O. y Computación, Universidad de La Laguna
E-38271 La Laguna, Canary Islands, Spain
{jrgonzal,cleon,casiano}@ull.es
http://nereida.deioc.ull.es

Abstract. This work presents a parallel skeleton for the Branch and Bound technique. The main contribution of the proposed skeleton is that it is fully distributed. The implementation has been written in MPI. The user interface is the same as the one provided by the combinatorial optimization library MaLLBa. Computational results for a heterogeneous Linux cluster of PC are presented.

1 Introduction

Branch and Bound (BnB) is a general purpose search method. The technique can be seen as the traverse of a graph, usually an acyclic one or a tree, to find an optimal solution of a given problem. The goal is to maximize (minimize) a function $f(x)$, where x belongs to a given feasible domain. To apply this algorithmic technique, it is required to have functions to compute the *lower* and *upper bounds* and methods to divide the domain and generate smaller subproblems (*branch*).

The technique starts considering the original problem inside the full domain (named *root problem*). The lower and upper bound procedures are applied to this problem. If both bounds produce the same value, an optimal solution has been found and the procedure finishes. Otherwise, the feasible domain is divided into two or more regions. Each of these regions is a section of the original one and defines the new search space of the corresponding *subproblem*. These subproblems are assigned to the children of the root node. The algorithm is recursively applied to the subproblems, generating this way a tree of subproblems. When an optimal solution of a subproblem is found it produces a feasible solution of the original problem and so, it can be used to prune the remaining tree. If the lower bound of the node is larger than the best feasible solution known then no global optimal solution exists inside the subtree associated with such node and so the node can be discarded from the subsequent search. The search through the nodes continues until all of them have been solved or pruned.

BnB can be regarded as an instance of the general problem of distributing a queue of tasks among processors. These tasks can generate new queues of tasks to be distributed. In this sense, solving BnB on a heterogeneous distributed platform constitutes

---

* This work has been supported by the EC (FEDER) and by the Spanish Ministry of Science and Technology contract number TIC2002-04498-C05-05.

D. Kranzlmüller et al. (Eds.): EuroPVM/MPI 2004, LNCS 3241, pp. 191–198, 2004.

an interesting laboratory to tackle the general work queue skeleton [7, 10] that have been solved by other authors for shared memory platforms.

BnB has been parallelized from different perspectives [3, 8, 11, 4] but there is lack of studies for the heterogeneous case. The work presented here uses the MaLLBa interface. The skeleton library MaLLBa [1] provides algorithmic skeletons for solving combinatorial optimization problems. More precisely, to solve a problem using the BnB technique the skeleton MaLLBa::BnB has to be instantiated [2]. This skeleton requires from the user the implementation of several C++ classes: a Problem class defining the problem data structures, a Solution class to represent the result and a SubProblem class to specify subproblems. In the SubProblem class is where the user has to provide the upper_bound(), lower_bound() and branch() methods corresponding to the upper and lower bounds and the dividing functions described in the previous paragraphs. MaLLBa::BnB provides a sequential resolution pattern, a MPI [9] message-passing master-slave resolution pattern for distributed memory machines and an *OpenMP* [6] based resolution pattern for shared memory machines. This work presents a new fully distributed parallel skeleton using the same user interface. A MPI asynchronous peer-processor implementation where all processors are peers and behave the same way (except during the initialization phase) and where decisions are taken based on local information has been developed.

The contents of the paper are as follows: The parallel MPI skeleton is explained in section 2. Computational results are discussed in section 3. Finally, the conclusions and future prospectives are commented in section 4.

2   The Algorithm of the Skeleton

The algorithm has been implemented using Object Oriented and Message Passing techniques. The search space is represented by a tree of *subproblems* where each node has a pointer to its father. Additionally, a queue with the nodes pending of being explored is kept. The search tree is distributed among the processors. The number of local children and the number of children sent to other processors (remote children) is registered in each node. Since the father of a node can be located in another processor, the rank of the processor owning the father is stored on each node. The algorithmic skeleton's implementation has been divided in three stages: An initialization phase, a resolution phase and a reconstruction phase.

2.1   The Initialization Phase

The data required to tackle the resolution of the problem is placed into the processors. Initially only the master processor (*master*) has the original problem and proceeds to distribute it to the remaining processors. Next, it creates the node representing the initial subproblem and inserts it into its queue. All the processors perform monitor and work distribution tasks. The state of the set of processors intervening in the resolution is kept by *all* the processors. At the beginning, all the processors except the *master* are idle. Once this phase is finished, all the processors behave the same way.

2.2   Resolution Phase

The goal of this stage is to find the optimal solution to the problem. The majority of the skeleton tasks concentrate in this phase. The main aim of the design of the skeleton has been to achieve a *balanced distribution* of the work load. Furthermore, such arrangement is performed in a *fully distributed* way using only asynchronous communication and information that is local to each processor, avoiding this way the use of a central control and synchronization barriers.

The developed control is based on a request-answer system. When a processor is idle it performs a request for work to the remaining processors. This request is answered by the remaining processors either with work or with a message indicating their inability to send work. Since the state of the processors has to be locally registered, each one sends its answer not only to the processor requesting work but also to the remaining others, informing of what the answer was (if work was sent or not) and to which processor it was. This allows to keep updated the state of *busy* or *idle* for each processor. According to this approach, the algorithm would finish when all the processors were marked as idle. However, since request and answer messages may arrive in any arbitrary order, a problem arises: the answers may arrive before the request or messages referring to different requests may be tangled. Additionally, it may also happen that a message is never received if the processors send a message corresponding to the resolution phase to a processor that has decided that the others are idle.

To attack these problems, a *turn of requests* is associated with each processor, assigning a rank or order number to the requests in such a way that it can be determined which messages correspond to what request. They also keep a *counter of pending answers*. To complete the frame, additional constraints are imposed to the issue of messages: (i) A processor that becomes idle, performs a single request until it becomes busy again. This avoids an excess of request messages at the end of the resolution phase. (ii) A processor does not perform any request while the number of pending answers to its previous request is not zero. Therefore, work is not requested until checking that there is no risk to acknowledge work corresponding to a previous request. (iii) No processor sends messages not related with this protocol while it is idle.

Though the former restrictions solve the aforementioned protocol problems, a new problem has been introduced: if a processor makes a work request and none sends work as answer, such processor, since it is not allowed to initiate new petitions, will remain idle until the end of the stage (*starvation*). This has been solved making the *busy* processors to check, when it is their turn to communicate, if there are idle processors without pending answers to their last petition and, if there is one and they have enough work, forcing the initiation of a new turn of request for such idle processor. Since the processor(s) initiating the new turn are working, such messages will be received before any subsequent messages of work request produced by it (them), so there is no danger of a processor finishing the stage before these messages are received and therefore they won't be lost.

The scheme in Figure 1 shows the five main tasks performed in this phase. The "conditional communication" part (lines 3-16) is in charge of the reception of all sort of messages and the work delivery. The communication is conditional since is not made per iteration of the main loop but when the condition time to communicate == TRUE is held. The goal of this is to limit the time spent checking for messages, assuring the fulfilment of a minimum work between communications. The current implementation performs this check every time the time that has passed is larger than a given

```
   ... // Resolution phase
 1 problemSolved = false;
 2 while (!problemSolved) {
 3   // Conditional communication
 4   if (time to communicate) {
 5     // Message reception
 6     while (pending packets) {
 7       inputPacket = packetComm.receive(SOLVING_TAG);
 8       switch (inputPacket.msgType) { // Messages types
 9         case NOT_WORKING_MSG:...
10         case CANT_SEND_WORK_MSG:...
11         case SEND_WORK_MSG_DONE:...
12         case SEND_WORK_MSG:...
13         case BEST_BOUND_MSG:...
14       } }
15     // Avoid starvation when a request for work was neglected
16   }
17   // Computing ...
18   // Broadcasting the best bound ...
19   // Work request ...
20   // Ending the resolution phase ...
21 } ...
```

Fig. 1. Resolution phase outline.

threshold. The threshold value has been established to a small value, since for larger values the work load balance gets poorer and the subsequent delay of the propagation of bounds leads to the undesired exploration of non promising nodes.

Inside the "message reception" loop (lines 6-14) the labels to handle the messages described in the previous paragraphs are specified. Messages of type BEST_BOUND _MSG are used to communicate the best bound among processors. Each processor has a *best bound* that is compared with the incoming one to keep it updated. There is another local variable storing the best bound known by the other processors, so that at any given time it is known if the local bound is the best and if it has to be sent or not. Only one of the processors keeps to true the variable that indicates that it owns the current optimal solution (if there are several, the one with smallest rank).

The "computing subproblems" part (line 17) is where the generic BnB algorithm is implemented. The best bound must be propagated (line 18) if the one recently got improves the one the other processors have. Each time the processor runs out of work, one and only one "work request" (line 19) is performed. This request carries the beginning of a new petition turn. To determine the "end of the resolution phase" (line 20) there is a check that no processor is working and that there are no pending answers. In such case the problem is solved and the solution found is optimal.

Figure 2 shows the computing task in more depth. First, the queue is checked and a node is removed if it is not empty. Its bounds are computed to determine if it is worth to explore and if so it is branched out. Between lines 7 and 15 is the code that deals with nodes that improve the knowledge of the previous best node and thus the former best node is substituted and the new one is made persistent. To make it persistent is to

```
1 if (!bbQueue.empty()) {    // Computing ...
2     node = bbQueue.remove();
3     sp = node.getSubProblem();
4     upper = sp.upper_bound(pbm);
5     if (upper > bestBound) {
6         lower = sp.lower_bound(pbm);
7         if (lower > bestBound) {
8             bestBound = lower;
9             if (bestNode != NULL) {
10                bestNode.setIsPersistent(false);
11                bbQueue.deleteTree(bestNode);
12            }
13            bestNode = node;
14            bestNode.setIsPersistent(true);
15        }
16        if (!sp.solve()) {
17            sp.branch(pbm, subPbms[]);
18            for (i = 1; i < subPbms[].size; i++) {
19                bbQueue.insert(bbNode.create(subpbms[i]), node);
20    } } }
21        bbQueue.deleteTree(node);
22 } ...
```

Fig. 2. BnB computing tasks.

give it a special mark so that the removal algorithm does not eliminate it. The removal function (deleteTree()) eliminates the nodes from the one that is its argument towards the root. It stops when it arrives at a node that has no explored children or that is marked as persistent. If a node has only "remote children" it is saved in a temporary queue to eliminate them at the end of the algorithm. They could be eliminated at the cost of sending a message to the processor owning the father of the nodes, with the subsequent benefit in memory management but the final design decision was to reduce communications.

2.3   Rebuilding the Best Solution

Once the problem has been solved, in order to recompose the solution we have to traverse in reverse order the corresponding path in the search tree. Since the search tree is distributed among the processors, the solution is distributed too. In this phase, the partial solutions are gathered to built the solution of the original problem (see Figure 3). In first place, the processor that owns the best solution performs a call to lower_bound() with the best node (bestNode) in order to complete the solution (sol) with the required values so that the target value coincides with the value returned as bound. The other processors call the function that removes the best node they have, to eliminate the part of the tree that is no longer needed (lines 1-2). Next, they go into the loop responsible for the rebuilding of the optimal solution (lines 4-13). There are two types of messages the processors expect: (i) Those of the type BEST_SOL_MSG (line 11) with a partially rebuilt optimal solution in which the next node that intervenes in the

```
       ... // Rebuilding the best solution phase
1      if (haveBestSol) { ... /* compute the best solution */}
2      else if (bestNode != NULL) { ... /* remove tree */ }
3      bestSolConstructed = false;
4      while (!bestSolConstructed) {
5        if (solReceived) {
6          ... Rebuild the optimal solution ...
7        }
8        else {
9          inputPacket = packetComm.receive(BUILDING_SOL_TAG);
10         switch (inputPacket.msgType) {
11           case BEST_SOL_MSG: ...
12           case END_MSG:...
13     } } } ...
```

Fig. 3. Outline of the rebuilding the best solution phase.

re-building is the local one. In such case, the re-building process continues locally. (ii) Those of type END_MSG (line 12) indicating that the optimal solution has been completely rebuilt and the execution can be finalized. The comment in line 6 stands for the code in charge of rebuilding the optimal solution through a bottom-up traverse in the tree, adding to the solution the decision represented by each node. Such rebuilding continues until the root of the tree is reached or until the next father node is in another processor. When the root node is reached, the optimal solution is complete and the execution is ended after sending the corresponding END_MSG messages to the remaining processors. Otherwise, not arriving to the root node means that the father of the node is in another processor; consequently, a message BEST_SOL_MSG is sent to such processor so that it continues with the rebuilding. After finishing with the partial rebuilding, the no longer needed part of the tree is erased using a call to deleteTree(). Storing the processor where the father of the node is, allows the straightforward rebuilding of the solution, even though the node may have travelled through several processors before being computed.

3   Computational Results

The experiments were performed instantiating our BnB skeleton with the methods described by Martello and Toth [5] for the 0-1 Knapsack. The results presented in this section correspond to a randomly generated set of problems. Since the difficulty of a Knapsack Problem is influenced by the relationship between profits and weights, we have observed those known as strongly correlated. The experiments were performed on a heterogeneous PC network, configured with: four 800 MHz AMD Duron processors and seven AMD-K6 3D 500 MHz processors, each one with 256 MBytes of memory and a 32 GBytes hard disk. The installed operating system was Debian Linux version 2.2.19 (herbert@gondolin), the C++ compiler used was GNU gcc version 2.95.4 and the message passing library was *mpich* version 1.2.0.

Figure 4 (a) shows the speedups obtained for a problem of size 50,000 that will be used throughout the section. The parallel times were the average of five executions.

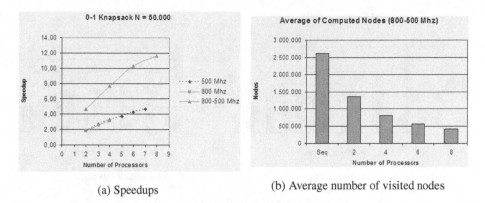

(a) Speedups   (b) Average number of visited nodes

Fig. 4. Results on a Heterogeneous cluster of PC.

The experiments labeled '500 Mhz' and '800 Mhz' were carried out on homogeneous set of machines of sizes seven and four respectively. Label '800-500 Mhz' depict the experiment on a heterogeneous set of machines where half the machines (four) were at 800 MHz and the other half were at 500 MHz. The sequential execution for the '800-500 Mhz' experiment was performed on a 500 MHz processor. To interpretate the '800-500 Mhz' line take into account that the ratio between the sequential executions on machines of both speeds was 3.81. For eight processors the maximum speed up expected will be 19.24, that is, 3.81×4 (fast processors)+4 (slow processors). Comparing the three experiments depicted we conclude that the algorithm does not experience any loss of performance due to the fact of being executed on a heterogeneous network. Figure 4 (b) represents the average number of visited nodes, for the experiment labelled '800-500 Mhz'. It is clear that an increase of the number of processors carries a reduction of the number of visited nodes. This evidences the good behaviour of the parallel algorithm.

A parameter to study is the load balance among the different processors intervening in the execution of the algorithm. Figure 5 (a) shows the per processor average of the number of visited nodes for the five executions. Observe how the slow processors examine less nodes than the faster ones. It is interesting to compare these results with those appearing in Figure 5 (b) corresponding the homogeneous executions. Both pictures highlight the fulfilled fairness of the work load distribution.

4 Conclusions

This work describes a parallel implementation of a skeleton for the Branch and Bound technique using the Message Passing paradigm. The main contribution of the algorithm is the achievement of a balanced work load among the processors. Furthermore, such arrangement is accomplished in a fully distributed way, using only asynchronous communication and information that is local to each processor. To this extent, the use of barriers and a central control has been avoided. The results obtained shows a good behaviour in the homogeneous and the heterogeneous case.

Ongoing work focuses on eliminate the replication of the global information in all processors and maintain it only in those ones which belong to a certain neighbourhood.

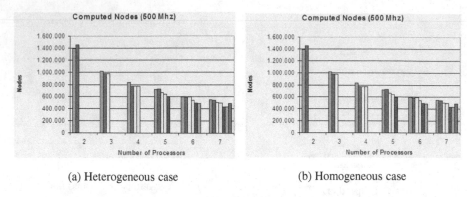

(a) Heterogeneous case   (b) Homogeneous case

Fig. 5. Load balancing.

Also, to achieve better results with problems with fine grain parallelism an algorithm mixing MPI and *OpenMP* is on the agenda.

References

1. E. Alba, F. Almeida, M. Blesa, J. Cabeza, C. Cotta, M. Díaz, I. Dorta J. Gabarró, C. León, J. Luna, L. Moreno, J. Petit, A. Rojas and F. Xhafa, *MaLLBa: A Library of skeletons for combinatorial optimisation*, Proceedings of the International Euro-Par Conference, Paderborn, Germany, LNCS 2400, 927–932, 2002.
2. I. Dorta, C. León, C. Rodríguez, and A. Rojas, *Parallel Skeletons for Divide-and-conquer and Branch-and-bound Techniques*, 11th Euromicro Conference on Parallel, Distributed and Network-based Processing, Geneva, Italy, 292–298, IEEE Computer Society Press, 2003.
3. B. Le Cun., C. Roucairol, The PNN Team, *BOB: a Unified Platform for Implementing Branch-and-Bound like Algorithms*, Rapport de Recherche n.95/16, 1999.
4. B. Di Martino, N. Mazzocca, S. Russo, *Paradigms for the Parallelization of Branch and Bound Algorithms.* 2nd International Workshop, PARA'95, Lyngby, Denmark, 141–150, LNCS-1041, 1996
5. S. Martello, P. Toth, *Knapsack Problems: Algorithms and Computer Implementations*, John Wiley & Sons Ltd, 1990.
6. OpenMP Architecture Review Board, *OpenMP C and C++ Application Program Interface*, Version 1.0, http://www.openmp.org, 1998.
7. S. Shah, G. Haab, P. Petersen, J. Throop, *Flexible Control Structures for Parallelism in OpenMP*, In 1st European Workshop on OpenMP, Lund (Sweden), September 1999.
8. Y. Shinano, M. Higaki, R. Hirabayashi, *A Generalized Utility for Parallel Branch and Bound Algorithms*, IEEE Computer Society Press, 392-401, 1995.
9. M. Snir, S.W. Otto, S. Huss-Lederman, D.W. Walker,J.J. Dongarra, *MPI: The Complete Reference*, The MIT Press, 1996.
10. Supercomputing Technologies Group, *Cilk-5.3 Reference Manual* June 2000. http://supertech.lcs.mit.edu/cilk
11. S. Tschöke, T. Polzer, *Portable Parallel Branch-and-Bound Library*, User Manual Library Version 2.0, Paderborn, 1995.

Parallelization of GSL: Architecture, Interfaces, and Programming Models*

J. Aliaga[1], F. Almeida[2], J.M. Badía[1], S. Barrachina[1], V. Blanco[2],
M. Castillo[1], U. Dorta[2], R. Mayo[1], E.S. Quintana[1], G. Quintana[1],
C. Rodríguez[2], and F. de Sande[2]

[1] Depto. de Ingeniería y Ciencia de Computadores
Univ. Jaume I, 12.071–Castellón, Spain
{aliaga,badia,castillo,mayo,quintana,gquintan}@icc.uji.es
[2] Depto. de Estadística, Investigación Operativa y Computación
Univ. de La Laguna, 38.271–La Laguna, Spain
{falmeida,vblanco,casiano,fsande}@ull.es

Abstract. In this paper we present our efforts towards the design and
development of a parallel version of the Scientific Library from GNU
using MPI and OpenMP. Two well-known operations arising in discrete
mathematics and sparse linear algebra illustrate the architecture and
interfaces of the system. Our approach, though being a general high-level
proposal, achieves for these two particular examples a performance close
to that obtained by an *ad hoc* parallel programming implementation.

1 Introduction

The GNU Scientific Library (GSL) [2] is a collection of hundreds of routines for
numerical scientific computations coded.Although there is currently no parallel
version of GSL, probably due to the lack of an accepted standard for developing parallel applications when the project started, we believe that with the
introduction of MPI and OpenMP the situation has changed substantially.

We present here our joint efforts towards the parallelization of of GSL using MPI and OpenMP. In particular, we plan our library be portable to several
parallel architectures, including distributed and shared-memory multiprocessors,
hybrid systems -consisting of a combination of both types of architectures-, and
clusters of heterogeneous nodes. Besides, we want to reach two different classes
of users: a programmer with an average knowledge of the C programming language but with no experience in parallel programming, that will be denoted as
user A, and a second programmer, or user B, that regularly utilizes MPI or
OpenMP. As a general goal, the routines included in our library should execute
efficiently on the target parallel architecture and, equally important, the library
should appear to user A as a collection of traditional serial routines. We believe
our approach to be different to some other existing parallel scientific libraries
(see, e.g., http://www.netlib.org) in that our library targets multiple classes of

---

* Supported by MCyT projects TIC2002-04400-C03, TIC2002-04498-C05-05.

D. Kranzlmüller et al. (Eds.): EuroPVM/MPI 2004, LNCS 3241, pp. 199–206, 2004.

architectures. Moreover, we offer the user a sequential interface while trying to avoid the usual loss of performance of high-level parallel programming tools.

In this paper we describe the software architecture of our parallel integrated library. We also explore the interface of the different levels of the system and the challenges in meeting the most widely-used parallel programming models using two classical numerical computations. Specifically, we employ the operation of sorting a vector and the USAXPY (unstructured sparse α times x plus y) operation [1]. Although we illustrate the approach with two simple operations, the results in this paper extend to a wide range of the routines in GSL. Our current efforts are focused on the definition of the architecture, the specification of the interfaces, and the parallelization of a certain part of GSL. Parallelizing the complete GSL can then be considered as a labor of software reusability. Many existing parallel routines can be adapted without too much effort to use our interfaces while, in some other cases, the parallelization will require a larger code re-elaboration.

The rest of the paper is structured as follows. In Section 2 we describe the software architecture of our parallel integrated library for numerical scientific computing. Then, in Sections 3–5, we describe the functionality and details of the different levels of the architecture from top to bottom. Finally, some concluding remarks follow in Section 6.

2 Software Architecture of the Parallel Integrated Library

Our library has been designed as a multilevel software architecture; see Fig. 1. Thus, each layer offers certain services to the higher layers and hides those layers from the details on how these services are implemented.

The *User Level* (the top level) provides a sequential interface that hides the parallelism to user A and supplies the services through C/C++ functions according to the prototypes specified by the sequential GSL interface (for example, a `gsl_sort_vector()` routine is provided to sort a `gsl_vector` data array).

The *Programming Model Level* provides a different instantiation of the GSL library for each one of the computational models: sequential, distributed-memory, shared-memory, and hybrid. The semantics of the functions in the Programming Model Level are those of the parallel case so that user B can invoke them directly from her own parallel programs. The function prototypes and data types in user A codes are mapped into the appropriate ones of this level by just a renaming procedure at compilation time. The Programming Model Level implements the services for the upper level using standard libraries and parallelizing tools like (the sequential) GSL, MPI, and OpenMP. In the distributed-memory (or message-passing) programming model we view the parallel application as being executed by p *peer* processes, P_0, P_1,...,P_{p-1}, where the same parallel code is executed by all processes on different data.

In the *Physical Architecture Level* the design includes shared-memory platforms, distributed-memory architectures, and hybrid and heterogeneous systems

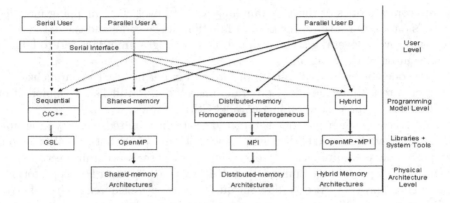

Fig. 1. Software architecture of the parallel integrated library for numerical scientific computing.

(clusters of nodes with shared-memory and processors with different capabilities). We map one process per processor of the target parallel system where, in order to balance the computational load, a process will carry out an amount of work that is proportional to the performance of the corresponding processor. The performance of the parallel routines will depend on the adequacy between the programming paradigm chosen by the user and the target architecture.

In the following sections we review the functionality, interfaces, and further details of the different levels of the software architecture, from top to bottom.

3   User Level

The basic purpose of this level is to present user A with the classical interface and interaction of the sequential GSL routines. To illustrate the challenges involved in this task, consider the following program which reads two vectors, computes their USAXPY, and outputs the result:

```
 1:  #include <gsl_sparse_vector.h>
 2:  void main (int argc, char * argv[]) {
     ...
 3:    scanf ("Value of nz %u", &nz);
 4:    y = gsl_vector_alloc (n);
 5:    x = gsl_sparse_vector_alloc (n, nz); // Allocate
 6:    gsl_vector_scanf (y, n);
 7:    gsl_sparse_vector_scanf (x, n, nz);
 8:    gsl_usaxpy (alpha, x, y);                 // USAXPY operation
 9:    printf ("Result y = alpha x + y\n"); // Output
10:    gsl_vector_printf (y, n);
11:    gsl_vector_free (y);                      // Deallocate
12:    gsl_sparse_vector_free (x); }
```

What a serial user in general expects is to compile this program using, e.g., the **make** utility, and execute the resulting runnable code from the command line. We offer the user the proper **Makefile** that at compilation time, depending on the programming model selected, maps (renames) the function prototypes into the appropriate ones of the Programming Model Level. This mapping includes user A's data I/O routines, the `main()` function in the code, and the data types. We also provide several `gslrun` scripts (one per programming model) to launch the execution of the program.

The code does not need to be transformed further in case the parallelization is targeted to a shared-memory programming model. Parallelism is exploited in this case inside the corresponding parallel GSL routines.

Nevertheless, when the code is targeted to the distributed-memory programming model, we still need to deal with a few problems. First, notice that following our *peer* processes approach, the execution of the user's program becomes the execution of p processes running in parallel, where user's data I/O is performed from a single process. Also, a different question to be considered is that of error and exception handling. Finally, some execution errors due to the parallel nature of the routines cannot be masked and must be reported to the end-user as such.

4   Programming Model Level

In this section we first review the interface of the routines in the parallel instantiations of the GSL library, and we then describe some details of the major parallelization approaches utilized in the library.

4.1   Programming Model Interface

All programming models present similar interfaces at this level. As an example, Table 1 relates the names of several sequential User Level routines with those of the different instantiations of the parallel library. The letters "sm", "dm", and "hs" after the GSL prefix ("gsl_") denote the programming model: shared-memory, distributed-memory, and hybrid systems, respectively. In the distributed-memory model, the following two letters, "rd" or "dd", specify whether the data are replicated or distributed.

Table 1. Mapping of User Level routines to the corresponding parallel routines.

User Level	Programming Model Level		
Sequential	Shared-memory	Distributed-memory	Hybrid
fscanf()	fscanf()	gsl_dmrd_fscanf() gsl_dmdd_fscanf()	gsl_hs_fscanf()
gsl_sort_vector()	gsl_sm_sort_vector()	gsl_dmrd_sort_vector() gsl_dmdd_sort_vector()	gsl_hs_sort_vector()

At the Programming Model Level the interface supplied to the User Level is also available as a parallel user-friendly interface to user B. The parallel routines can thus be employed as building blocks for more complex parallel programs.

A sorting routine is used next to expose the interface of the distributed-memory programming model:

```
1: #include <mpi.h>
2: #include <gsl_dmdd_sort_vector.h>
3: void main (int argc, char * argv []) {
   ...
4:    MPI_Init (& argc, & argv);
5:    gsl_dmdd_set_context (MPI_COMM_WORLD); // Allocate
6:    gsl_dmdd_scanf ("Value of n %u", &n);  // Read
7:    v = gsl_dmdd_vector_alloc (n, n);      // Block-Cyclic Allocation
8:    gsl_dmdd_vector_scanf (v, n);
9:    status = gsl_dmdd_sort_vector (v);     // Sorting operation
10:   printf ("Test sorting: %d\n", status); // Output
11:   gsl_dmdd_vector_free (v);              // Deallocate
12:   MPI_Finalize (); }
```

Here the user is in charge of initializing and terminating the parallel machine, with the respective invocations of routines MPI_Init() and MPI_Finalize(). Besides, as the information about the parallel context is needed by the GSL kernel, the user must invoke routine gsl_dmdd_set_context to transfer this information from the MPI program to the kernel and create the proper GSL context. The MPI program above assumes the vector to sort to be distributed among all processes so that, when routine gsl_dmdd_vector_alloc(n, cs) is invoked, the allocation for the n elements of a gsl_vector is distributed among the whole set of processors following a block-cyclic distribution policy with cycle size cs. In the case of heterogeneous systems, the block sizes assigned depend on the performance of the target processors. The call to gsl_dmdd_sort_vector sorts the distributed vector following the PSRS algorithm [3] described later.

4.2   Implementation in the Distributed-Memory Programming Model

Our library currently supports two data distributions: In the replicated layout a copy of the data is stored by all processes. In the distributed layout the data are partitioned into a certain number of blocks and each process owns a part of these blocks; in heterogeneous systems the partitioning takes into consideration the different computational capabilities of the processors where the processes will be mapped.

All I/O routines in the distributed-memory programming model (e.g., routine gsl_dmdd_fscanf) perform the actual input from P_0 and any value read from the input is then broadcasted to the rest of processes; analogously, any data to be sent to the output is first collected to P_0 from the appropriate process. While scalar data can be easily replicated using the policy just described, replication of GSL arrays has to be avoided in order to minimize memory requirements. This implies parallelizing the routines of GSL which deal with these derived data types. Notice that data I/O performed by user B in her program directly refers to the routines in the stdio library and therefore is not mapped to the I/O routines in the distributed-memory programming model.

A sorting routine is used next to illustrate the parallelization in the distributed-memory programming model. For generality, we have chosen the well-known Parallel Sort by Regular Sampling (PSRS) algorithm, introduced in [3]. This algorithm was conceived for distributed-memory architectures with homogeneous nodes and has good load balancing properties, modest communication requirements, and a reasonable locality of reference in memory accesses.

The PSRS algorithm is composed of the following five stages:

1. Each process sorts its local data, chooses $p-1$ "pivots", and sends them to P_0. The stride used to select the samples is, in the case of heterogeneous contexts, different on each processor and is calculated in terms of the size of the local array to sort.
2. Process P_0 sorts the collected elements, finds $p-1$ pivots, and broadcasts them to the remaining processes. Again, for the heterogeneous systems, the pivots are selected such that the merge process in step 4 generates the appropriate sizes of local data vectors, according to the computational performance of the processors.
3. Each process partitions its data and sends its i-th partition to process P_i.
4. Each process merges the incoming partitions.
5. All processes participate in redistributing the results according to the data layout specified for the output vector.

The case where the vector to sort is replicated poses no special difficulties: a simple adjustment of pointers allows the processes to limit themselves to work with their corresponding portions of the vector. Only stage 5 implies a redistribution. When the output vector is replicated, each process has to broadcast its chunk. Redistribution is also required even for a vector that is distributed by blocks, since the resulting chunk sizes after stage 4 in general do not fit into a proper block distribution. The average time spent in this last redistribution is proportional to the final global imbalance.

4.3   Implementation in the Shared-Memory Programming Model

In this subsection we explore the parallelization on a shared-memory parallel architecture of the USAXPY operation using OpenMP. A simplified serial implementation of this operation is given by the following loop:

```
1:    for (i = 0; i < nz; i++) {
2:      iy = indx [i];
3:      y [iy] += alpha * valx [i];  }
```

As specified by the BLAS-TF standard [1], sparse vectors are stored using two arrays: one with the nonzero values and the other an integer array holding their respective indices (valx and indx in our example).

The USAXPY operation, as many others arising in linear algebra, is a typical example of a routine that spends much of its time executing a loop. Parallel shared-memory architectures usually reduce the execution time of these routines by the executing iterations of the loops in parallel across multiple processors. In particular, the OpenMP compiler generates code that makes use of *threads* to execute the iterations concurrently.

The parallelization of such a loop using OpenMP is quite straight-forward. We only need to add a **parallel for** compiler directive, or *pragma*, before the loop, and declare the scope of variable iy as being **private**. As all iterations perform the same amount of work, a *static schedule* will produce an almost perfect load balancing with a minimum overhead.

Table 2. Characteristics of the platforms handled in the experimental evaluation.

Type	Manufacturer/ Architecture	Processor Frequency	Memory size (RAM/Cache)	#Proc.	Commun. Network
PC cluster distributed-memory	Intel Pentium Xeon	2.4GHz	1GB/512KB L2	34	Gigabit Ethernet Myrinet
Heterogeneous Cluster	Intel Pentium Xeon AMD (Duron) AMD-K6	1.4 GHz 800 MHz 500 MHz	2GB/512 L2 256MB/64 L2 256MB/64 L2	4 4 6	Fast Ethernet
CC-NUMA SGI Origin 3800	SGI MIPS R14000	500MHz	1GB/16MB L2	160	Crossbar Hypercube
Shared-memory	Intel Pentium Xeon	700MHz	1GB/1MB L2	4	System bus

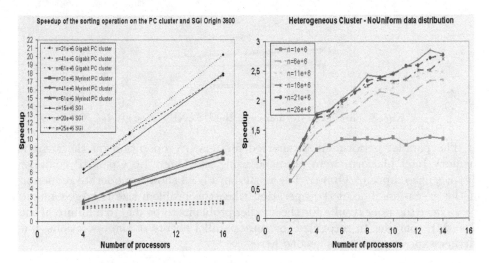

Fig. 2. Parallel performance of the Sorting case studies.

5   Physical Architecture Level

In this section we report experimental results for the parallel versions of the sorting and USAXPY operations on several platforms; see Table 2.

The parallel PSRS algorithm is evaluated on three different platforms (see Fig. 2): a PC cluster (using a distributed data layout), a SGI Origin 3800 (with a replicated data layout), and a heterogeneous cluster (with a nonuniform data distribution). In the PC cluster we used MPICH 1.2.5 for the Gigabit Ethernet and GMMPI 1.6.3 for the Myrinet network. The native compiler was used for the SGI Origin 3800. The MPICH 1.2.5 implementation was also employed for the heterogeneous cluster. The use of the Myrinet network instead of the Gigabit Ethernet in the PC cluster achieves a considerable reduction in the execution time. The parallel algorithm presents acceptable parallel performances. As could be expected, the execution time is reduced in both architectures when the number of processors is increased. We observed super-linear speed-ups in the SGI platform that are a consequence of a better use of the cache memory in the parallel algorithms. Due to the heterogeneous data distribution, better speed-ups are also achieved in the heterogeneous cluster.

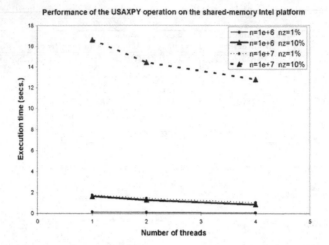

Fig. 3. Parallel performance of the USAXPY case studie.

The parallel performances reported for USAXPY operation on the shared-memory Intel platform were obtained using the Omni 1.6 OpenMP compiler (http://phase.hpcc.jp/Omni). The results in Fig. 3 show a moderate reduction in the execution time of the operation when the problem size is large enough. We believe the poor results for the smaller problems to be due to a failure of the OpenMP compiler to recognize the pure parallel nature of the operations, but further experimentation is needed here.

6   Conclusions

We have described the design and development of an integrated problem solving environment for scientific applications based on the GNU Scientific Library. Our library is portable to multiple classes of architectures and targets also a class of users with no previous experience in parallel programming.

Two simple operations coming from sparse linear algebra and discrete mathematics have been used here to expose the architecture and interface of the system, and to report preliminary results on the performance of the parallel library.

References

1. I.S. Duff, M.A. Heroux, and R. Pozo. An overview of the sparse basic linear algebra subprograms. *ACM Trans. Math. Software*, 28(2):239–267, 2002.
2. M. Galassi, J. Davies, J. Theiler, B. Gough, G. Jungman, M. Booth, and F. Rossi. *GNU scientific library reference manual*, July 2002. Ed. 1.2, for GSL Version 1.2.
3. X. Li, P. Lu, J. Schaeffer, J. Shillington, P.S. Wong, and H. Shi. On the versatility of parallel sorting by regular sampling. *Parallel Computing*, 19(10):1079–1103, 1993.

Using Web Services
to Run Distributed Numerical Applications

Diego Puppin, Nicola Tonellotto, and Domenico Laforenza

Institute for Information Science and Technologies
ISTI - CNR, via Moruzzi, 56100 Pisa, Italy
{diego.puppin,nicola.tonellotto,domenico.laforenza}@isti.cnr.it

Abstract. MPI is a *de facto* standard for high performance numerical applications on parallel machines: it is available, in a variety of implementations, for a range of architectures, ranging from supercomputers to clusters of workstations. Nonetheless, with the growing demand for distributed, heterogeneous and Grid computing, developers are hitting some of its limitations: e.g. security is not addressed, and geographically distributed machines are difficult to connect.
In this work, we give an example of a parallel application, implemented with the use of Web Services. Web Services represent an emerging standard to offer computational services over the Internet. While this solution does not reach the same performance of MPI, it offers a series of advantages: high availability, rapid design, extreme heterogeneity.

1   Introduction

The emergence of computational Grids is shifting the interest of the computational scientists. Their applications are not anymore (or not only) CPU-bound numerical kernels, but can be seen as collections of services. This causes a growing need for ways to connect heterogeneous tools, machines and data repositories. In this direction, there is a growing demand for a standard, cross-platform communication and computational infrastructure.

At the present day, MPI is a *de facto* standard for the development of numerical kernels: parallel applications have been using MPI for years; several MPI implementations are available for a variety of machines; binding to many programming languages are available. Nonetheless, MPI incurs in communication problems, very difficult to overcome, when the application is spread over remote machines, when firewalls are present, when machine configurations are different. There have been some experimental works to overcome these limitations, but so far there is no agreement on a common solution.

On the other side, there is a growing interest toward the Service Oriented Architecture (SOA) [1], a computing paradigm that considers services as building blocks for applications, and Web Services (WSs) are one of its implementations. A WS is a specific kind of service, identified by a URI, whose description and transport utilize open Internet standards. When wrapped as http data, requests to and responses from any Web Service can cross, without problems, firewalls,

D. Kranzlmüller et al. (Eds.): EuroPVM/MPI 2004, LNCS 3241, pp. 207–214, 2004.

differences in implementation languages or in operating systems. This is why we discuss about the opportunity of using Web Services to implement parallel applications, as an alternative to MPI. Clearly, it is not our intention to state that Web Services are candidates for substituting MPI for scientific applications: rather, we want to show that WSs can be fruitfully used in an area where typically MPI is used, when there is a stronger need for flexibility and interoperability among application components. We want to face this comparison from many different points of view: ease of programming, performance, availability.

The rest of the paper is structured as follows. In the next section, we give an overview of related work. Then, we introduce Web Services and their features. In section 4, we compare these two approaches by using a very simple benchmark application. Lastly, we conclude by showing some open research directions.

2   Related Work

There is an emerging attention to the convergence of MPI parallel software and the Service-Oriented Architecture. An interesting approach to this is taken in [2], where authors show an effective way to include a legacy MPI algorithm into a larger application using the OGSA standard, by creating a surrounding Grid Service. Also, whole parallel applications are wrapped as Web Services in recent works [3, 4].

Our work differs in that we are trying to use WSs as workers for a numerical application, rather than wrapping the whole application to be used as a service.

On the other side, there is interest in overcoming some limitations of MPI so to use it to program distributed applications:

- the internal nodes of a private cluster often do not have a static public IP, and cannot connect directly to the nodes outside the cluster; this way, they cannot participate to the *MPI world* of an application running remotely;
- if a machine is behind a firewall, MPI messages could be filtered out by it.

Several solutions have been proposed. In order to connect distinct private machine clusters, PACX-MPI [5] uses a two-level communication: MPI-based within the cluster, and TCP/IP-based across clusters. Two daemons, sitting on publicly visible nodes, mediate the two types of communications. MPICH-G2 [6] uses Globus services to cross firewall boundaries, so to connect clusters within different, secure administrative domains. An extension to MPICH-G2, called MPICH-GP [7] and still under development, uses the Linux NAT address translation, along with a user-level proxy, to make two firewall-protected clusters, with private IP address, to cooperate as a single MPI machine, with limited overhead.

3   Web Services

Web Services (WS) are a way to make data and computational services available on the Internet: using the WS standard, machines connected through the Internet can interact with each other to perform complex activities. A simple http

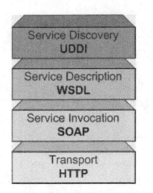
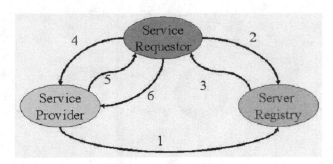

Fig. 1. A logical representation of messages hierarchy for Web Services (left). Web Service life cycle (right).

connection is usually enough to connect to a server and use the services it offers. One of the goals of this architecture is to give a standard interface to new or existing applications, i.e. to offer a standard description to any computational service, which can be stored in public repositories.

The WS standard is designed around three main types of messages, based on XML (see Figure 1): *Web Service Description Language (WSDL)* [8] is used to describe the service implementation and interface; *Simple Object Access Protocol (SOAP)* [9], used in the communication and data exchange between a WS and a client; *Universal Description, Discovery and Integration (UDDI)* [10], used to register and discover public services.

Figure 1 shows the typical scenario of Web Service utilization. The Web Service is first published using a UDDI repository (1). When a Web Service consumer is looking for a specific service, it submits a query to the repository (2), which answers back with the URL of the service that best matches the query (3). By using this URL, the Consumer asks for the WSDL describing the needed service (4); then, by using this WSDL description (5), it invokes the service (with a SOAP message) (6). The Web Service Consumer and Web Service provider have to share only the description of the service in a WSDL standard format. So, Internet access and support for SOAP message are all that is needed to use a Web Service.

The use of a WS is simplified by WS-oriented frameworks, which implement the infrastructure to communicate over the Internet with SOAP messages, and offer a *procedure-call abstraction* to service invocation (see Figure 2).

4   Experimental Results

We compared MPI and WSs with a simple client/server (task farm) application: workers perform an intensive computational loop on floating-point values, sent in blocks (with varying granularity) by the master, which then collects the results. Clearly, this limited experimentation is not exhaustive of the matter.

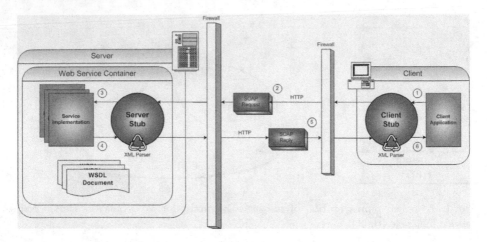

Fig. 2. Web Service architecture.

Rather, it wants to be an initial evaluation of the difference of the two discussed methodologies.

We performed two experiments. First, to understand the overhead related to the use of WSs, our applications were tested on our local cluster, using its internal private network. It is built up of 8 nodes, each of which with a dual 2 GHz Intel Xeon processor, 1 GB RAM memory. On the internal network, all nodes have a unique static IP that can be used by applications.

Second, we launched our WS-based application across the Internet: we installed our services on remote machines in San Diego. They have a (slower) 1.4 GHz AMD Athlon K6 processor, 1 GB RAM memory. They are not accessible with MPI because they are sitting behind firewalls.

MPI Implementation. Our farm application was implemented very simply with MPI: a master is responsible for distributing sequences of floating-point numbers among slaves, which are running on the internal nodes; then, it waits for answer, and gives new data to the idle workers. Our tests were performed using MPICH-1.2.5. Its implementation is a single C file, 218-line long, and can be run with a single shell command.

WS Implementation. To test the WS implementation, we installed a web server on each of the internal nodes of the cluster, and we used their internal IP to connect to them. We also linked our service to the web servers running on the remote machines in San Diego.

We used Apache Tomcat (V4.0) and Apache Axis. Tomcat is an open-source Web Application container developed with Java. Axis allows the developer to use Java classes as WS, by intercepting SOAP messages and by offering a procedure-call abstraction.

The master, running on the cluster front-end, invokes the services of the farm workers: each worker appears as a WS running on a different machine. The

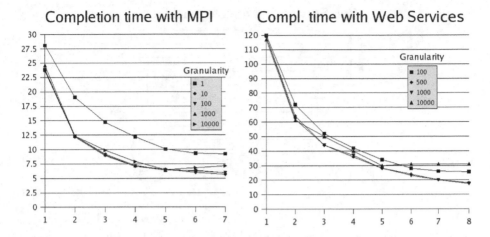

Fig. 3. Completion time of a MPI farm (left) and a WS farm (right), with different granularity, varying the number of workers, on our local cluster. Time in seconds.

master, with the roles of dispatcher and collector, is implemented as a client application. The master spawns a thread for each worker, with the responsibility of waiting for the responses and submitting new values to it.

Creating a WS is relatively simple: a Java class has to be developed, with the code implementing the numerical kernel. Then, it has to be *deployed* into Tomcat. We developed a set of scripts that perform most of this task automatically, using *ssh*. An MPI programmer interested in using WSs to implement his/her application, should focus mainly on translating its algorithm so to use a client/server (procedure-call) paradigm. We are interested in using one-way WS communications to mimic MPI messages: in the future, porting an MPI applications to WSs should be much simpler.

The WS implementation of our benchmark is composed of three Java files: the implementation of the master (318 lines), the worker interface (6 lines), and the worker implementation (20 lines). Its length is motivated by the verbosity of Java, and by the need of creating new threads for each worker. Also, the information about the topology of the computing environment is encoded within the master, while, in the MPI application, it is part of the environment configuration.

4.1   Programming Complexity and Performance

The MPI application behaves as expected, with good scalability, as shown in Figure 3(left). The best speedup is obtained with granularity 100.

The WS-based version, running on the local cluster, has an overhead ranging from 3x to 4x. This is due to a number of factors, including difference in performance from Java to C, overhead in message marshalling and unmarshalling (heavier for SOAP than MPI), overhead of the Web Service container (communication is not mediated in MPI). Also, the use of the procedure-call paradigm can

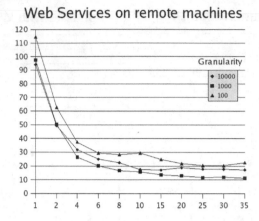

Fig. 4. Completion time of a WS farm when using remote machines. Time in seconds.

slow down the control flow, because there is the need for multiple threads, waiting for replies from the services. Recent advancements in RPC-based scheduling (e.g. [11]) could be useful in this context too.

Our WS-based application was able to use the computing services offered by a set of remote machines located in San Diego. We were able to use a pool of up to 35 workers. Using a total of 25 workers, with the best choice of granularity, we completed our application in 11.59 seconds, that is about twice as much as the best timing we had running MPI locally (5.87 seconds, with 7 workers).

In this example, the overhead related to moving from an MPI implementation to a WS implementation is somewhat limited, and can be acceptable if firewalls and difference in operating systems or programming languages prevent the use of MPI. Using WSs, we could connect to a set of (slower) remote machines, not reachable with MPI, and had an acceptable overall performance (2x slower).

4.2   Communication Overhead

MPI messages are very compact, and highly optimized: the MPI envelope is very small compared to the payload, as it stores very limited information, such as sender's and receiver's ID, message tag and MPI communicator number, totaling 16 bytes of data. Also, the array of double-precision numbers is sent in a very compact binary way, 8 bytes per element (see Table 1(a)).

On the other side, SOAP has to pay a high cost in order to be inter-operable: the payload is not sent in a binary form, but in a more portable, verbose way – doubles are converted to ASCII, and then converted back by the receiver. This enlarges greatly the message size (see Table 1(b)), and also is cause of a big performance overhead (discussed in detail in [12]).

Many researchers are exploring ways to reduce the overhead introduced by SOAP. The most interesting effort in this direction is being performed by W3C,

Table 1. Message size (bytes) for MPI (a) and WS (b), varying granularity.

Granularity	In data	Out Data	MPI request	MPI reply	Overhead
1	8	8	24	24	200%
10	80	80	96	96	20%
100	800	800	816	816	2%
1000	8000	8000	8016	8016	0.2%
10000	80000	80000	80016	80016	0.002%

(a)

Granularity	In data	Out Data	SOAP Req.	SOAP Resp.	Overhead
100	800	800	2692	4181	236% - 422%
500	4000	4000	11992	18981	199% - 374%
1000	8000	8000	23494	37482	193% - 368%
10000	80000	80000	230496	370483	188% - 363%

(b)

which is investigating the XML-binary Optimized Packaging protocol (XOP), which allows binary transmission of data between services when possible [13].

5   Conclusions and Future Work

While MPI is a strong, efficient, very accepted standard for parallel applications, there is a growing need for more general solutions, especially when distributed/Grid applications are to be developed. A variety of MPI extensions have been proposed, each addressing a specific MPI limitation, but so far there is no general agreement on a standard solution.

On the other side, WSs are emerging as a standard for distributed, Internet-oriented applications. Legacy applications can be easily wrapped into a WS interface, and made available to any client on the Internet. In this work, we showed how a computationally intensive application can be performed by using WSs as workers. While the performance is clearly lower, it can be a choice when firewalls are present, when operating systems and programming languages are different.

Also, for a variety of frameworks (.NET, J2EE...), it is very easy to wrap an existing application into a WS: this can be a way to do rapid prototyping of the parallel/distributed version of legacy software. Here we showed that a simple Java class of 20 lines, implementing a numerical kernel, can be simply transformed into a WS, and then used by a client to perform a distributed computation. We could run our application on our private cluster, and then across the Internet with no modification.

Future work includes a comparison of some of the available MPI extensions with WSs in a real distributed application, a detailed analysis of the overhead introduced by the use of WSs and of the suitable granularity for applications built as collections of services. Also, we are interested in implementing the MPI protocols over WSs: the MPI world could be emulated by a pool of coordinated services, communicating directly with one another.

Acknowledgements

We want to thank Massimo Serranò for his help with the experimental setting, and the group led by Dean Tullsen at UCSD for letting us run our tests on their machines. This work has been partially supported by the MIUR GRID.it project (RBNE01KNFP) and the MIUR CNR Strategic Project L 499/97-2000.

References

1. Papazoglou, M.P., Georgakopoulos, D., eds.: Service-Oriented Computing. Volume 46 (10) of Communications of ACM. (2003)
2. Floros, E., Cotronis, Y.: Exposing mpi applications as grid services. In: Proceedings of EuroPar 2004, Pisa, Italy (2004) To appear.
3. Gannon, D.: Software component architecture for the grid: Workflow and cca. In: Proceedings of the Workshop on Component Models and Systems for Grid Applications, Saint Malo, France (2004)
4. Balis, B., Bubak, M., Wegiel, M.: A solution for adapting legacy code as web services. In: Proceedings of the Workshop on Component Models and Systems for Grid Applications, Saint Malo, France (2004)
5. Beisel, T., Gabriel, E., Resch, M.: An extension to mpi for distributed computing on mpps. In: Recent Advances in PVM and MPI, LNCS (1997) 75–83
6. Karonis, N., Toonen, B., Foster, I.: MPICH-G2: A Grid-Enabled Implementation of the Message Passing Interface. JPDC **63** (2003) 551–563
7. Kwon, O.Y.: Mpi functionality extension for grid. Technical report, Sogang University (2003) Available at: http://gridcenter.or.kr/ComputingGrid/ file/gfk/Oh-YoungKwon.pdf.
8. Christensen, E., Curbera, F., Meredith, G., Weerawarana, S.: Web services description language (wsdl) 1.1. Technical report, W3C (2003) Available at: http://www.w3.org/TR/wsdl.
9. Box, D., Ehnebuske, D., Kakivaya, G., Layman, A., Mendelsohn, N., Nielsen, H.F., Thatte, S., Winer, D.: Simple object access protocol (soap) 1.1. Technical report, W3C (2003) Available at: http://www.w3.org/TR/SOAP/.
10. Bryan, D., *et al.*: Universal description, discovery and integration (uddi) protocol. Technical report, W3C (2003) Available at: http://www.uddi.org.
11. Gautier, T., Hamidi, H.R.: Automatic re-scheduling of dependencies in a rpc-based grid. In: Proceedings of the 2004 International Conference on Supercomputing (ICS), Saint Malo, France (2004)
12. Chiu, K., Govindaraju, M., Bramley, R.: Investigating the limits of soap performance for scientific computing. In: Proceedings of HPDC 11, IEEE (2002) 246
13. Web Consortium (W3C): The xml-binary optimized protocol (2004) Available at: http://www.w3.org/TR/xop10/.

A Grid-Based Parallel Maple

Dana Petcu[1,2], Diana Dubu[1,2], and Marcin Paprzycki[3,4]

[1] Computer Science Department, Western University of Timişoara, Romania
[2] Institute e-Austria, Timişoara, Romania
[3] Computer Science Department, Oklahoma State University, USA
[4] Computer Science Department, SWPS, Warsaw, Poland
{petcu,ddubu}@info.uvt.ro, marcin.paprzycki@swps.edu.pl

Abstract. Popularity and success of computational grids will depend, among others, on the availability of application software. Maple2g is a grid-oriented extension of Maple. One of its components allows the access of grid services within Maple, while another one use of multiple computational units. The latter component is discussed in this paper. It is based on a master-slave paradigm, and it is implemented using Globus Toolkit GT3, mpiJava and MPICH-G2. Preliminary experiments are reported and discussed. These are proving that a reasonable time reduction of computational-intensive applications written in Maple can be obtained by using multiple kernels running on different grid sites.

1 Introduction

Computer algebra systems (CAS) can be successfully used in prototyping sequential algorithms for symbolical or numerical solution of mathematical problems as well as efficiently utilized as production software in large domain of scientific and engineering applications. It is especially in the latter context, when computationally intensive problems arise. Obviously, such applications of CAS can become less time consuming if an efficient method of utilizing multiple computational units is available. This can be achieved in a number of ways: (a) parallelisation provided "inside" of a CAS or (b) parallelization facilitated by the environment in which multiple copies of the CAS are executing; furthermore, utilization of multiple computational units can occur through (c) computations taking place on a parallel computer, or (d) utilization of distributed resources (i.e. grid based computing). Since coarse grain parallelism has been proved to be efficient in an interpreted computation environment such as the CAS, in this paper we are particularly interested in the case of running multiple copies of the CAS working together within a grid. To be able to facilitate this model of parallelism, a CAS interface to a message-passing library is needed and in our research we have developed such an interface for Maple.

Maple is a popular mathematical software that provides an easy to use interface to solve complicated problems and to visualize the results of computations. Our main reason for choosing Maple is that, despite its robustness and user friendliness, we were not able to locate efforts to link Maple with grids. Second, it is well known that Maple excels other CAS in solving selected classes of

D. Kranzlmüller et al. (Eds.): EuroPVM/MPI 2004, LNCS 3241, pp. 215–223, 2004.

problems like systems of nonlinear equations or inequalities [11]. Furthermore, Maple has already a socket library for communicating over the Internet. Finally, distributed versions of Maple have been recently reported in [6] and [9].

We intend to provide an environment where utilization of multiple computational units is possible within the Maple environment such that the programmer does not have to leave the familiar CAS interface. While several parallel or distributed versions of Maple, mentioned in Section 2, were developed for clusters of computers, our goal is to "port" Maple to the computational grid. Success of our project will allow multiple grid users with Maple installed on their computers, to pool their resources and form a distributed Maple environment.

We have therefore proceeded to develop Maple2g: the grid-wrapper for Maple. It consists of two parts: one which is CAS-dependent and another, which is grid-dependent. In this way, any change in the CAS or the grid needs to be reflected only in one part of the proposed system. The CAS-dependent part is relatively simple and can easily be ported to support another CAS or a legacy code. Maple2g should therefore be portable to any new commercial version of Maple, including the Socket package. The system only relies on basic interfaces to the Maple kernel. More details about Maple2g architecture are given in Section 3. We continue by describing, in Section 4, the proposed approach to distributed computing. Finally, results of our experiments are presented in Section 5, while conclusions and future improvements are enumerated in Section 6.

2   Parallel and Distributed Versions of Maple

With the increasing popularity of parallel and distributed computing, researchers have been attempting at building support for parallel computing into Maple. We are aware of the following attempts to supplement Maple with parallel and/or distributed computation features.

‖Maple‖ is a portable system for parallel symbolic computations built as an interface between the parallel programming language Strand and Maple [10]. Sugarbush combines the parallelism of C/Linda with Maple [2]. Maple was ported also to the Intel Paragon architecture [1]. Five message passing primitive were added to the kernel and used to implement a master-slave relationship amongst the nodes. The manager could spawn several workers and asynchronously await the results. Finally, FoxBox provides an MPI-compliant distribution mechanism allowing parallel and distributed execution of FoxBox programs; it has a client/server style interface to Maple [3].

All these attempts took place in the 1990th and are all but forgotten. In recent years we observe a renewed interest parallel/distributed Maple.

Distributed Maple is a portable system for writing parallel programs in Maple, which allows to create concurrent tasks and have them executed by Maple kernels running on separate networked computers. A configurable program written in Java starts and connects external computation kernels on various machines and schedules concurrent tasks for execution on them. A small Maple package implements an interface to the scheduler and provides a high level parallel programming model for Maple [9].

Parallel Virtual Maple (PVMaple) was developed to allow several independent Maple kernels on various machines connected by a network to cooperate in solving a problem. This is achieved by wrapping Maple into an external system which takes care of the parallel execution of tasks: a special binary is responsible for the message exchanges between Maple processes, coordinates the interaction between Maple kernels via PVM daemons, and schedules tasks among nodes [6]. A Maple library implements a set of parallel programming commands in Maple making the connections with the command messenger. The design principles are very similar to those of the Distributed Maple.

The above mentioned parallel or distributed Maple projects make use of message-passing for interprocessor communication and provide message-passing interfaces to the user. Commands, like *send* and *receive*, are available so that a user can write a parallel Maple program using the message-passing programming model. These commands are implemented either in user written subroutines callable by Maple or in script files written in Maple's native programming language. They utilize low level message-passing routines from the standard MPI/PVM libraries, simple communication functions, or file synchronization functions (in the case of a communication via a shared file system). Actually, existing parallel/distributed versions Maple can be regarded as a *message-passing extension of Maple*. A recent more comprehensive description of the available parallel and distributed versions of Maple can be found in [9].

None of the attempts at adding parallel/distributed features to Maple, that we were able to located tried to introduce Maple to the grids. It is the latter idea that became focus of our work. Our experiments with PVMaple showed sufficient efficiency in solving large problems to follow this design paths in Maple2g development. As will be seen below, the later has similar facilities with PVMaple.

3   Maple2g Architecture

Rewriting a CAS kernel in order to supplement its functionality with grid capabilities can be a complicated and high-cost solution. Wrapping the existing CAS kernel in an interface between the grid, the user and the CAS can be done relatively easily as an added functionality to the CAS. In addition, it can also be adapted on-the-fly when new versions of the CAS in question become available. In this way Maple2g is a prototype grid-enabling wrapper for Maple.

Maple2g allows the connection between Maple and computational grids based on the Globus Toolkit. The prototype consists of two parts. A CAS-dependent part (*m2g*) is the Maple library of functions allowing the Maple user to interact with the grid or cluster middleware. A grid-dependent part (*MGProxy*) is the middleware, a package of Java classes, acting as interface between m2g and the grid environment. The m2g functions are implemented in the Maple language, and they call MGProxy which accesses the Java CoG API [5]. A preliminary description of fundamental Maple2g concepts is present in [7].

Maple2g has three operating modes: *user mode*, for external grid-service access, *server mode*, for exposing Maple facilities as grid services, and *parallel mode* for parallel computations in Maple using the grid.

In the current version of Maple2g we have implemented a minimal set of functions allowing the access to the grid services:

m2g_connect(): connection via Java COG to the grid;
m2g_getservice(c, l): search for a service c and retrieve its location l;
m2g_jobsubmit(t, c): job submission on the grid based of the command c;
m2g_results(t): retrieve the results of the submitted job labeled t.

More details about the implementation of the grid services access procedures from Maple illustrated by several examples can be found in [8]. Let us now proceed to present some details of utilizing Maple2g in parallel on the grid.

4   Coupling Maple Kernels over Grid - Maple2g Approach

The computational power of a CAS can be augmented by using several other CAS kernels (the same or different CASs) when the problem to be solved can be split between these kernels or a distributed-memory parallel method is used in order to solve it. The usage of a standard message-passing interface for inter-kernel communication allows the portability of the parallel version of a CAS in particular an easy deployment on clusters and grids (Figure 1).

The two extreme approaches to design the interaction with the message-passing interface are minimal, respectively full, access to the functions of the message-passing interface. In the first case the set of functions is restricted to those allowing to send commands and receive results from the remote kernels. In the second case it is possible to enhance the CAS with parallel or distributed computing facilities, allowing the access of the CAS to other parallel codes than the ones written in the CAS language (the message-passing interface can be used as interpreter between parallel codes written in different languages). The first approach has been followed in our Maple2g prototype.

Parallel codes using MPI as the message-passing interface can be easily ported to grid environments due to the existence of the MPICH-G2 version which runs on top of the Globus Toolkit. On other hand, the latest Globus Toolkit GT3 is built in Java, and the Java clients are easier to write. This being the case, we selected mpiJava as the message-passing interface between Maple kernels.

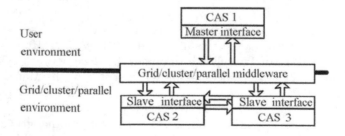

Fig. 1. Coupling CAS kernels over the grid using a master-slave approach.

Table 1. Maple2g functions/constants for remote process launch/communications.

Function/const.	Description
m2g_maple(p)	Starts p processes MGProxy in parallel modes
m2g_send(d, t, c)	Send at the destination d a message labeled t containing the command c; d, t are numbers, c, a string; when d is "all", c is send to all kernels
m2g_recv(s, t)	Receive from the source s a message containing the results from the a previous command labeled t; when s is 'all', a list is returned with the results from all kernels which have executed the command t
m2g_rank	MGProxy rank in the MPI World, can be used in a command
m2g_size	Number of MGProxy processes, can be used in a command

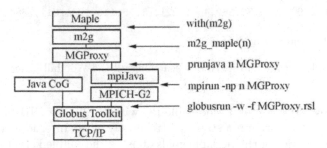

Fig. 2. From a m2g command to a grid request.

In Maple2g a small number of commands have been implemented and made available to the user, for sending commands to other Maple kernels and for receiving their results (Table 1).

Maple2g facilities are similar to those introduced in the PVMaple [6]. The user's Maple interface is seen as the master process, while the other Maple kernels are working in a slave mode. Command sending is possible not only from the user's Maple interface, but also from one kernel to another (i.e. a user command can contain inside a send/receive command between slaves).

Figure 2 shows how the m2g_maple command is translated in a grid request.

MGProxy is activated from user's Maple interface with several other MGProxy copies by m2g_maple command. The copy with the rank 0 enters in user mode and normally runs in the user environment, while the others enter in server mode. Communication between different MGProxy copies is done via mpiJava.

5   Test Results

We have tested the feasibility of Maple2g approach to development of distributed Maple applications on a small grid based on 6 Linux computers from two locations: 4 PCs located in Timisoara, Romania; each with a 1.5 GHz P4 processor and 256 Mb of memory, connected via a Myrinet switch at full 2Gb/s and 2 computers located at the RISC Institute in Linz[1]; one with a P4 processor running

---

[1] In the frame of the IeAT project supported by Austrian Ministries BMBWK project no. GZ 45.527/1-VI/B/7a/02, BMWA project GZ no. 98.244/1-I/18/02.

```
>with(m2g); m2g_MGProxy_start();
[m2g_connect,m2g_getservice,m2g_jobstop,m2g_jobsubmit,m2g_maple,m2g_rank,
m2g_recv,m2g_results,m2g_send,m2g_size,m2g_MGProxy_end,m2g_MGProxy_start]
        Grid connection established
>p:=4: a:=1: b:=2000: m2g_maple(n);
        Connect kernel 1: successful
        Connect kernel 2: successful
        Connect kernel 3: successful
        Connect kernel 4: successful
>m2g_send("all",1,cat("s:=NULL:a:=",a,":b:=",b,": for i from a+m2g_rank"
  ," to b by m2g_size do if isprime(i*2^i-1) then s:=s,i fi od: s;")):
>m2g_recv("all",1);
        [[81,249],[2,6,30,362,462,822],[3,75,115,123,751],[384,512]]
>m2g_MGProxy_end();
        Grid connection closed
```

Fig. 3. Maple2g code and its results searching all the Woodall primes in $[a, b]$.

at 2.4 GHz and 512 Mb, and a portable PC with a 1.2 GHz PIII processor and 512 Mb, connected through a standard (relatively slow) Internet connection.

Note that the one of the possible goals of using the Maple kernels on the grid is to reduce the computation time (it is **not** to obtain an optimal runtime, like on a cluster or a parallel computer. When a grid user executes a parallel Maple program, other computers typically become available as "slaves". Taking into account the possible relative slowness of the Internet (and unpredictable connection latency), it would be costly to pass data frequently among the computational nodes. This being the case, the best possible efficiency for embarrassingly parallel problems; for example, when each slave node receives a work package, performs the required computation, and sends the results back to the master. In what follows we present two examples of such computations, which therefore can be treated as the "best case" scenarios.

There are several codes available on Internet to solve in parallel open problems like finding prime numbers of specific form [4]. For example, currently the Woodall numbers, the primes of the form $i2^i - 1$, are searched in the interval $[10^5, 10^6]$. Figure 3 presents the Maple2g code and its results searching the Woodall primes in a given interval $[a, b]$ using $p = 4$ Maple computational units.

A second example involves graphical representation of results of a computation. Given a polynomial equation, we count the number of Newton iterations necessary to achieve a solution with a predescribed precision and starting from a specific value on the complex plane. If we compute these numbers for the points of a rectangular grid in a complex plane, and then we interpret them as colors, we may obtain a picture similar to that from Figure 4. The same figure displays the Maple2g code in the case of using 4 kernels; vertical slices of the grid are equally distributed among these kernels.

We have run our experiments on two basic combinations of available computers. First, using 4 machines clustered in Timişoara and, second, using 2 machines in Timişoara and 2 in Linz. We have experimented with a number of possible

```
>with(m2g): m2g_MGProxy_start(); no_procs:=4;
>m2g_maple(no_procs): d:="all";
>m2g_send(d,1,"f:=x->x^7+x^6+5*x^5+3*x^4+87*x^3
            +231*x^2+83*x+195:"):
>m2g_send(d,2,"newton:=proc(x,y) local z,dif,m;
        dif:=1; z:=evalf(x+y*I);
        for m to 30 while abs(dif)>0.1*10^(-8) do
        dif:=f(z)/D(f)(z); z:=z-dif od; m end:");
>m2g_send(d,3,"plot3d(0,-5+10*m2g_rank/m2g_size..
            -5+10*(m2g_rank+1)/m2g_size, -5..5, grid=[160/m2g_size,160],
            style=patchnogrid,orientation=[90,0],color='newton');"):
>plots[display3d](m2g_recv(d,3)); m2g_MGProxy_end();
```

Fig. 4. Maple2g code and the graphical result in measuring the levels of Newton iterations to solve in the complex plane a polynomial equation of degree seven.

approaches to the solution of the two problems, where both problems come in two different sizes representing a "small" and a "large" problem. Table 2 summarizes the results and the notations used there refers to the following case studies:

Sequential: the Woodall prime list and the plot were constructed without any splitting technique and the results come form one of the PC's in Timişoara.

Ideal: the maximum possible reduction of the time using p processors;

Cycle: the Woodall prime list or the plot were constructed in a sequential manner, but in a cycle with p steps. The time per step is variable. We registered the maximum of the time value of each step (on a cluster's PC).

MPI-cluster: The codes from the Figs. 3 and 4 are used on p processors of the cluster; here mpiJava is installed over MPICH version; Globus is not used.

G2-cluster: Same codes were used on p processors of the cluster in Timişoara, here mpiJava is installed over MPICH-G2 using Globus Toolkit 3.0.

G2-net: Same codes were used on p different processors running the mpiJava based on MPICH-G2: in the case of $p = 2$, one PC in Timişoara and the faster machine in Linz are used; in the case of $p = 4$, two PC's in Timişoara and two computers in Linz are used. The parallel efficiency is computed.

Overall, it can be said that a reasonable parallel efficiency for the larger problem has been achieved: for the Woodall primes: 73% for 2 processors and 44% for 4 processors; for Newton iteration visualization: 75% for 2 processors and 40% for 4 processors. As expected, efficiency is improving as the problem size is increasing.

At the same time the results are somewhat disturbing when one considers the current state of grid computing. On the local cluster, the results based on mpiJava and MPICH are substantially better than these obtained when the mpiJava and MPICH-G2 are used. This indicates a considerable inefficiency in the MPICH-G2 package. Furthermore, our results indicate that currently, realistic application of grids over the Internet makes sense only for very large and easy to parallelize problems (like seti@home). For instance, when machines residing at two sites were connected then the efficiency dropped by about 18%.

Table 2. Time results.

Problem	Woodall primes		Newton iterations	
p Implementation	$[a,b]=[1,2000]$	$[a,b]=[1,4000]$	grid=160×160	grid=300×300
1 Sequential	236 s	3190 s	208 s	1804 s
2 Ideal	118 s	1595 s	104 s	902 s
Cycle	122 s	1643 s	105 s	911 s
MPI-cluster	135 s	1725 s	123 s	1020 s
G2-cluster	153 s	1846 s	138 s	1071 s
G2-net	185 s	2197 s	160 s	1199 s
4 Ideal	59 s	797 s	52 s	451 s
Cycle	65 s	885 s	55 s	473 s
MPI-cluster	79 s	1027 s	73 s	654 s
G2-cluster	107 s	1263 s	94 s	784 s
G2-net	160 s	1831 s	138 s	1129 s

Obviously, in this case this is not the problem with the grid tools, but with the Internet itself. However, since the grid is hailed as the future computational infrastructure, and since our two problems represented the best case scenario, it should be clear to everyone that, unfortunately, we are far away from the ultimate goal of the grid paradigm.

6   Conclusions and Future Developments

At this stage, the proposed extension of Maple exists as a demonstrator system. Maple2g preserves the regular Maple instruction set and only add several new instructions. Further work is necessary to make it a more comprehensive package and to compare it with similar tools build for clusters. In this paper we have shown that utilizing Maple2g allows developing grid-based parallel applications. Our initial test have also indicated satisfactory efficiency of Maple2g, especially when native MPI tools are used (instead of their Globus based conuterparts). In the near future we plan intensive tests on grids on a large domain of problems to help guide further development of the system. Among others, the master-slave relationship between nodes will be extended to allow slaves to become masters themselves and thus facilitate the development of hierarchical grid applications.

References

1. Bernardin, L.: Maple on a massively parallel, distributed memory machine. In Procs. 2nd Int. Symp. on Parallel Symbolic Computation,Hawaii (1997), 217-222.
2. Char B. W.: Progress report on a system for general-purpose parallel symbolic algebraic computation. In ISSAC '90, ACM Press, New York (1990).
3. Diaz A., Kartofen E.: FoxBox: a system for manipulating symbolic objects in black box representation. In ISSAC '98, ACM Press, New York (1998).
4. Internet-based Distributed Computing Projects, www.aspenleaf.com/distributed/.
5. Java CoG Kit, http://www-unix.globus.org/cog/java/.

6. Petcu D.: PVMaple – a distributed approach to cooperative work of Maple processes. In LNCS 1908, eds. J.Dongarra et al., Springer (2000), 216–224
7. Petcu D., Dubu D., Paprzycki M.: Towards a grid-aware computer algebra system, In LNCS 3036, eds. M.Bubak, J.Dongarra, Springer (2004), 490–494.
8. Petcu D., Dubu D., Paprzycki, M.: Extending Maple to the grid: design and implementation. Procs. ISPDC'2004, Cork, July 5-7, 2004, IEEE series, in print.
9. Schreiner W.,Mittermaier C.,Bosa K.: Distributed Maple–parallel computer algebra in networked environments.J.Symb.Comp.35(3),Academic Pr.(2003),305–347.
10. Siegl K.: Parallelizing algorithms for symbolic computation using ‖Maple‖. In Procs. 4th ACM SIGPLAN Symp. ACM Press, San Diego (1993), 179–186.
11. Wester M.: A critique of the mathematical abilities of CA systems. In CASs - A Practical Guide, ed. M.Wester, J.Wiley (1999), math.unm.edu/~wester/cas_review.

A Pipeline-Based Approach
for Mapping Message-Passing Applications
with an Input Data Stream*

Fernando Guirado[1], Ana Ripoll[2], Concepció Roig[1], and Emilio Luque[2]

[1] Univ. de Lleida, Dept. of CS. Jaume II 69, 25001 Lleida, Spain
{fernando,roig}@eup.udl.es
[2] Univ. Autònoma de Barcelona, Dept. of CS. 08193 Bellaterra, Barcelona, Spain
{ana.ripoll,emilio.luque}@uab.es

Abstract. Pipeline applications simultaneously execute different instances from an input data set. Performance parameters for such applications are latency (the time taken to process an individual data set) and throughput (the aggregate rate at which data sets are processed). In this paper, we propose a mapping algorithm that improves activity periods for processors by maximizing throughput and maintaining latency. The effectiveness of this mapping algorithm is studied for a representative set of message-passing pipeline applications having different characteristics.

1 Introduction

Many applications in image processing, signal processing and scientific computing are naturally expressed as collections of message-passing pipelined tasks [1] [2] [3]. A pipeline allows different instances of data from an input stream to be processed simultaneously. In this model, each task repeatedly receives input from its predecessor tasks, performs its computation and sends output to its successor tasks.

An important factor in the performance of pipeline applications on a parallel system is the manner in which the computational load is mapped onto the system processors. Ideally, to achieve maximum parallelism, the load must be evenly distributed across processors. The problem of statically mapping the workload of a parallel application onto processors in a distributed memory system has been studied under different problem models [4] [5] [6]. These static mapping policies do not model applications consisting of a sequence of tasks where the output of a task becomes input for the next task in sequence.

The mapping mechanism for a pipeline application has to exploit both parallelism (spatial concurrency) and pipelining (temporal concurrency). Parallelism is used to allow several processors to execute tasks concurrently, and pipelining allows a set of tasks to be divided into stages, with each stage handling results

---

* This work was supported by the MCyT under contract 2001-2592 and partially sponsored by the Generalitat de Catalunya (G. de Rec. Consolidat 2001SGR-00218).

D. Kranzlmüller et al. (Eds.): EuroPVM/MPI 2004, LNCS 3241, pp. 224–233, 2004.

obtained from the previous stage. Response time (or latency) is defined as the time required to execute all the tasks for a single input data. The iteration period is defined as the interval of time needed to execute consecutive data from the input stream. Parallelism helps to reduce the response time of a single data, while pipelining is used to reduce the iteration period (i.e., increase throughput) whenever an input data stream is processed in sequence.

The solution to the mapping problem is known to be NP-complete when it tries to minimize response time for a single data. Further, since our objective is to maximize throughput, rather than just minimize response time, we also need to bear pipelining in mind. Exploiting parallelism with pipelining makes the scheduling process much more difficult.

The system is assumed to be dedicated and deterministic. The application has a stable behavior so that the set of tasks and their computational load is known in advance. Input of the mapping algorithm is a description of the application in terms of a directed task graph with known computation and communication times. The output generated by the algorithm is the assignment of tasks to processors.

A feasible mapping for pipeline applications should meet two constraints: the time constraint requires schedules to achieve desired throughput. The resource constraint requires schedules to use available resources only. These two constraints are related to the following mapping problems:

- The resource optimization problem. Given a task graph and a desired iteration period, find a mapping with this iteration period using the minimum number of processors.
- The throughput optimization problem. Given a task graph and a machine architecture, find a mapping that achieves maximum throughput for this graph with the given architecture.

In this paper, we adapt the resource optimization problem and propose an algorithm that finds the minimum number of processors allowing a maximum throughput and minimum latency to be achieved.

The mapping of pipeline applications is mainly solved in the literature for applications that can be modelled as one-way linear structure [7] [8] [9] or a set of homogeneous tasks at each stage [10]. In this work, we generalize this previous mapping research and deal with non-homogeneous pipeline applications that can be composed of several stages with different tasks at each stage. The proposed mapping algorithm proceeds by defining synchronous stages and allocating tasks to processors, based on these stages. This allows the pipelining to be exploited by minimizing the number of required processors.

The remaining sections are organized as follows: Section 2 presents the mapping methodology that we propose in this paper. Section 3 shows the experimental results that were obtained for a set of PVM applications with a pipeline structure running and a cluster of workstations. Finally, Section 4 outlines the main conclusions to our work.

2   Mapping Methodology

Pipeline applications can be structured in a pipeline with multiple ways or lines, because different functions are applied to the same data, or because tasks are replicated in order to process split portions of data in a data-parallel fashion.

In this work, we deal with pipeline applications that exhibit a non-homogeneous behavior. An application is composed of a sequence of computation stages defined by the user, that perform in disjoint sets of input data. Each stage is composed of an arbitrary number of tasks, where each task Ti has a computation time $\mu(Ti)$ and a communication time $c(Ti, Tj)$ with its successor (adjacent) task Tj. The first stage receives the external data, while the last stage produces the results. Figure 1 illustrates the model of pipeline applications under consideration.

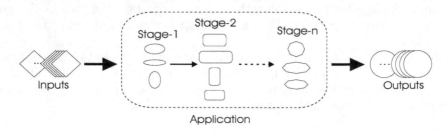

Fig. 1. Task model for a non-homogeneous pipeline application.

In the literature, there is a great amount of algorithms that solve the mapping problem of this task model with heuristics that work with a Directed Acyclic Graph (DAG). In [11] there is an exahustive comparison of the most relevant DAG heuristics. However, the main goal of DAG mapping policies is to minimize the execution time for only one data (i.e. to minimize the latency), and the iterative behaviour of a pipeline application is not considered in this mapping model.

The mapping algorithm that we propose in this paper is based on the one hand on improving the application throughput by defining pipeline stages with similar computation time, that can run synchronously for diffentent instances of the input stream and to minimize latency for one data. On the other hand, it reduces the number of required processors and as a consequence their activity periods are improved.

The strategy of the algorithm is based on forming clusters of tasks, where each cluster will be assigned to the same processor. The algorithm evolves through the three following steps, which are set out following the example of the 3-way pipeline application shown in Figure 2. For sake of illustration, we assume the communication times of adjacent tasks to be zero. Note that throughput is only affected by the computation time of tasks because communications are overlapped with computations [12].

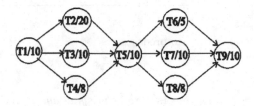

Fig. 2. Example of a non-homogeneous pipeline application.

Step 1. Identification of all different paths in the pipe.

Starting from the task graph of the application, all the different paths are identified. For each path, the corresponding *path_time* value is calculated, which corresponds to the accumulated computation time of all the tasks in the path.

$$path\_time = \sum_{Ti \in path} \mu(Ti)$$

All the paths are ordered in a list in increasing order of *path_time*. For example, the first path of the list will be T1 → T4 → T5 → T6 → T9 with *path_time*=43.

Step 2. Compose synchronous stages.

In order to have an execution that is as synchronous as possible, this step identifies those tasks included in each synchronous stage, called *syn_stage*. Each synchronous stage includes task clusters that allow equilibration of the computation time among processors without decreasing maximum throughput. In the specific case of a one-way pipeline, each stage will be composed of only one task. In the general case of an n-way pipeline, the algorithm proceeds as follows:

The first synchronous stage, *syn_stage*(1), is defined with the tasks that do not have precedence relationships. A stage number i, *syn_stage*(i), is defined with the two next phases:

1. The preliminary stage, *pre_stage*, is defined with the tasks that are successor to the tasks in *syn_stage*(i − 1). The *max_time_stage* value is calculated as the maximum computation time for these tasks.

$$max\_time\_stage = max_{Ti \in pre\_stage} \mu(Ti)$$

2. All the paths of the list created in Step 1 are examined in order to include task clusters that equilibrate the computation in the *syn_stage*(i). For each path that contains a task of the pre_stage, it is assessed whether there are successor tasks not already assigned in which the accumulated computation time, *ac_comput*, is less or equal than *max_time_stage*. In these case, they are joined in a cluster that is added to the *syn_stage*(i). In the case of the evaluated task not being able to form a cluster, it will be included alone in *syn_stage*(i).

```
N = set of tasks
syn_stage(1)=tasks with no predecessors
N=N-syn_stage(1)
i=2
while N ≠ ∅
    pre_stage=set of successors of tasks in syn_stage(i − 1)
    max_time_stage = max_{Tj∈pre_stage}μ(Tj)
    for each path in the list containing Tj ∈ pre_stage
        ac_comput = μ(Tj)
        cluster=Tj
        pre_stage=pre_stage-Tj
        N=N-Tj
        for each successor task Tk of Tj in the path with Tk ∈ N
            if (ac_comput + μ(Tk))≤ max_time_stage
                cluster = cluster ∪ Tk
                ac_comput = ac_comput + μ(Tk)
                N = N-Tk
            end_if
        end_for
        syn_stage(i) = syn_stage(i)∪ cluster
    end_for
end_while
```

Fig. 3. Algorithm for defining synchronous stages (Step 2).

It has to be noted that the paths are examined in increasing order of their *path_time* value. Thus, the paths with a lower value of *path_time* have more priority for having task clusters within them. The reason for this criterion is based on the goal of not increasing the latency of the application. When a task is included on a cluster, it may be included in a path where it did not belong. In this case, its *path_time* and, consequently, its global latency, are incremented. Thus, by prioritizing clusters in the least loaded paths, negative effects on latency will mainly be avoided.

Figure 3 shows the pseudocode corresponding to Step 2 of the mapping algorithm. Table 1 shows the development of this step for the example of Figure 2. As can be observed, all the stages are composed of the tasks of the pre_stage except in *syn_stage(2)*, where we identified the cluster (T4,T5). It has to be noted that in *syn_stage(2)*, cluster (T3,T5) was also possible because it has an accumulated computation time of 20. But the algorithm first made cluster (T4,T5) as this belongs to a line with lower *path_time*. Thus, this cluster is discarded as T5 was already assigned.

Step 3. Definition of clusters inter-stages.

In order to equilibrate the computation among the processors, this step identifies task clusters that can be composed between stages. This step starts with

Table 1. Composition of the stages of the pipe.

	pre_stage	max_time_stage	clusters
$syn\_stage(1)$	T1	$\mu(T1)=10$	T1
$syn\_stage(2)$	T2,T3,T4	$\mu(T2)=20$	T2,T3,(T4,T5)
$syn\_stage(3)$	T6,T7,T8	$\mu(T7)=10$	T6,T7,T8
$syn\_stage(4)$	T9	$\mu(T9)=10$	T9

the clustered graph obtained in Step 2. Thus, a cluster, CLi, refers here to both a single task or a cluster of tasks obtained in the previous step.

The list of paths in Step 1 is recalculated, taking into account the clusters obtained in Step 2. All the paths are evaluated from the initial to the end cluster. For each cluster CLi in the list path, we compute the successor clusters for which the accumulated computation time is less or equal to $\mu(CLi)$. In this case, the clusters will be grouped into a new one. Figure 4 shows the algorithm pseudocode solving Step 3.

```
for each path in the list
    CLi=initial cluster of the path
    while CLi ≠ final cluster of the path
        CLj=successor cluster of CLi
        ac_comput=μ(CLj)
        cluster=CLj
        while ac_comput ≤ μ(CLi) and CLj has successors
            CLj=successor cluster of CLj
            ac_comput=ac_comput+μ(CLj)
            if ac_comput ≤ μ(CLi) cluster=cluster ∪ CLj
        end_while
        update cluster in the paths
        CLi=cluster
    end_while
end_for
```

Fig. 4. Algorithm for defining clusters of tasks between stages (Step 3).

Table 2 shows the evaluation of the first two paths for the graph corresponding to the example. As can be observed, in the first path we found a new cluster with (T6,T9), because their accumulated computation time of 15 is less than that of (T4,T5). In the second path in the table, it was not possible to find a new cluster. Finally with the evaluation of all the paths, no additional new clusters were found. Figure 5 shows the task graph of the example with the clusters obtained after applying the proposed new algorithm and the resulting mapping of tasks to processors.

Table 2. Development of Step 3 for the two first paths of the list.

Path	CLi	μ(CLi)	CLj	μ(CLj)	ac_comput	cluster
T1-(T4,T5)-T6-T9	T1	10	(T4,T5)	18	18	
	(T4,T5)	18	T6	5	5	
			T9	10	15	(T6,T9)
T1-T3-(T4,T5)-(T6-T9)	T1	10	T3	10	10	
	T3	10	(T4,T5)	18	18	
	(T4,T5)	18	(T6,T9)	15	15	

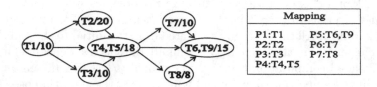

Fig. 5. Clustered task graph after applying the mapping algorithm and the resulting mapping.

3   Experimental Results

In this section, we conducted an experimentation process aimed at evaluating the quality of the mapping solution proposed in this paper. We experimented with a set of message-passing pipeline applications having different characteristics for task interaction pattern and task computation times. For each mapping, we measured latency, throughput and periods of processor inactivity in order to analyze the effectiveness of our mapping in the system utilization and in application performance.

The execution of each application was carried out with the simulation framework DIMEMAS [13] that works with message-passing applications. The underlying system was modelled by defining a set of homogeneous nodes fully-connected by a 100 Mbps Fast Ethernet. The execution of the applications was carried out with a dedicated system. The set of applications that we experimented with can be classified into two categories: (a) a set of synthetic applications and, (b) an image-processing application. The results obtained for each of these are reported below.

a) Pipeline synthetic applications.

Due to the lack of accepted benchmarks, we conducted an experiment with a set of five synthetic applications. We compared the effectiveness of our mapping with the assignment of one-task-per-processor and with the optimum assignment. Each application had eight tasks only in order to be able to generate all possible combinations for finding the optimum assignment that provides minimum latency for a single data and maximum throughput for an input data stream, respectively. For the same reason, the use of a simulation framework facilitated execution of all possible combinations for the optimum.

Table 3. Performance results for the mappings: one-task-per-processor, proposed mapping algorithm and optimum mapping.

	latency (seconds)			throughput (data/sec.)		
	1t_1p	mapping	optimum	1t_1p	mapping	optimum
pipe_1 (3)	90	85	80	0.032	0.025	0.025
pipe_2 (4)	80	75	65	0.014	0.019	0.019
pipe_3 (4)	90	80	75	0.0327	0.0328	0.0328
pipe_4 (5)	90	85	75	0.0390	0.0391	0.0282
pipe_5 (4)	95	90	85	0.0652	0.0649	0.0495

The execution of these applications was carried out for a stream of 100 input data sets. Table 3 shows the obtained values of latency and average throughput that were obtained with the different assignments. For each application, the number of processors that were used by the mapping and optimum columns in the table is shown in parentheses.

As can be observed by applying the mapping algorithm, on average we reduced by 50% the number of processors with respect to the one-task-per-processor assignment that needed eight processors. In all cases, this reduction yielded improvements in latency and throughput values close to those of the optimum assignment; in certain cases, the results are actually equal to optimum.

b) Image processing application.

We experimented with an image-processing application. In this case, we simulated the application known as BASIZ (Bright And Saturated Image Zones), which processes a stream of images from a video sequence in order to identify those zones most sensitive to human vision. This application was implemented in the PVM framework and was adapted from URT modules [14]. It is composed of 39 tasks arranged in seven pipeline stages that perform the following functions: color separation, gaussian blur, adding blurred images, color merging, image conversion, threshold and image generation.

The tasks are distributed in nine internal task parallel ways, where each is divided into three ways that execute in a data parallel fashion. This results in a 27-way pipeline application, whose tasks have different computation times ranging from 10 to 570 seconds. The execution of BASIZ was simulated in DIMEMAS for an input data stream of 100 image frames. Application behavior was synthesized from an execution trace.

The obtained mapping for BASIZ application used 29 processors, giving a latency of 1.11 seconds and an average throughput of 0.01719 frames/sec. We compared these results with the assignment of one-task-per-processor (i.e. 39 processors) that yielded a latency of 1.032 seconds and an average throughput of 0.01720 frames/sec. As can be observed, latency increased only slightly, and throughput is nearly the same. However, the differences existing in the execution times of tasks yielded considerable periods of processor inactivity during execution. This is illustrated in Figure 6, which shows the average time in seconds of CPU utilization for both executions and waiting periods. As can be observed,

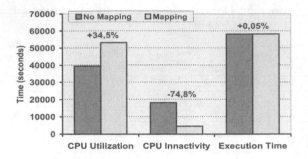

Fig. 6. Average utilization of the CPU and execution time for BASIZ application.

there is a significant reduction in CPU inactivity when we used only 29 processors with the assignment provided by the mapping algorithm. It can also be observed that using ten fewer processors did not imply an increase in global execution time for the application, as the difference is only one of 0.05%.

The results obtained in this experimentation process indicate that the use of the proposed mapping algorithm provides processor utilization of greater efficiency in the execution of pipeline applications having different characteristics. This improvement in system utilization has been achieved in every case by maintaining maximum throughput for the application and with very low increase in application latency.

4   Conclusions

Efficient utilization of processors in the execution of pipeline applications requires an allocation of task-to-processor in such a way that the different stages of the application can run synchronously. In this work, we have proposed a mapping algorithm that carries out the allocation of tasks-to-processors for pipeline applications that do not exhibit a homogeneous behavior. The goal of the algorithm is to join tasks into clusters capable of running synchronously. This allows reducing the number of processors as well as their inactivity periods, during execution.

The effectiveness of the proposed algorithm was proven with a representative set of pipeline applications. Results show that significant reductions in the number of processors and great improvement in their utilization are both obtained. These improvements have been achieved in all cases by maintaining maximum throughput for the application, and with very low increase in latency.

References

1. A. Coudhary, W. K. Liao, D. Weiner, P. Varshwey, R. Linderman and M. Linderman: Design Implementation and Evaluation of Parallel Pipelined STAP on Parallel Computers. Proc. 12th Int. Parallel Processing Symposium. Florida. pp: 220-225. April 1998.

2. J. W. Webb: Latency and Bandwidth Consideration in Parallel Robotics Image Processing. Proc. Supercomputing'93. pp: 230-239. Portland. Nov. 1993.
3. M. Yang, T. Gandhi, R. Kasturi, L. Coraor, O. Camps and J. McCandless: Real-Time Obstacle Detection System for High-Speed Civil Transport Supersonic Aircraft. Proc. IEEE Nat'l Aeroscope and Electronics Conference. Oct. 2000.
4. M. Berger and S. Bokhari: A partitioning Strategy for Nonuniform Problems on Multiprocessors. IEEE Trans. on Computers. Vol 36. N. 5. Pp: 570-580. 1987.
5. F. Berman and L. Snyder: On Mapping Parallel Algorithms into Parallel Architectures. Journal of Parallel and Distributed Computing. Vol 4. Pp: 439-458. 1987.
6. C. Roig, A. Ripoll, M. a. Senar, F. Guirado and E. Luque: A New Model for Static Mapping of Parallel Applications with Task and Data Parallelism. IEEE Proc. Conf. IPDPS 2002. ISBN: 0-7695-1573-8. Florida. April 2002.
7. J. Subhlok and G. Vondram: Optimal Use of Mixed Task and Data Parallelism for Pipelined Computations. J. of Parallel and Distr. Compt. 60. pp. 297-319. 2000.
8. A. Choudhary, B. Narahari, D. Nicol and R. Simha: Optimal Processor Assignment for a Class of Pipelined Computations. IEEE Trans.. Parallel and Distributed Systems. Vol. 5,. N. 4. Pp: 439-445. April 1994.
9. C. L. Wang, P. B. Bhat and V. K. Prasanna: High Performance Computing for Vision. Proc. of the IEEE. Vol. 84. N. 7. July 1996.
10. M. Lee, W. Liu and V. K. Prasanna: a Mapping Methodology for Designing Software Task Pipelines for Embedded Signal Processing. Proc. Workshop on Embedded HPC Systems and Applications of IPPS/SPDP 1998. pp: 937-944.
11. Kwok Y.K. and Ahmad I.: Benchmarking and Comparison of the Task Graph Scheduling Algorithms. J. Parallel and Distributed Computing, vol 59, pp. 381-422, 1999.
12. M-T. Yang, R. Kasturi and A. Sivasubramaniam: A Pipeline-Based Approach for Scheduling Video Processing Algorithms on NOW. IEEE Trans, on Parallel and Distributed Systems. Vol. 14. N. 2. Feb. 2003.
13. J. Labarta and S. Girona: Analysing Scheduling Policies Using DIMEMAS. Parallel Computing. Vol. 23. Pp: 23-34. 1997.
14. The Utah Raster Toolkit. Utah University http://www.utah.com

Parallel Simulations of Electrophysiological Phenomena in Myocardium on Large 32 and 64-bit Linux Clusters

Paweł Czarnul and Krzysztof Grzęda

Faculty of Electronics, Telecommunications and Informatics
Gdansk University of Technology, Poland
pczarnul@eti.pg.gda.pl, kgrzeda@biomed.eti.pg.gda.pl
http://fox.eti.pg.gda.pl/~pczarnul

Abstract. Within this work we have conducted research and performed simulations on electrophysiological phenomena in the myocardium using a custom-built parallel code using MPI. We have been able to investigate and implement efficient code improvements and refinements leading to good speed-ups as well as to test the performance of the latest Linux based clusters powered by 32 and 64-bit Intel processors. Our work should be thought of as research on and a parallel implementation of a well known numerical approach on large top-of-the-line clusters. We have aimed at and implemented a portable parallel MPI code which will be used as a testbed for testing various partitioning and allocation algorithms.

1 Introduction

Although new powerful supercomputers ([1], [2]) and clusters ([3], [4]) allow to simulate and understand more and more complex phenomena by using higher spatial and time resolution, still new challenges like climate, ocean modeling, electromagnetic and CFD simulations, ray tracing ([5]) call for more. Traditionally, such applications are coded as parallel SPMD programs implemented with the use of MPI ([6]) or PVM ([7]).

2 Problem Formulation

The living myocardium has unique electrical properties, different from other materials commonly known in engineering. In fact, the bidomain nature of the myocardium coming from its histological structure, and non-linear properties of the cellular membrane make the myocardium an excitable medium.

A big class of electrophysiological problems can be modeled using the simplified (by using only local dependencies between nodes) bidomain model and Finite Difference Method as an equivalent electric circuit, presented in Fig. 1 and described by the following large system of ordinary differential equations:

$$\begin{cases} \frac{dV_{m[j]}}{dt} = \frac{\sum_k g_{j,k} \cdot (V_{m[k]} - V_{m[j]}) + I_{stim[j]} + i_{ion}(V_{m[j]}, \mathbf{U}_{[j]})}{C_0} \\ \frac{d\mathbf{U}_{[j]}}{dt} = \mathbf{u}(V_{m[j]}, \mathbf{U}_{[j]}) \end{cases} \tag{1}$$

D. Kranzlmüller et al. (Eds.): EuroPVM/MPI 2004, LNCS 3241, pp. 234–241, 2004.

where j iterates through all nodes of the FDM grid, k iterates through neighbor nodes of the j-th node, $V_{m[j]}$, $\mathbf{U}_{[j]}$ are unknowns attached to the j-th node of the FDM grid, $g_{j,k}$ is conductance of the branch directly connecting neighbor j-th and k-th nodes (assumed as zero for the other pairs of nodes), C_0 is capacitance attached to each node, $I_{stim[j]}$ is the stimulus current introduced to the j-th node and $\mathbf{u}$, i_{ion} are some given, usually non-linear functions. The structure of vector $\mathbf{U}$ and the form of $\mathbf{u}$, i_{ion} is determined by model of the cellular membrane being used.

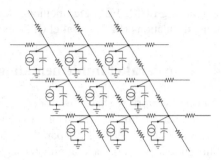

Fig. 1. Equivalent Electric Circuit

In present work, for modeling the membrane kinetics we used the Drouhard-Roberge modification ([8]) of the Beeler-Reuter model ([9], BRDR model), adopted by Skouibine et al. ([10]) to handle large, non-physiological values of transmembrane potentials. In this case, vector $\mathbf{U}$ contains 6 real components, representing intracellular calcium ion concentration and five gating variables.

3 Related Work

Simulations of phenomena occurring in the heart is an important investigation technique for studying cardiac arrhythmias ([11], [12]) and modeling electrotherapy procedures. Since solving bidomain equations governing the electrophysiological phenomena in the heart with sufficient time and spatial resolution requires a lot of computational effort, many proposals to improve this procedure were presented.

FDM is widely used for simulation of phenomena occurring just in the heart, like arrhythmias, where uniform spatial resolution is acceptable. In opposite, the Finite Element Method is very useful for static simulations performed to find out how the current applied by the medical devices distributes in the patient body. FEM is also used in other cases requiring modeling organs outside the heart, where different spatial resolution for different organs is required ([12]). A wide range of time constants appearing in various states of the myocytes, from under $20\,\mu s$ to milliseconds led to attempts at using an adaptive time step ([13]).

Parallel computing was used to solve bidomain equations as well. Porras et al. presented results of simulations performed on eight processors using PVM ([14]). In [11] a Silicon Graphics Origin2000 with 38 processors was used, but the paper was related to medical interpretation and no data about efficiency was presented.

Apart from static schemes for parallelization, there are many available dynamic load balancing algorithms and strategies ([15]). There are tools available for easy parallelization of e.g. parallel image processing (Parallel Image Processing Toolkit, PIPT - [16]), divide and conquer (DAMPVM /DAC - [17] for C/C++, Satin for Java – [18]) and others as well as general load balancing frameworks like Zoltan ([19]).

Our work should be regarded as coupling of the FDM method and code parallelization used for solving biomedical problems. Parallelization of other real world applications using FDM, FEM etc. includes electromagnetic simulations ([20], [21]), electric field simulations ([22]), parallel CFD and others ([5]).

4   Our Parallel Implementation

Within this work we have developed a parallel, GNU GPL licensed, C++/MPI code for inter-process communication easy to be modified for particular needs.

As in other SPMD MPI parallel codes ([7], [20], [21]) using FDM, the code applies updates in iterations running through time which are interleaved with inter-process communication needed to exchange the boundary data. Processes are assigned disjoint subdomains of the initial 1-3–dimensional domain. We present code and results for statically partitioned domains, however, we plan on experimenting with dynamic load balancing techniques as discussed below.

In the initial version of the communication code, developed for verification purposes rather than tuned for achieving high speed-ups, we used a sequenced series of blocking send and receive operations: in step 0 subdomains send data to subdomains with larger ids, in step 1 receive data from subdomains with larger ids using blocking communication. The persistent mode was then used for better performance.

In this work, we focus on later versions of the partitioning/communication code in which non-blocking persistent mode communication functions are used:

Ver. 1: Stripe partitioning of the domain in which successive nodes are assigned to subdomains in the following fashion (node (x, y, z) is assigned to subdomain):

```
  subdomain=0; count=0; subdomainnodecount=SIZE_X*SIZE_Y*SIZE_Z/PROC_COUNT;
2 for( x,y,z from 0 through DOMAIN_SIZE_{X|Y|Z} )
    if (( count++)<subdomainnodecount ) subdomain[x][y][z]=subdomain;
4   else { count=0; subdomain++; }
```

Ver. 2: Partitioning has been optimized by cutting the whole domain in 3 dimensions into identical rectangular prisms to reach the global minimum of the sum of their (prisms) surface areas. This approximates communication costs to be minimized. Although non-blocking MPI functions were used, the computation and communication phases are strictly separated.

Ver. 3: In this case, version 2 has been augmented with non-blocking sends and receives invoked (persistent communication calls) after the nodes on the boundaries of each subdomain have been updated. The computations of the interior part of each subdomain follow in every process and the iteration is finalized by calling MPI_Waitall completing both pending send and receive operations. This enables to overlap ([6]) the computations of the interior part of the domain with the communication of the boundary node values.

4.1   Proposed Data Representation and Node Mapping

The code has been designed in such a way that the initial assignment of nodes to successive subdomains is defined in a file. For each node, its unique (within the entire domain) id, (x, y, z) coordinates are given as well as the unique ids of the nodes it is connected to as well as link conductances (see Fig. 1). However, only the unique id is used in the assignment of the node to a subdomain, requiring that all the nodes of a subdomain have successive unique ids. The number of dimensions is not hard-coded in the code with the exception of a tuned version described below. On the one hand, it makes the definition of each subdomain easy as one defines the starting id of each subdomain and the number of nodes. If one wanted to partition the domain into disjoint rectangular subdomains, the approach requires that continuous regions of nodes are assigned successive unique ids. We have developed an external geometry generator which easily produces any assignment (either continuous subdomains or disjoint areas assigned to one process).

Versions 1 and 2 use the notion of unique node ids only, not the actual coordinates of the nodes. Communication between processors is then determined by the code by reading only links, not coordinates of, between nodes. The code itself handles arrays of indexes which nodes belong to which processors and fills up necessary communication buffers. This approach will allow us to investigate dynamic load balancing techniques in future versions. Dynamic repartitioning is reasonably easy with this approach as a global data exchange (e.g. with MPI_Alltoall) is necessary, remapping of node links between the nodes as well as possibly different ranges of unique ids.

Version 3 of the code, however, has been optimized for overlapping communication and computations which means that it needs to know which node belongs to the boundary of each processor and which belongs to the interior part. This can only be achieved by getting to know the coordinates of each node. This version is activated by setting a proper #define variable. The original update code in versions 1 and 2 is fast since it updates nodes in a simple loop:

```
  for (long i = 0; i < nX; i++) {
2   nodeArray[i] . updateNode ();
  };
```

In version 3, however, the loop needs to be split into several loops which are put into two methods - one updating the boundaries of each processor and the other the interior nodes.

We have experimented with various data representations. A special class Domain PartitioningAlgorithm has been implemented which provides methods both for generating optimized input files for versions 2 and 3 linked with the input generator program as well as methods linked with the actual simulation code. In the latter case the methods prepare (before the main simulation loop starts) one-dimensional arrays for storing indexes to nodes on all boundaries (6 rectangular boundaries in 3D) and the interior part. This data representation proved to be the fastest among the tested ones, especially on the IA64 holk cluster. In particular, a 3D array was slower as in the former case the code needs to fetch successive memory locations only, as shown below.

```
  i =0; max=nDomainSizeX*nDomainSizeY;
2 for ( i =0; i<max ; i++)
    nodeArray [ dpaAlgorithm−>nLocalIndexesXYMin[ i ]] . calcLocalDependentVars ();
4 i =0; max=(nDomainSizeX −2)*(nDomainSizeY −2)*(nDomainSizeZ −2);
  for ( i =0; i<max ; i++)
6   nodeArray [ dpaAlgorithm−>nLocalIndexesInterior [ i ]] . calcLocalDependentVars ();
```

5   Experimental Results

5.1   Exemplary Simulations

For the first verification tests of our software we chose a simulation demonstrating the induction of a spiral wave in a 2D excitable medium (in practice, both 2D and 3D problems are modeled). We model a square sheet of the myocardium, composed of 80×80 nodes. The membrane capacitance attached to each node is set to $0.5\,\mu F$ and the conductance between neighbor nodes is set to $0.1315\,mS$.

The myocardium is excited by two stimuli: S1 applied to the left edge of the sheet and S2 applied to the bottom edge. At the beginning of the experiment the whole myocardium has a negative transmembrane potential about $-85\,mV$, marked as black in Fig. 2. This is the resting state. Stimulus S1 excited the myocardium and the transmembrane potential rose above zero (in terms of electrophysiology a depolarization occurs). The wave of depolarization starts to propagate through the myocardium from its origin. After the wave has passed, the myocardium restores its resting potential (a repolarization).

Stimulus S2 initiates the second wavefront moving in a perpendicular direction to the first one. By interaction of the two waves a spiral wave initiates; the name „spiral" comes from the shape of the wavefront. A spiral wave can circulate in the medium theoretically for infinite time around a core but in our simulation properties of the spiral wave are strongly distorted by the geometry of the stimuli and the sheet being modeled.

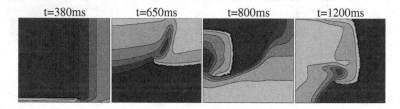

t=380ms   t=650ms   t=800ms   t=1200ms

Fig. 2. Slides From Test Simulation. Gray levels denote values of transmembrane potentials from $-100\,mV$ (black) to $+50\,mV$ (white)

5.2   Parallel Clusters Used

For simulations we used two Linux-based clusters located at the Academic Computer Center in Gdansk, Poland, galera and holk. The simulation code provides a direct performance comparison between the two systems, especially important as the latter cluster has just been deployed and put into operation.

galera ([4]) is built using 32 4-processor SMP nodes using IA32 Intel Pentium III Xeons (700 MHz, 1 MB Cache L2) for a total of 128 processors with 16GBs of RAM. The nodes are connected using the SCI ([23]) communication infrastructure which connects each network card with 4 other cards forming a 2-dimensional torus, reducing the latency by eliminating the software runtime layers compared to standard Ethernet interconnects. The system runs Linux Debian 3.0. We used the g++ compiler with the -O3 optimization option linked with the SCALI MPI.

holk ([3]) is a recently deployed IA-64 cluster containing 256 Intel Itanium II 1.3 GHz, 3 MB Cache L3, EPIC processors (2-processor nodes) with 256GBs of RAM running Linux Debian 3.0. Nodes are connected using single links to Gbit Dlink switches. We used the ecc compiler with the -O3 optimization option linked with LAM MPI.

5.3   Results

As a test case for parallel runs to measure speed-ups in large configurations, we used a 3D domain of $100 \times 100 \times 100$ nodes and ran all three versions of the code on both systems. For real biomedical problems, we will use larger domains which are expected to give even better benefits from parallelization. In particular, we wanted to see the performance comparison between the two clusters. Execution times and speed-ups are shown in Fig. 3.

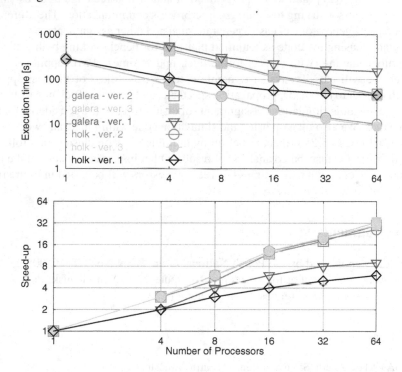

Fig. 3. Execution Time and Speed-up

It can be noticed that, as predicted, versions 2 and 3 produce significantly better times than version 1. holk is 5.11 times faster on 1 processor and 4.68 times faster on 64 processors (both compared to galera). Overlapping communication and computations showed better times and thus speed-ups on both systems, slightly better on galera though which we attribute to the SCI communication interconnect.

Regarding the data representation in version 3: on galera representing the boundary and interior indexes both in a 3D and 1D array were faster than version 2 while on holk only the latter gave an advantage over version 2.

6   Summary and Future Work

A big challenge is to improve our software to handle the true bidomain system of equations. The approach proposed in [24] may be very useful here.

Another area for improvement is to replace the BRDR model of membrane kinetics with one of Luo-Rudy models ([25]) which are more accurate. Since in our software the model of membrane kinetics is encapsulated in two particular classes, its replacement is trivial from a programmer's point of view. However, the Luo-Rudy models are more complex than the BRDR model. Therefore we can expect changes in the relationship between communication and computation times.

We also consider the implementation of an adaptive time step, but using it in the parallel code will make load balancing more difficult. In this case, apart from repartitioning schemes ([21]) and tools ([19]), disjoint and/or non-rectangular areas can be assigned to processors trying to minimize interprocess communication. The aforementioned geometry generator can easily generate such and other assignments.

We plan on using the code as a unified platform for benchmarking both static and dynamically changing (with respect to the node update time) SPMD applications. In particular, we will classify sets of dynamic patterns (regardless of the real application) with respect to the shape, propagation speed etc. and propose efficient partitioning schemes. In dynamic simulations, assignments of disjoint spaces are considered for performance runs. We also plan on using the simulation code within the framework of the KBN No. 4 T11C 005 25 project to be run by multiple users on clusters and followed by collaborative and distributed analysis of results. Other improved versions of the metrics minimizing communication times between processors will be tested in future runs, especially for various domain sizes.

Acknowledgments

Work partially sponsored by the Polish National Grant KBN No. 4 T11C 005 25. Calculations were carried out at the Academic Computer Center in Gdansk, Poland. The source code is available on request from the authors.

References

1. (C) JAMSTEC / Earth Simulator Center: (Earth Simulator)
   http://www.es.jamstec.go.jp/esc/eng/ES/index.html.
2. TOP500 Supercomputer Sites: (http://www.top500.org/)
3. holk Itanium2 Cluster: (holk.task.gda.pl) www.task.gda.pl/kdm/holk/.
4. galera IA32 Cluster: (galera.task.gda.pl) www.task.gda.pl/klaster/.
5. Buyya, R., ed.: High Performance Cluster Computing, Programming and Applications. Prentice Hall (1999)
6. Dongarra, J., et al.: Mpi: A message-passing interface standard. Technical report, University of Tennessee, Knoxville, Tennessee, U.S.A. (1995) Nonblocking communication, http://www.mpi-forum.org/docs/mpi-11-html/node44.html.
7. Wilkinson, B., Allen, M.: Parallel Programming: Techniques and Applications Using Networked Workstations and Parallel Computers. Prentice Hall (1999)

8. Drouhard, J., Roberge, F.: Revised formulation of the hodgkin-huxley representation of the sodium current in cardiac cells. Computers and Biomedical Research **20** (1987) 333–50
9. Beeler, G., Reuter, H.: Reconstruction of the action potential of ventricular myocardial fibres. Journal of Physiology **268** (1977) 177–210
10. Skouibine, K., Trayanova, N., Moore, P.: Anode/cathode make and break phenomena in a model of defibrillation. IEEE Transactions on Biomedical Engineering **46** (1999) 769–77
11. Virag, N., Vesin, J., Kappenberger, L.: A computer model of cardiac electrical activity for the simulation of arrhythmias. Pacing and Clinical Electrophysiology **21** (1998) 2366–71
12. Clayton, R., Holden, A.: Computational framework for simulating the mechanisms and ecg of re-entrant ventricular fibrillation. Physiological Measurements **23** (2002) 707–26
13. Quan, W., Evans, S., Hastings, H.: Efficient integration of a realistic two-dimensional cardiac tissue model by domain decomposition. IEEE Transactions on Biomedical Engineering **45** (1998) 372–85
14. Porras, D., Rogers, J., Smith, W., Pollard, A.: Distributed computing for membrane-based modeling of action potential propagation. IEEE Transactions on Biomedical Engineering **47** (2000) 1051–7
15. Willebeek-LeMair, M., Reeves, A.: Strategies for dynamic load balancing on highly parallel computers. IEEE Transactions on Parallel and Distributed Systems **4** (1993) 979–993
16. Squyres, J.M., Lumsdaine, A., Stevenson, R.L.: A Toolkit for Parallel Image Processing. In: Proceedings of SPIE Annual Meeting Vol. 3452, Parallel and Distributed Methods for Image Processing II, San Diego (1998)
17. Czarnul, P.: Programming, Tuning and Automatic Parallelization of Irregular Divide-and-Conquer Applications in DAMPVM/DAC. International Journal of High Performance Computing Applications **17** (2003) 77–93
18. van Nieuwpoort, R.V., Kielmann, T., Bal, H.E.: Satin: Efficient Parallel Divide-and-Conquer in Java. In: Euro-Par 2000 Parallel Processing, Proceedings of the 6th International Euro-Par Conference. Number 1900 in LNCS (2000) 690–699
19. Devine, K., Hendrickson, B., Boman, E., St.John, M., Vaughan, C.: Zoltan: A Dynamic Load-Balancing Library for Parallel Applications; User's Guide. Technical Report SAND99-1377, Sandia National Laboratories, Albuquerque, NM (1999)
20. Sarris, C.D., Tomko, K., Czarnul, P., Hung, S.H., Robertson, R.L., Chun, D., Davidson, E.S., Katehi, L.P.B.: Multiresolution Time Domain Modeling for Large Scale Wireless Communication Problems. In: Proceedings of the 2001 IEEE AP-S International Symposium on Antennas and Propagation. Volume 3. (2001) 557–560
21. Czarnul, P., Venkatasubramanian, S., Sarris, C.D., Hung, S.H., Chun, D., Tomko, K., Davidson, E.S., Katehi, L.P.B., Perlman, B.: Locality Enhancement and Parallelization of an FDTD Simulation. In: Proceedings of the 2001 DoD/HPCMO Users Conference, U.S.A. (2001) http://www.hpcmo.hpc.mil/Htdocs/UGC/UGC01/.
22. Trinitis, C., Schulz, M., Karl, W.: A Comprehensive Electric Field Simulation Environment on Top of SCI. In Kranzlmüller, D., Kacsuk, P., Dongarra, J., Volkert, J., eds.: Recent Advances in Parallel Virtual Machine and Message Passing Interface, 9th European PVM/MPI Users' Group Meeting, Linz, Austria, September 29 - October 2, 2002, Proceedings. Volume 2474 of Lecture Notes in Computer Science., Springer (2002) 114–121
23. Buyya, R., ed.: High Performance Cluster Computing, Architectures and Systems. Prentice Hall (1999)
24. Vigmond, E., Aguel, F., Trayanova, N.: Computational techniques for solving the bidomain equations in three dimensions. IEEE Transactions on Biomedical Engineering **49** (2002) 1260–9
25. Luo, C., Rudy, Y.: A model of the ventricular cardiac action potential. depolarization, repolarization, and their interaction. Circulation Research **68** (1991) 1501–26

MPI I/O Analysis and Error Detection
with MARMOT

Bettina Krammer, Matthias S. Müller, and Michael M. Resch

High Performance Computing Center Stuttgart
Allmandring 30, D-70550 Stuttgart, Germany
{krammer,mueller,resch}@hlrs.de

Abstract. The most frequently used part of MPI-2 is MPI I/O. Due
to the complexity of parallel programming in general, and of handling
parallel I/O in particular, there is a need for tools that support the application development process. There are many situations where incorrect
usage of MPI by the application programmer can be automatically detected. In this paper we describe the MARMOT tool that uncovers some
of these errors and we also analyze to what extent it is possible to do so
for MPI I/O.

1   Introduction

The Message Passing Interface (MPI) is a widely used standard [12] for writing
parallel programs. The availability of implementations on essentially all parallel
platforms is probably the main reason for its popularity. Yet another reason is
that the standard contains a large number of calls for solving standard parallel
problems in a convenient and efficient manner. However, a drawback of this is
that the MPI-1.2 standard, with its 129 calls, has a size and complexity that
makes it possible to use the MPI API incorrectly.

Version 2 of the MPI standard [13] extends the functionality significantly,
adding about 200 functions. This further increases the API's complexity and
therefore the possibilities for introducing mistakes. Several vendors offer implementations of MPI-2 and there are already open source implementations [2, 3, 6],
which cover at least some of the new features. Due to the demand for I/O support on the one hand and the availability of the free ROMIO implementation [15]
on the other hand, MPI I/O is probably the most widely used part of MPI-2.
The MPI I/O chapter describes 53 different calls.

2   Related Work

Finding errors in MPI programs is a difficult task that has been addressed in
various ways by existing tools. The solutions can be roughly grouped into four
different approaches: classical debuggers, special MPI libraries and other tools
that may perform a run-time or post-mortem analysis.

D. Kranzlmüller et al. (Eds.): EuroPVM/MPI 2004, LNCS 3241, pp. 242–250, 2004.

1. Classical debuggers have been extended to address MPI programs. This is done by attaching the debugger to all processes of the MPI program. There are many parallel debuggers, among them the very well-known commercial debugger Totalview [1]. The freely available gdb debugger currently has no support for MPI; however, it may be used as a back-end debugger in conjunction with a front-end that supports MPI, e.g. mpigdb. Another example of such an approach is the commercial debugger DDT by the company Streamline Computing, or the non-freely available p2d2 [7, 14].

2. The second approach is to provide a debug version of the MPI library (e.g. mpich). This version is not only used to catch internal errors in the MPI library, but also to detect some incorrect uses of MPI calls by the programmer, e.g. a type mismatch between sending and receiving message pairs [5].

3. Another possibility is to develop tools and environments dedicated to finding problems within MPI applications. Three different message-checking tools are under active development at present: MPI-CHECK [11], Umpire [16] and MARMOT [8]. MPI-CHECK is currently restricted to Fortran code and performs argument type checking or finds problems such as deadlocks [11]. Like MARMOT, Umpire [16] uses the MPI profiling interface. These three tools all perform their analysis at runtime.

4. The fourth approach is to collect all information on MPI calls in a trace file, which can be analyzed by a separate tool after program execution [10]. A disadvantage with this approach is that such a trace file may be very large. However, the main problem is guaranteeing that the trace file is written in the presence of MPI errors, because the behavior after an MPI error is implementation defined.

The approach taken by tools such as MARMOT has the advantage that they are able to combine extensive checking with ease of use. However, since they offer specific support for MPI problems, they have to be extended to cover the new functionality.

3   Short Description of MARMOT

MARMOT uses the MPI profiling interface to intercept MPI calls and analyze them, as illustrated in Fig. 1. It issues warnings if the application relies on non-portable MPI constructs, and error messages if erroneous calls are made. MARMOT can be used with any standard-conforming MPI implementation and it may thus be deployed on any development platform available to the programmer. The tool has been tested on Linux Clusters with IA32/IA64 processors, IBM Regatta and NEC SX systems. It currently supports the MPI-1.2 interface, although not all possible checks (such as consistency checks) have been implemented yet. Functionality and performance tests have been performed with test suites, microbenchmarks and real applications [8, 9].

Local checks including verification of arguments such as tags, communicators, ranks, etc. are performed on the client side. For all tasks that cannot be handled within the context of a single MPI process, e.g. deadlock detection, an

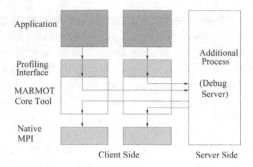

Fig. 1. Design of MARMOT.

additional MPI process is added. Information is transferred between the original MPI processes and this additional process using MPI. An alternative approach would be to use a thread instead of an MPI process and shared memory communication instead of MPI [16]. The advantage of the approach taken here is that the MPI library does not need to be threadsafe.

In order to ensure that the additional debug process is transparent to the application, we map MPI_COMM_WORLD to a MARMOT communicator that contains only the application processes. Since all other communicators are derived from MPI_COMM_WORLD they will also automatically exclude the debug server process. A similar strategy is used to map all other MPI resources, such as groups, datatypes, etc. to the corresponding MARMOT groups, datatypes, etc., and vice versa. Thus, MARMOT is able to keep track of the proper construction, usage and freeing of these resources.

4   Short Description of MPI I/O

Version 2 of MPI [13] provides a high-level interface to parallel I/O that enables the creation of portable applications. To support high performance I/O and a convenient interface for the user, MPI-2 not only includes Unix/Posix-like read and write operations, but also collective I/O operations and operations with strided memory and/or file access. Among other things, it also introduces the concept of split collective operations. I/O access is based on the concept of MPI predefined or derived datatypes.

The MPI-2 standard distinguishes between MPI calls for file manipulation, e. g. MPI_File_open (see Table 1), calls for file view such as MPI_File_set_view (see Table 2), calls for data access with many different read/write options (see Table 3) and calls for file pointer manipulation (see Table 4), calls for file interoperability such as MPI_File_get_type_extent and calls for consistency such as MPI_File_sync.

5   Possible Checks

File management errors are common, and MPI-2 therefore also includes error handling. According to the MPI standard, some of the errors related to these I/O calls are to be flagged in error classes: for example, errors related to the access mode in MPI_File_open belong to the class MPI_ERR_AMODE. However, the goal of MARMOT is not to detect errors of this kind. Our attention is focussed on the kind of errors that stem from erroneous use of the interface, and these are often not detected by the implementation itself. The question, of course, is to what extent such errors can be uncovered by verification tools such as MARMOT. Since the MPI standard also allows implementation dependent behavior for I/O calls, tools like MARMOT may help the user detect non-portable constructs.

5.1 Classification of I/O Calls and General Checks

In addition to their grouping according to functionality, several other criteria can be used to classify MPI calls. One possibility is to distinguish between collective and non-collective calls. As with all other collective routines, care has to be taken to call the collective functions in the same order on each process and, depending on the call, to provide the same value for certain arguments. For the numerous data access routines, two additional criteria are positioning and synchronism (see Table 3). The first of these distinguishes calls based on their use of explicit offsets, individual file pointers and shared file pointers. The second one groups the calls into blocking, nonblocking and split collective routines.

In the following subsections, we consider the possible checks for these classes of calls in more detail.

5.2 File Manipulation

MPI-2 contains several functions for creating, deleting and otherwise manipulating files. Table 1 classifies calls as either collective or non-collective. For the collective calls, the general checks described in Section 5.1 are applied. All query functions are non-collective. The only non-collective function that is not a query function is MPI_File_delete. If it is invoked when a process currently has the corresponding file open, the behavior of any accesses to the file (as well as of any outstanding accesses) is implementation dependent. MARMOT must therefore keep track of the names of all currently open files. For the sake of simplicity, we assume a global namespace when comparing the filenames on different processes.

Table 1. Classification of File Manipulation functions.

Non-collective	Collective
MPI_File_delete	MPI_File_open
	MPI_File_close
	MPI_File_preallocate
MPI_File_get_size	MPI_File_set_size
MPI_File_get_group	
MPI_File_get_amode	
MPI_File_get_info	MPI_File_set_info

There is not much that needs to be verified for query functions. One exception is the group that is created in calls to MPI_File_get_group. MARMOT issues a warning if the group is not freed before MPI_Finalize is called.

For MPI_File_get_info and MPI_File_set_info a portability warning will be issued if a key value is used that is not reserved by the standard. In addition, for the appropriate keys an error is flagged if the value provided is not the same on all processes. For the reserved keys, we can also check whether the value given has the appropriate data type (integer, string, comma separated list, ...).

The two calls MPI_File_open and MPI_File_close require more attention. The user is responsible for ensuring that all outstanding non-blocking requests and split collective operations have been completed by a process before it closes the file. MARMOT must therefore keep track of these operations and verify this condition when the file is closed. The user is also responsible for closing all files before MPI_Finalize is called. It is thus necessary to extend the checks already implemented for this call in MARMOT to determine whether there are any open files.

The access mode argument of MPI_File_open has strong implications. First, it is illegal to combine certain modes (e.g. MPI_MODE_SEQUENTIAL with MPI_-MODE_RDWR). Second, certain modes will make specific operations on that file illegal. If the file is marked as read only, any write operation is illegal; a read operation is illegal in write-only mode. If MPI_MODE_SEQUENTIAL was specified in MPI_File_open, the routines with explicit offset, the routines with individual pointers, MPI_File_seek, MPI_File_seek_shared, MPI_File_get_position, and MPI_File_get_position_shared must not be called.

5.3   File Views

Table 2 lists the calls related to setting/getting file views. For the collective call MPI_File_set_view(MPI_File fh, MPI_Offset disp, MPI_Datatype etype, MPI_Datatype filetype, char *datarep, MPI_Info info), the values for the datarep argument and the extents of etype must be identical in all processes in the group, whereas the values for the disp, filetype and info arguments may vary. However, the range of values allowed for disp depends on the mode chosen in MPI_File_open. Calling MPI_File_set_view is erroneous if there are non-blocking requests or split collective operations on the file handle fh that have not completed.

Table 2. Classification of File View functions.

noncollective	collective
MPI_File_get_view	MPI_File_set_view

Special attention has to be paid to verifying the correct usage of the datatypes etype and filetype in MPI_File_set_view and MPI_File_get_view. However these are based on MPI predefined or derived datatypes, so MARMOT can determine whether the datatypes are properly constructed, committed or freed in just the same way that it already handles similar MPI-1.2 calls. Absolute addresses must not be used in the construction of the `etype` and `filetype`. Moreover, the displacements in the typemap of the `filetype` are required to be non-negative and monotonically non-decreasing, the extent of any hole in the `filetype` must be a multiple of the `etype`'s extent and neither the `etype` nor the `filetype` is permitted to contain overlapping regions once the file is opened for writing.

5.4   Data Access

Table 3 shows that the MPI standard is very flexible with regard to data access, where it provides 30 different calls to meet the user's requirements. There are again collective and non-collective calls, which can be grouped into synchronous, i. e. blocking, and asynchronous, i. e. non-blocking and split collective calls, and according to positioning, i. e. into calls using explicit offsets, individual file pointers or shared file pointers.

Table 3. Classification of Data Access routines of MPI-2 [13].

positioning	synchronism	coordination	
		noncollective	*collective*
explicit offsets	*blocking*	MPI_File_read_at MPI_File_write_at	MPI_File_read_at_all MPI_File_write_at_all
	nonblocking & split collective	MPI_File_iread_at MPI_File_iwrite_at	MPI_File_read_at_all_begin MPI_File_read_at_all_end MPI_File_write_at_all_begin MPI_File_write_at_all_end
individual file pointers	*blocking*	MPI_File_read MPI_File_write	MPI_File_read_all MPI_File_write_all
	nonblocking & split collective	MPI_File_iread MPI_File_iwrite	MPI_File_read_all_begin MPI_File_read_all_end MPI_File_write_all_begin MPI_File_write_all_end
shared file pointer	*blocking*	MPI_File_read_shared MPI_File_write_shared	MPI_File_read_ordered MPI_File_write_ordered
	nonblocking & split collective	MPI_File_iread_shared MPI_File_iwrite_shared	MPI_File_read_ordered_begin MPI_File_read_ordered_end MPI_File_write_ordered_begin MPI_File_write_ordered_end

For all calls, the offset arguments must never be negative and the type signatures of the datatype argument must match the signature of some number of contiguous copies of the etype argument of the current view. The datatype for reading must not contain overlapping regions.

As buffering of data is implementation dependent, the only way to guarantee that data has been transferred to the storage device is to use the MPI_File_sync routine. Non-blocking I/O routines follow the naming convention of the MPI 1.2 standard and are named `MPI_File_iXXX`. Just as with the well-known MPI-1.2 non-blocking calls, it is necessary to insert calls like `MPI_Test` or `MPI_Wait` etc. that request completion of operations before an attempt is made to access the data. The split collective calls are named `MPI_File_XXX_begin` and `MPI_File_XXX_end`, respectively. They must be inserted in matching begin/end pairs of calls; no other collective I/O operation is permitted on their file handle between them. A file handle may have at most one active split collective operation at any time in an MPI process.

The data access routines that accept explicit offsets are named `MPI_File_XXX_at` when they are non-collective and `MPI_File_XXX_at_all_YYY` when they are collective. The routines with shared pointers are named `MPI_File_XXX_shared` when they are non-collective and `MPI_File_XXX_ordered_YYY` when they are collective. They may only be used when all processes share the same file view.

File Pointer Manipulation. Table 4 gives an overview of routines for pointer manipulation. With the exception of `MPI_File_get_byte_offset`, it is erroneous to use them if `MPI_MODE_SEQUENTIAL` is set. The seek functions are permitted to have a negative offset, i.e. to seek backwards; however, the user must ensure that the seek operation does not reach a negative position in the view.

Table 4. Classification of File Pointer Manipulation functions.

individual file pointer	shared file pointer	view-relative offset
MPI_File_seek	MPI_File_seek_shared	
MPI_File_get_position	MPI_File_get_position_shared	
		MPI_File_get_byte_offset

5.5   File Interoperability

MPI-2 guarantees that a file written by one program can be read by another program, independent of the number of processes used. However, this is only true if both programs make use of the same MPI implementation. In order to allow file sharing between different implementations, several conditions have to be fulfilled. First, the data has to be written using the data format "external 32". This must be specified in the `MPI_File_open` call. Second, the datatypes for etypes and filetypes have to be *portable*. A datatype is *portable*, if it is a predefined datatype, or is derived from a predefined datatype using only the type constructors `MPI_Type_contiguous`, `MPI_Type_vector`, `MPI_Type_indexed`, `MPI_Type_indexed_block`, `MPI_Type_create_subarray`, `MPI_Type_dup`, and `MPI_Type_Create_Darray`. Other datatypes may contain platform-dependent byte displacements for padding.

5.6   Consistency

The consistency semantics of MPI-2 describes the validity and outcome of *conflicting* data access operations, which occur when two data access operations overlap and at least one of them is a write access. The user is able to control the consistency behavior by setting the function MPI_File_set_atomicity when desired. If atomic mode is thus set, MPI guarantees that the data written by one process can be read immediately by another process. A race condition may exist only if the order of the two data access operations is not defined. More care has to be taken in the default, non-atomic mode, where the user is responsible for ensuring that no write *sequence* in a process is concurrent with any other *sequence* in any other process. In this context, a *sequence* is a set of file operations bracketed by any pair of the functions MPI_File_sync, MPI_File_open, and MPI_File_close [4].

To find potential race conditions, MARMOT has to analyze which data is written in which order by all participating processes. For all conflicting data access operations, it is necessary to verify that they are separated by an appropriate synchronization point. We are currently investigating strategies for doing this in a portable and efficient way within the limitations imposed by the design of MARMOT.

There are other potential problems related to the contents of the file. For example, the value of data in new regions created by a call to MPI_File_set_size is undefined, and it is therefore erroneous to read such data before writing it.

6   Conclusions and Future Work

In this paper, we analyzed the MPI I/O Interface with regard to potential errors that can be made by the user. Altogether more than 50 different error types have been identified. We found that in most cases these errors can be detected by tools like MARMOT following the approach taken for MPI-1. The rules for data consistency are an exception, and new solutions have to be considered for them. This is currently work in progress. Since the implementation is not yet completed, another open question is the performance impact MARMOT will have on the application.

Acknowledgments

The development of MARMOT is partially supported by the European Commissionion through the IST-2001-32243 project "CrossGrid".

References

1. WWW. http://www.etnus.com/Products/TotalView.
2. Greg Burns, Raja Daoud, and James Vaigl. LAM: An Open Cluster Environment for MPI. In *Proceedings of Supercomputing Symposium*, pages 379–386, 1994.
3. W. Gropp, E. Lusk, N. Doss, and A. Skjellum. A high-performance, portable implementation of the MPI message passing interface standard. *Parallel Computing*, 22(6):789–828, September 1996.

4. William Gropp, Ewing Lusk, and Rajeev Thakur. *Using MPI-2: Advanced Features of the Message-Passing Interface*. MIT Press, 1999.
5. William D. Gropp. Runtime Checking of Datatype Signatures in MPI. In Jack Dongarra, Peter Kacsuk, and Norbert Podhorszki, editors, *Recent Advances in Parallel Virtual Machine and Message Passing Interface*, volume 1908 of *Lecture Notes In Computer Science*, pages 160–167. Springer, Balatonfüred, Lake Balaton, Hungary, Sept. 2000. 7th European PVM/MPI Users' Group Meeting.
6. William D. Gropp and Ewing Lusk. *User's Guide for* mpich, *a Portable Implementation of MPI*. Mathematics and Computer Science Division, Argonne National Laboratory, 1996. ANL-96/6.
7. Robert Hood. Debugging Computational Grid Programs with the Portable Parallel/Distributed Debugger (p2d2). In *The NASA HPCC Annual Report for 1999*. NASA, 1999. http://hpcc.arc.nasa.gov:80/reports/report99/99index.htm.
8. Bettina Krammer, Katrin Bidmon, Matthias S. Müller, and Michael M. Resch. MARMOT: An MPI Analysis and Checking Tool. In *Proceedings of PARCO 2003*, Dresden, Germany, September 2003.
9. Bettina Krammer, Matthias S. Müller, and Michael M. Resch. MPI Application Development Using the Analysis Tool MARMOT. In M. Bubak, G. D. van Albada, P. M. Sloot, and J. J. Dongarra, editors, *Computational Science — ICCS 2004*, volume 3038 of *Lecture Notes in Computer Science*, pages 464–471, Krakow, Poland, June 2004. Springer.
10. D. Kranzlmueller, Ch. Schaubschlaeger, and J. Volkert. A Brief Overview of the MAD Debugging Activities. In *Fourth International Workshop on Automated Debugging (AADEBUG 2000)*, Munich, 2000.
11. Glenn Luecke, Yan Zou, James Coyle, Jim Hoekstra, and Marina Kraeva. Deadlock Detection in MPI Programs. *Concurrency and Computation: Practice and Experience*, 14:911–932, 2002.
12. Message Passing Interface Forum. *MPI: A Message Passing Interface Standard*, June 1995. http://www.mpi-forum.org.
13. Message Passing Interface Forum. *MPI-2: Extensions to the Message Passing Interface*, July 1997. http://www.mpi-forum.org.
14. Sue Reynolds. System software makes it easy. *Insights Magazine*, 2000. NASA, http://hpcc.arc.nasa.gov:80/insights/vol12.
15. Rajeev Thakur, Robert Ross, Ewing Lusk, and William Gropp. *Users Guide for ROMIO: A High-Performance, Portable MPI-IO Implementation*. Argonne National Laboratory, January 2002. Technical Memorandum ANL/MCS-TM-234.
16. J.S. Vetter and B.R. de Supinski. Dynamic Software Testing of MPI Applications with Umpire. In *Proceedings of the 2000 ACM/IEEE Supercomputing Conference (SC 2000)*, Dallas, Texas, 2000. ACM/IEEE. CD-ROM.

Parallel I/O in an Object-Oriented Message-Passing Library*

Simon Pinkenburg and Wolfgang Rosenstiel

Wilhelm-Schickard-Institut für Informatik
Department of Computer Engineering, University of Tübingen
Sand 13, 72076 Tübingen, Germany
{pinkenbu,rosen}@informatik.uni-tuebingen.de
http://www-ti.informatik.uni-tuebingen.de

Abstract. The article describes the design and implementation of parallel I/O in the object-oriented message-passing library TPO++. TPO++ is implemented on top of the message passing standard MPI and provides an object-oriented, type-safe and data-centric interface to message-passing. Starting with version 2, the MPI standard defines primitives for parallel I/O called MPI-IO. Based on this layer, we have implemented an object-oriented parallel I/O interface in TPO++. The project is part of our efforts to apply object-oriented methods to the development of parallel physical simulations. We give a short introduction to our message-passing library and detail its extension to parallel I/O. Performance measurements between TPO++ and MPI are compared and discussed.

1 Introduction

The work is part of a government funded project, which is done by collaboration of physicists, mathematicians and computer scientists to develop large-scale physical simulations for massive parallel computers. Our group focuses on the development of the runtime environments and libraries to parallelize these simulations efficiently. Recently, the computation of our massive parallel application showed a lack in performance due to sequential I/O. Therefore parallel I/O systems are required.

With MPI-IO [2], the I/O part of the widely accepted message passing standard MPI-2 [6], a portable interface for parallel I/O has been developed. MPI provides support for object-oriented languages by defining C++ bindings for all MPI interface functions. But these are only wrapper functions which do not fit well into the object-oriented concepts. Moreover, MPI does not support the transmission of objects or standardized library data structures such as the Standard Template Library (STL) containers. To shorten the gap between the

---

* This project is funded by the DFG within the Collaborative Research Center (CRC) 382: Verfahren und Algorithmen zur Simulation physikalischer Prozesse auf Höchstleistungsrechnern (Methods and algorithms to simulate physical processes on supercomputers).

D. Kranzlmüller et al. (Eds.): EuroPVM/MPI 2004, LNCS 3241, pp. 251–258, 2004.

object-oriented software development and the procedural communication, we developed a communication library for parallelizing object-oriented applications called TPO++ [4] which offers the functionality of MPI 1.2. It includes a type-safe interface with a data-centric rather than a memory-block oriented view and concepts for inheritance of communication code for classes. Other goals were to provide a light-weight, efficient and thread-safe implementation, and, since TPO++ is targeted to C++, the extensive use of all language features that help to simplify the interface. A distinguishing feature compared to other approaches [5, 3] is the tight integration of the STL. TPO++ is able to communicate STL containers and adheres to STL interface conventions.

Since parallel I/O in MPI is handled like sending or receiving messages, the concepts for mapping objects into simpler communication structures should be reused for the implementation of parallel I/O. To preserve the high portability of TPO++ and the compatibility to existing applications, the design of the parallel I/O interface is very important.

This article is organized as follows. In section 2, we state the design goals followed by the arising problems in section 3. Section 4 gives details of the implementation and provides some examples. In section 5, we present and discuss the performance measurements of our message-passing library supporting parallel I/O, which lines out the great efficiency. The conclusion is drawn in section 6 including an outlook on our future work.

2   Design Goals

In this section we give an overview of the design goals which are most important on integrating the object-oriented parallel I/O interface into TPO++:

MPI-IO Conformity. The design should conform to the MPI-IO interface, naming conventions and semantics as closely as possible without violating object-oriented concepts. It helps migrating from C bindings and eases porting of the existing C or C++ code.

Efficiency. The implementation should not much differ from MPI in terms of transfer rates and memory efficiency. Therefore it should be as lightweight as possible, allowing the C++ compiler to statically remove most of the interface overhead. Data transfer should be done directly via MPI calls and, if feasible, with no additional buffers. This saves memory and enables the underlying MPI implementation to optimize the data transfer.

C++ Integration. An implementation in C++ should take into account all recent C++ features, like the usage of exceptions for error handling and, even more important, the integration of the Standard Template Library by supporting the STL containers as well as adopting the STL interface conventions.

3   Problems

In this section we adress the problems arising on implementing an object-oriented parallel I/O interface.

Transfering Objects. A first approach for transfering objects is to overload the corresponding method. This works for built-in types and known library types such as STL containers. Obviously, this cannot work for the user-defined types since the user would have to overload methods of our message-passing classes. A possible solution is to provide an abstract base class from which the user types could inherit marshalling methods, then, overload the write and read methods of this base class and use virtual marshalling methods to get data in and out from the user-defined types. This is certainly an approach for dynamic user-defined data structures. However, the drawback is the additional overhead of virtual method calls and the need for the user to implement serialize and deserialize methods for *each* class.

Recovering Dynamic Objects. On the termination of an application the data structure information of each object gets lost. Therefore, the library has to provide mechanisms which either make the object's structure persistent or explicitly rebuild it on restart of the application, e.g. setting up the data structure in the object's constructor. While the former reduces the I/O performance through transfering type information to disk, the latter reduces the compute performance.

File Views. are a powerful functionality of MPI-IO, which allows the user to access noncontiguous data from a file into memory with only one single I/O call. Therefore the user has to create a `file view` and commit it to the system [6]. This view represents the data which can be accessed by the process. This ability is very important, because noncontiguous accesses are very common in parallel applications [9]. The conformity of object-oriented concepts demands extending the granularity from a single concatenation of data types to object level. However, accessing single members of an object is only possible through breaking up the structure of the object and violating object-oriented principles.

Collective I/O. Another key feature of MPI-IO are collective I/O functions, which must be called by all processes, that opened the file together. This enables the MPI-IO implementation to analyze and merge small noncontigous requests of different processes into large contigous requests leading to a significant improvement in performance [12]. Since file views are fundamental for integrating collective I/O, extending TPO++ to collective I/O through implementing views becomes possible.

4   Implementation and Examples

In this section, we address the issues raised in section 3. In an object-oriented parallel I/O system one would ideally like to have a simple interface providing a single write and a single read method to which every object could be passed in a type-safe manner and without having the user to give any information about the objects to be transfered. TPO++ already preserves this functionality for sending and receiving objects. Therefore, on extending the library to parallel I/O we reuse the serialization and deserialization mechanisms already implemented in TPO++ to make objects and containers persistent. The new implemented class File represents the interface to object-oriented writing and reading. All data transfer methods provide the same orthogonal interface for specifying the data objects to read or write. The user has two options: provide a single datatype (basic or object) or a range of data elements by using a pair of STL iterators.

Transfering Predefined C++ Types. Predefined C++ types are library and basic types and can be read and written very simple (left code, **read** analog). STL containers can be transfered using the same overloaded method (right code, **read** analog). The STL conventions require two iterators specifying begin and end of a range, which also allows to transfer subranges of containers:

```
double d;                        vector<double> vd;
File fh;                         File fh;
fh.open(...);                    fh.open(...);
fh.write(d);                     fh.write(vd.begin(), vd.end());
fh.close();                      fh.close();
```

The same overloaded methods can be used to specify an offset in the file. The user simply has to add a third parameter in the call containing the offset.

Conforming to the STL interface, the data can be read into a full container, where the previous data will be overwritten, or an empty container. Note that the user has to resize the empty container to the amount of data he wants to read, since the number of bytes to be read depends on the size of allocated memory for the container:

```
Status status;
vector<double> vd1(10);
vector<double> vd2;
File fh;

fh.open(...);
status=fh.read(vd1.begin(), vd1.end()); /* read 10 doubles */
status=fh.read(vd2.begin(), vd2.end()); /* read 0  doubles! */
vd2.resize(10); /* resize container */
status=fh.read(vd2.begin(), vd2.end()); /* read 10 doubles */
fh.close();
```

User-Defined Types. When user-defined types were declared for transmission through adding serialize and deserialize methods to the classes they can be written to or read from file exactly like basic datatypes or STL containers of basic datatypes. The code to marshall an object can be reused in derived classes and our library is able to handle arbitrary complex and dynamic datatypes.

The problem of memory allocation on reading still resides: On reading a user-defined object or container of these objects, the user has to ensure that the object's structure in memory is set up correct before reading data into this region. This can easily be done within the objects constructor.

Recovering Objects. Since the size of an object is not known when reading it for the first time, we generate a meta file when the object is written. This file contains some minimal type information which is needed to successfully restore the object. The saved information is kept as small as possible due to performance reasons. In addition, a seperation of meta information and raw data enables the system to cache the small meta file leading to a minimum of loss in performance.

File Views. Implementing views in TPO++ is done by defining a STL vector containing boolean values[1]. Due to the granularity on object level it is only possible to select or deselect the whole object. The object is selected by setting the boolean value TRUE and deselected by setting it FALSE. The following code gives an example on how to define a view in TPO++ which allows reading every second Circle from the file:

```
Circle c;
vector<Circle> vc(20);
File fh;
vector<bool> mask; // file type
int disp = 0; // offset in file
mask.push_back(TRUE);
mask.push_back(FALSE);

fh.open(...);
fh.set_view(disp, c, mask, "native", TPO_INFO_NULL);
fh.write(vc.begin(), vc.end());
```

The example shows that after defining the file type with push_back, no explicitly commit by the user has to be done. The implementation of set_view implicitly generates and commits the file type on basis of MPI_BYTE and defines the view.

Collective I/O. Using the combination of serialize/deserialize mechanism and the implementation of views enables collective I/O to TPO++. The view on the object is internally transformed into a byte stream view using MPI_BYTE as elementary datatype within the set_view command of MPI. Then, collective TPO++ calls can profit from the MPI internal optimization of noncontiguous collective accesses.

5   Performance Measurements

The comparison of MPI-IO and the object-oriented parallel I/O part of TPO++ has been measured on our Kepler-Cluster, a self-made clustered supercomputer based on commodity hardware [13]. Its architecture consists of two parts: The

---

[1] The STL type bitset is not sufficient since its extent has to be defined on compile time.

first diskless part with 96 nodes each running with two Pentium III processors at 650 MHz and having 1 GB of total memory, or 512 MB per processor, and the newer second part with 32 nodes each running with two AMD Athlon processors at 1.667 GHz, sharing 2 GB of total memory, and using a 80 GB disk. The whole system has two interconnects, a fast ethernet for booting the nodes, administration purposes and as storage area network (SAN), and a Myrinet network for the communication between parallel applications. Both have a multi-staged hierarchical switched topology organized as a fat tree. The Myrinet has a nominal bandwidth of 133 MB/s which is the maximum PCI transfer rate. Measurements give about 115 MB/s effective bandwidth and $7\mu s$ latency. The nominal transfer rate of Fast Ethernet is 12.5 MB/s and about 11 MB/s effective due to TCP/IP overhead. The parallel I/O architecture consists of the MPI-IO implementation ROMIO [11], which is set up on top of the file system using an abstract device interface for I/O (ADIO) [10]. The underlying parallel file system is PVFS [1] which is configured as follows: The 32 disks of the AMD nodes act as I/O nodes and the whole 128 nodes are PVFS clients. The SAN is used for striping the data onto these disks.

We compared the efficiency of our library on this platform using simple read/write tests as well as collective read/write tests and measured the achieved aggregated bandwidths of MPI and TPO++. Figure 1 shows the whole set of our measurements on Kepler. The overall limit is determined by the 32 I/O nodes using PVFS over the SAN. The I/O nodes are divided into two parts connected hierarchically to a Fast Ethernet backplane. Measurements give an effective aggregated bandwith of about 180 MB/s. The results on top are simple read (left) and write (right) tests where each processor reads and writes to a single file on PVFS in parallel. Using collective I/O, all processors read and write single blocks to and from specific positions within the same file in parallel. While single accesses are limited to hardware restrictions independent of the number of processors, the collective performance declines when increasing the amount of processors due to the collective overhead. Independent measurements with "The Effective I/O Bandwidth Benchmark" [7] confirmed the validity of the results.

The differences seen in all charts between MPI and TPO++ represent the loss in performance due to the object-oriented abstraction. Evidently, object-oriented parallel I/O using TPO++ achieves almost the same bandwidth as MPI-IO.

Application

Within a project funded by the DFG called "Gene sequence analysis on high performance parallel systems" we developed a software tool for searching motifs with structural and biochemical properties in biological sequences [8]. The search is very complex because of using regular expressions for defining the input. Hence, the search time can increase quickly. In addition, the whole sequence data typically reaches hundreds of megabyte. Therefore, we incorporated an object-oriented parallel search which uses our parallel I/O and implemented it on our

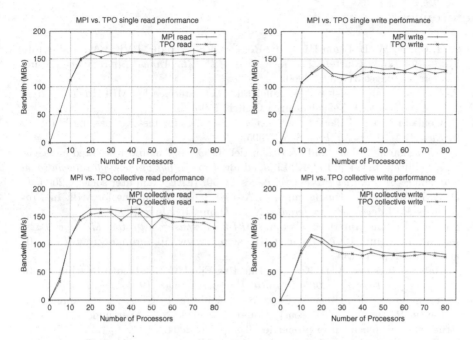

Fig. 1. Comparison of parallel I/O using MPI and TPO++.

Cluster. Due to algorithmic constraints we could not use collective I/O or file views and therefor the performance improved only about 30%.

6 Conclusion

In this paper we have discussed our implementation of an object-oriented parallel I/O system. Our system exploits object-oriented and generic programming concepts, allows easy reading and writing of objects and makes use of advanced C++ techniques and features as well as supporting these features, most notably it supports STL datatypes. The system introduces object-oriented techniques while preserving MPI-IO semantics and naming conventions as far as possible and reasonable. This simplifies the transition from existing code. While providing a convenient user interface its design is still efficient. The code to marshall an object can be reused in derived classes and our library is able to handle arbitrary complex and dynamic datatypes. The interface can be implemented with only small loss in performance compared to MPI. However, the advantages of object-oriented programming definitly outweigh this loss in performance.

While individual file pointers and explicit offsets are supported, we focus next on integrating shared file pointers. Another lack of integration lies in nonblocking calls (`MPI_File_ixxx`) as well as non-blocking collective calls (split collective I/O). Finally, the interface shall be integrated in other object-oriented applications to increase their performance and functionality.

References

1. P. H. Carns, W. B. Ligon III, R. B. Ross, and R. Thakur. PVFS: A Parallel File System for Linux Clusters. In *Proceedings of the 4th Annual Linux Showcase and Conference*, pages 317–327, 2000.
2. P. Corbett et al. MPI-IO: A Parallel File I/O Interface for MPI. In *Tech. Rep. NAS-95-002*. NASA Ames Research Center, 1995.
3. O. Coulaud and E. Dillon. Para++: C++ bindings for message-passing libraries. In *EuroPVM Users Meeting*, Sept. 1995.
4. T. Grundmann, M. Ritt, and W. Rosenstiel. TPO++: An object-oriented message-passing library in C++. In D. J. Lilja, editor, *Proceedings of the 2000 International Conference on Parallel Processing*, pages 43–50. IEEE Computer society, 2000.
5. D. G. Kafura and L. Huang. mpi++: A C++ language binding for MPI. In *Proceedings MPI developers conference*, Notre Dame, IN, June 1995.
6. Message Passing Interface Forum. *MPI-2: Extensions to the Message Passing Interface*, July 1997. Online. URL: http://www.mpi-forum.org/docs/mpi-20-html/mpi2-report.html.
7. R. Rabenseifner and A. E. Koniges. Effective File-I/O Bandwidth Benchmark. In *Proceedings of Euro-Par 2000 – Parallel Processing*, pages 1273–1283, August 2000.
8. M. Schmollinger, I. Fischer, C. Nerz, S. Pinkenburg, F. Götz, M. Kaufmann, K.-J. Lange, R. Reuter, W. Rosenstiel, and A. Zell. Parseq: Searching motifs with structural and biochemical properties. To appear 2004.
9. E. Smirni, R. Aydt, A. Chien, and D. Reed. I/O Requirements of Scientific Applications: An Evolutionary View. In *Proceedings of the Fifth IEEE International Symposium on High Performance Distributed Computing*, pages 49–59, 1996.
10. R. Thakur, W. Gropp, and E. Lusk. An Abstract-Device Interface for Implementing Portable Parallel-I/O Interfaces. In *Proc. of the 6th Symposium on the Frontiers of Massively Parallel Computation*, pages 180–187. Argonne National Lab., 1996.
11. R. Thakur, W. Gropp, and E. Lusk. Users Guide for ROMIO: A High-Performance, Portable MPI-IO Implementation. In *Technical Memorandum ANL/MCS-TM-234*, volume (28)1, pages 82–105. Mathematics and Computer Science Division, Argonne National Laboratory, 1998.
12. R. Thakur, W. Gropp, and E. Lusk. Optimizing Noncontiguous Accesses in MPI-IO. In *Parallel Computing*, volume (28)1, pages 82–105, 2002.
13. University of Tübingen. *Kepler cluster website*, 2001. Online. URL: http://kepler.sfb382-zdv.uni-tuebingen.de.

Detection of Collective MPI Operation Patterns

Andreas Knüpfer[1], Dieter Kranzlmüller[1,2], and Wolfgang E. Nagel[1]

[1] Center for High Performance Computing
Dresden University of Technology, Germany
{knuepfer,nagel}@zhr.tu-dresden.de
http://www.tu-dresden.de/zhr
[2] GUP – Institute of Graphics and Parallel Processing
Joh. Kepler University Linz, Austria/Europe
kranzlmueller@gup.jku.at
http://www.gup.uni-linz.ac.at/

Abstract. The Message Passing Interface standard MPI offers collective communication routines to perform commonly required operations on groups of processes. The usage of these operations is highly recommended due to their simplified and compact interface and their optimized performance. This paper describes a pattern matching approach to detect clusters of point-to-point communication functions in program traces, which may resemble collective operations. The extracted information represents an important indicator for performance tuning, if point-to-point operations can be replaced by their collective counterparts. The paper describes the pattern matching and a series of measurements, which underline the feasibility of this idea.

1 Motivation

Message-passing programming is accepted as the most important programming paradigm for todays High Performance Computers (HPC) [1, 4], which includes multiprocessor machines, constellations, clusters of workstations, and even grid computing infrastructures. The success and acceptance of message passing on all these different architectures can partially be attributed to the availability of the Message Passing Interface standard MPI [5]. This concerted effort of the MPI forum delivered a "practical, portable, efficient, and flexible standard for writing message-passing programs".

Yet, as experienced in many of our MPI training courses, the rich functionality is seldom exploited. Instead, novice as well as experienced programmers tend to utilize only a relatively small set of functions throughout their codes, which seems sufficient for all their purposes. Furthermore, this "initial set" of functions is often limited to point-to-point communication functionality, where messages are communicated between two endpoints. More advanced routines, such as collective operations - e.g. broadcast, scatter, and gather - are widely ignored.

Consequently, when supporting users during performance optimizations of MPI programs with our program analysis tools Vampir [2] and MAD [10], we

D. Kranzlmüller et al. (Eds.): EuroPVM/MPI 2004, LNCS 3241, pp. 259–267, 2004.

observe a strong preference for point-to-point communication over collective operations, even though collective operations offer a series of advantages over their manually constructed counterparts. On the one hand, the usage of collective operations instead of groups of point-to-point communications is much more elegant and usually delivers code, that is easier to read and maintain. On the other hand, the implementation of the MPI standard offers the HPC vendors enough freedom and flexibility to optimize their MPI implementation for a particular hardware architecture or even with dedicated hardware support [15]. This enables the development of highly efficient MPI implementations.

The approach presented in this paper attempts to provide an improvement to this situation. Our hypothesis is as follows: If we are able to automatically detect occurrences of groups of point-to-point messages, which can possibly be replaced with a collective operation, we would provide the users with important suggestions for improving their code in terms of performance and readability. For this reason, we developed a proof-of-concept tool that performs pattern matching on program traces and presents the detected patterns to the user.

The existence of patterns in the execution of parallel programs is a well-known fact. In [8], [3] and [6] patterns are exploited for automatic or semi-automatic parallelization, while [7] describes a pattern-oriented parallel debugger. Pattern matching based on static source code analysis is described in [14], while pattern matchin on program traces is introduced in [13] and [16]. First ideas of our pattern analysis approach are discussed in [11] and [9].

The usage of our prototype tool is explained in the following section. Section 3 discusses the actual recognition of message patterns, while Section 4 shows the performance advantages achievable by using collective operations over point-to-point messages. Finally, Section 5 summarizes our results and provides an outlook on future goals in this project.

2   Overview: The Pattern-Matching Tool

The pattern-matching tool presented in this paper operates on Vampir trace files. The tool itself is implemented as a command-line application, which reads a given trace file and presents its results either as a textual list of detected patterns (see Figure 1) or as a color coded space-time diagram (see Figure 2).

The patterns derived by the tool are candidates for performance optimization. With the information shown in Figures 1 and 2, programmers may replace clusters of point-to-point communication operations with corresponding collective MPI operations. Obviously, the tool's findings are intended as hints, while the final decision for the substitution is solely up to the application developer.

Reasons for not performing the suggested modification may be that the detected pattern is a *pseudo pattern* or that the substitution would disturb the general conception of a program, i.e., its modular structure. Pseudo patterns are communication patterns that are not present in the general conception of a parallel program but arise by coincidence, e.g. only with a specific number of processors. Furthermore, these pseudo patterns may be detected because essential computation takes place in-between the send/receive events. This is invisible

```
in function 'alltoall_own': pattern(s) GATHER SCATTER ALL-TO-ALL
    1 --> 0 len 1048576 comm 135018888 tag 1              S A
    1 --> 2 len 1048576 comm 135018888 tag 1              S A
    1 --> 3 len 1048576 comm 135018888 tag 1              S A
    0 --> 1 len 1048576 comm 135018888 tag 0           G  A
    2 --> 1 len 1048576 comm 135018888 tag 2           G  A
    3 --> 1 len 1048576 comm 135018888 tag 3           G  A

in function 'alltoall_own': pattern(s) ALL-TO-ALL
    1 --> 0 len 1048576 comm 135018888 tag 1                 A
    1 --> 2 len 1048576 comm 135018888 tag 1                 A
    0 --> 1 len 1048576 comm 135018888 tag 0                 A
    1 --> 3 len 1048576 comm 135018888 tag 1                 A
    2 --> 1 len 1048576 comm 135018888 tag 2                 A
    3 --> 1 len 1048576 comm 135018888 tag 3                 A
```

Fig. 1. List of detected communication patterns as textual output. All patterns (GATHER, SCATTER, ALL-TO-ALL) are reported with the name of the most low-level function completely containing them (`alltoall_own`) and a list of associated messages. For every message a single letter code G, S and A determines which pattern it belongs to in detail. It is possible to provide timestamps to the reported patterns as well, however, this would enforce to print repetitive patterns multiple (probably very many) times.

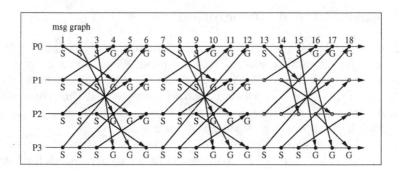

Fig. 2. Message graph with resulting patterns. The single send/receive events are shown in logical global time and connected by arrows. Send events that belong to a SCATTER pattern are marked with S, receive events belonging to a GATHER pattern are marked with G and messages that are part of an ALL-TO-ALL pattern are shown as bold arrows. (Usually, these properties are visualized with different colors.) In this figure, three successive ALL-TO-ALL patterns are shown. Please note, that ALL-TO-ALL does not necessarily consist of SCATTER and GATHER.

to the detection algorithm and prevents the modification of the communication scheme. However, all spots that may be suitable for optimizing the MPI communication are reported by our tool.

Another difficulty arises from the different kinds of possible patterns. The MPI standard defines several types of collective communication operations, such as broadcast, gather, scatter, gather-to-all, all-to-all, reduce, all-reduce, reduce-scatter, scan and their variations [15]. Each of them can be manually recreated by a pattern of point-to-point messages. Unfortunately, some of these operations deliver similar patterns, and with the mere trace information, it is quite difficult to decide the exact match without knowledge about the actual message data. For example, a series of point-to-point messages from a single sender to all other processes could be converted into both, a broadcast operation - if all messages contain the same data - or a scatter operation otherwise.

At present, our tool focuses on three classes of detectable patterns:

- Collection operations (`GATHER`): gather, reduce
- Distribution operations (`SCATTER`): scatter, broadcast
- All-to-all operations (`ALL-TO-ALL`): all-to-all, all-gather

All remaining types of collective operations can be composed from the previous mentioned ones, and are therefore not treated separately.

Each replacement should result in some performance profit compared to the 'self-made' counterpart, with a rule of thumb that the more restrictive operations deliver better performance results. For example, the more restrictive broadcast operation is usually advantageous over the more general scatter (See measurements of Section 4.).

3   Message Pattern Recognition

One of the main challenges of automatic performance analysis is the fact that it represents a performance critical task by itself. Thus, when designing the pattern detection algorithm, the computational effort should not be neglected. Even though our tool has to perform its job without any real-time constraints, a quick answer would be desirable. For very huge traces, as frequently encountered in HPC nowadays, appropriate scaling is a basic necessity.

In order to meet these requirements, we chose an advanced approach to handle trace data efficiently. In a two stage algorithm we first search for patterns local to each process, before trying to combine these local patterns into final global patterns. The principle of this technique is as follows:

1. In the first stage we try to detect local sequences of message events, that fulfill certain necessary conditions for the existence of patterns.
2. In the second stage only candidates identified above are taken into account, trying to combining them into larger global patterns and checking the necessary global conditions.

3.1   Local Patterns in CCG

The first stage of our pattern detection algorithm relies on a sophisticated data structure called *Complete Call Graph* (CCG)[12, 9]. CCGs basically are call trees containing all observed events as nodes. They are not reduced to mere relations of *"function A has been called by B, C and D"* like in common call graph structures. Furthermore, CCGs can be dramatically compressed by abandoning the tree-property and mapping repetitive subtrees (i.e. call sequences) onto one another. This results in so called *compressed Complete Call Graphs* (cCCG). (For more details, please refer to [9].)

The pattern detection within the cCCG works by traversing the recorded call hierarchy. The algorithm searches for a set of message events that satisfy necessary conditions for the existence of patterns. Only non-trivial message events are recognized, i.e., with sender $\neq$ receiver.

Let S and R be the sets of send and receive events, respectively, and P_S resp. P_R the sets of peer processes, i.e.

$$\exists s \in S \text{ with } p = \text{receiver}(s) \Longrightarrow p \in P_S \tag{1}$$

$$\exists r \in R \text{ with } p = \text{sender}(r) \Longrightarrow p \in P_R. \tag{2}$$

Then the necessary conditions for the three kinds of patterns are in detail:

GATHER: $|P_R| = N - 1$
SCATTER: $|P_S| = N - 1$
ALL-TO-ALL: $|P_R| = N - 1$ and $|P_S| = N - 1$

where N is the number of overall processes.

The call hierarchy is traversed in a bottom-up scheme in order to check for these properties. The traversal is stopped as soon as the check has been positive. Otherwise it is continued for the parent node with sets S' and R' resp. P'_S and P'_R. There is $S \subseteq S'$ and $R \subseteq R'$ and thus $P_S \subseteq P'_S$ and $P_R \subseteq P'_R$.

After a positive check, it is known that there is a set of consecutive send and receive operations to $N - 1$ distinctive peers, i.e., all other processes. Finally, a consecutive (=uninterrupted) sequence of $N - 1$ of these send resp. receive operations is sought.

For the GATHER and SCATTER patterns the presence of such an uninterrupted sequence is regarded as sufficient condition for their existence. For ALL-TO-ALL patterns further checks are necessary.

3.2   Global Patterns

For the second stage of the detection algorithm inter-process relationships have to be taken into account. At first, corresponding send and receive events are identified. For the subsequent recognition algorithms only candidates identified by the first stage are accepted, all others are ignored.

After the mapping of send and receive events it is possible to interchange the information about local patterns between pairs of events. For every event it is known whether itself and its peer event belongs to a local GATHER, SCATTER or ALL-TO-ALL.

With this extra information it is now possible to test for the sufficient condition of the global ALL-TO-ALL pattern. A locally detected candidate for an ALL-TO-ALL pattern is accepted as global ALL-TO-ALL pattern, if for all send/receive events the corresponding peer events are part of a local ALL-TO-ALL candidate, too. The results of this search are displayed in textual form or as a message graph as shown in Figures 1 and 2.

For the current implementation, all patterns found in the cCCG are reported only once in the output list. That means if such a pattern occurs multiple times in the trace and has been mapped to a single sub-structure of the cCCG, then all the identical cases are reported as a single entry. Therefore, no timestamps are provided for the single message events (compare Figure 1).

Another option concerns ALL-TO-ALL patterns. An ALL-TO-ALL pattern may consist of GATHER and SCATTER patterns, but doesn't need to (see Figures 1 and 2). Thus, it might be considered unnecessary to report GATHER and SCATTER inside of an ALL-TO-ALL pattern.

4   Performance Gain

The arguments described above are only applicable, if the actual performance improvements are achievable on a given hardware platform. For this reason, we performed a number of benchmarks for different kinds of collective operations. In particular, we compared gather, scatter, broadcast, and all-to-all from available MPI libraries with 'self-made' counterparts constructed from point-to-point operations. The implementation has been straight-forward, without any optimizations (such as tree hierarchies).

The results of our experiments are shown in Figures 3 and 4. The run-time ratio $time_{self-made}/time_{built-in}$ is plotted over the message length for several numbers of participating MPI processes. All experiments have been run multiple times, the run-time values were taken as the minima over all repetitions. These measurements have been performed on JUMP [17], a cluster of IBM p690 nodes connected by a federation switch at the John von Neumann Institute for Computing (NIC) of the Research Center Jülich, Germany.

For gather and scatter the performance advantage of built-in operations ranges up to 200 % (= twice as fast) and 300 %, respectively. As expected, broadcast can profit most (up to 750 %) from using built-in operations compared to the self-made counterparts. This applies particularly to large messages and higher process counts. All-to-all achieves an acceleration of over 400 % with 32 processes.

To our surprise, several instances of the experiments revealed that built-in operations as provided by vendors today are sometimes *slower* than the self-made counterparts! (Similar effects as on the IBM have been observed on the SGI O3K platform). This has been unexpected, especially since the built-in operations could operate similar to the straightforward self-made implementations. While one might consider a minor performance disadvantage for very small messages acceptable, middle sized messages with notable losses are definitely not what users expect to see! (We are in discussion with the concerned vendors and intend to do a more detailed investigation of this matter.)

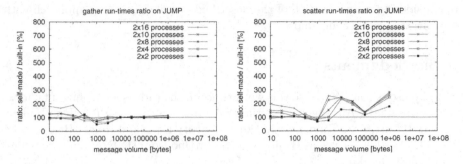

Fig. 3. Comparison of built-in vs. self-made gather (left) and scatter operations (right).

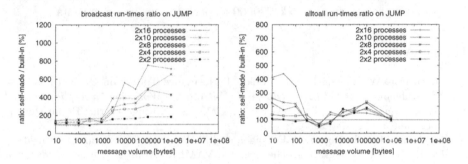

Fig. 4. Comparison of built-in vs. self-made broadcast (left) and all-to-all operations (right).

5   Conclusions and Future Work

Program development for HPC machines can be a difficult and tedious activity, especially for the less experienced users. For this reason, many program development and analysis tools try to offer sophisticated support for various subtasks of software engineering.

The approach described in this paper focuses on the application of collective operations in message passing parallel programs. Today, many users tend to implement their own collective operations instead of using optimized functionality from a given MPI library. Since this practice leads to suboptimal performance, our pattern-matching tool tries to identify such clusters of point-to-point messages. As a result, the user is informed about possible replacement operations, that may lead to increased runtime performance.

The current version of the pattern-matching tool delivers a subset of possible patterns, based on straight-forward implementations. Over time, more and even optimized patterns (e.g. by using tree hierarchies for broadcasting) will be added to the pool of detectable patterns, thus increasing the capabilities of the tool. However, as mentioned above more research is needed to distinguish similar communication patterns with different semantic functionality. At present, we are considering a combination of the described pattern matching approach with an arbitrary source code analyzer.

Acknowledgments

Several persons contributed to this work through their ideas and comments, most notably Bernhard Aichinger, Christian Schaubschläger, and Prof. Jens Volkert from GUP Linz, Axel Rimnac from the University Linz, and Holger Brunst from ZHR TU Dresden, as well as Beniamino Di Martino from UNINA, Italy, to whom we are most thankful.

We would also like to thank the John von Neuman Institute for Computing (NIC) at the Research Center Juelich for access to their IBM p690 machine Jump under project number #k2720000 to perform our measurements.

References

1. H. Brunst, W. E. Nagel, and S. Seidl. Performance Tuning on Parallel Systems: All Problems Solved? In *Proceedings of PARA2000 - Workshop on Applied Parallel Computing*, volume 1947 of *LNCS*, pages 279–287. Springer-Verlag Berlin Heidelberg New York, June 2000.
2. H. Brunst, H.-Ch. Hoppe, W.E. Nagel, and M. Winkler. Performance Otimization for Large Scale Computing: The Scalable VAMPIR Approach. In *Proceedings of ICCS2001, San Francisco, USA, LNCS* 2074, p751ff. Springer-Verlag, May 2001.
3. B. Di Martino and B. Chapman. Program Comprehension Techniques to Improve Automatic Parallelization. In *Proceedings of the Workshop on Automatic Data Layout and Performance Prediction*, Center for Research on Parallel Computation, Rice University, 1995.
4. I. Foster. Designing and Building Parallel Programs. Addison-Wesley, 1995.
5. W. Gropp, E. Lusk, and A. Skjellum. UsingMPI - 2nd Edition. MIT Press, November 1999.
6. B. Gruber, G. Haring, D. Kranzlmüller, and J. Volkert. Parallel Programming with CAPSE - A Case Study. in *Proceedings PDP'96, 4th EUROMICRO Workshop on Parallel and Distributed Processing*, Braga, Portugal, pages 130–137, January 1996.
7. A.E. Hough and J.E. Cuny. Initial Experiences with a Pattern-Oriented Parallel Debugger. In *Proceedings of the ACM SIGPLAN/SIGOPS Workshop on Parallel and Distributed Debugging*, Madison, Wisconsin, USA, SIGPLAN Notices, Vol. 24, No. 1, pp. 195–205, January 1989.
8. Ch. Kessler. Pattern-driven Automatic Parallelization. Scientific Programming, Vol. 5, pages 251–274, 1996.
9. A. Knüpfer and Wolfgang E. Nagel. Compressible Memory Data Structures for Event Based Trace Analysis. Future Generation Computer Systems by Elsevier, January 2004. [accepted for publication]
10. D. Kranzlmüller, S. Grabner, and J. Volkert. Debugging with the MAD Environment. *Parallel Computing*, Vol. 23, No. 1–2, pages 199–217, April 1997.
11. D. Kranzlmüller. Communication Pattern Analysis in Parallel and Distributed Programs. In *Proceedings of the 20th IASTED Intl. Multi-Conference Applied Informatics (AI 2002)*, International Association of Science and Technology for Development (IASTED), ACTA Press, Innsbruck, Austria, p153–158, February 2002.

12. D. Kranzlmüller, A. Knüpfer and Wolfgang E. Nagel. Pattern Matching of Collective MPI Operations. In *Proceedings of The 2004 International Conference on Parallel and Distributed Processing Techniques and Applications*. Las Vegas, Juli 2004.
13. T. Kunz and M. Seuren. Fast Detection of Communication Patterns in Distributed Executions. In *Proceedings of the 1997 Conference of The Centre for Advanced Studies on Collaborative Research*, IBM Press, Toronto, Canada, 1997.
14. B. Di Martino, A. Mazzeo, N. Mazzocca, U. Villano. Parallel Program Analysis and Restructuring by Detection of Point-To-Point Interaction Patterns and Their Transformation into Collective Communication Constructs. In *Science of Computer Programming*, Elsevier, Vol. 40, No. 2-3, pp. 235–263, July 2001.
15. M. Snir, S. Otto, S. Huss-Lederman, D. Walker, J. Dongarra. MPI: The Complete Reference. MIT Press, September 1998.
16. F. Wolf and B. Mohr. EARL - A Programmable and Extensible Toolkit for Analyzing Event Traces of Message Passing Programs. Technical report, Forschungszentrum Jülich GmbH, April 1998. FZJ-ZAM-IB-9803.
17. U. Detert. Introduction to the Jump Architecture. Presentation, Forschungszentrum Jülich GmbH, 2004. http://jumpdoc.fz-juelich.de/

Detecting Unaffected Race Conditions in Message-Passing Programs*

Mi-Young Park and Yong-Kee Jun**

Dept. of Computer Science, Gyeongsang National University
Chinju, 660-701 South Korea
{park,jun}@race.gsnu.ac.kr

Abstract. Detecting unaffected race conditions before which no other races causally happened is important to debugging message-passing programs effectively, because such a message race can affect other races to occur or not. The previous techniques to detect efficiently unaffected races do not guarantee that all of the detected races are unaffected. This paper presents a novel technique that traces the states of the locally-first race to occur in every process, and then visualizes effectively the affect-relations of all the locally-first races to detect unaffected races.

1 Introduction

In asynchronous message-passing programs, *a message race* [2, 7, 10, 11, 17] occurs toward a receive event if two or more messages are sent over communication channels on which the receive listens and they are simultaneously in transit without guaranteeing the order of arrival of them. Message races should be detected for debugging a large class of message-passing programs [1, 4, 18] effectively, because nondeterministic order of arrival of the racing messages causes unintended nondeterminism of programs [7, 9, 11, 13, 14, 16]. Especially, it is important to detect efficiently *unaffected races* before which no other races causally happened, because such races may make other affected races appear or be hidden. Debugging message races is difficult, because simple errors in message-passing can produce timing-dependent message races that can take days to months to track down.

The previous techniques [2, 7, 8, 11, 13, 14, 16, 17, 19] to detect unaffected races do not guarantee that all of the detected races are unaffected. The only efficient technique [16] trying to detect unaffected races in its two monitored executions detects racing messages by halting at the receive event of the locally-first race to occur in each process. However, if a process halts at the racing receive, the process cannot send any message thereafter and then does hide chains of

---

* This work was supported in part by Grant No. R05-2003-000-12345-0 from the Basic Research Program of the Korea Science and Engineering Foundation.
** *Corresponding author.* Also involved in Research Institute of Computer and Information Communication (RICIC), Gyeongsang National University.

D. Kranzlmüller et al. (Eds.): EuroPVM/MPI 2004, LNCS 3241, pp. 268–276, 2004.

affect-relations among those races. This previous technique therefore does not guarantee that all of the detected races are unaffected.

In this paper, we present a novel technique that traces the states of the locally-first race to occur in every process, and then visualizes effectively the affect-relations of all the locally-first races to detect unaffected races. To detect the locally-first race to occur in each process, we monitor the program execution to determine if each receive event is involved in a race and if the race occurs first in the process. During the other execution, we trace the states of the race and events appeared in each process into two trace files respectively. After the execution, we visualize the locally-first races to occur in all processes and the affect-relations among those races of all processes to detect unaffected races.

In the following section, we explain the importance of detecting unaffected races for debugging message-passing programs and the problem of the previous technique to detect unaffected races. In section 3, we present a novel technique which captures affect-relations among processes by tracing the state transitions of the locally-first races. In section 4, we explain how to visualize the affect-relations among the detected races and then show that our technique is effective in detecting unaffected races using a published benchmark program. Finally, we conclude it with future work.

2   Background

We model asynchronous message-passing [1, 11, 18, 19] between processes as occurring over *logical channels* [16], and assume that each send or receive event specifies a set of logical channels over which it operates to send copies of one message or to receive one message from the channels. If more than one channel have a message available, the receive event nondeterministically chooses a channel among them to receive one message. We assume that any message sent over a channel is received by exactly one receive event, and all messages sent during program execution are eventually received at the corresponding receive events. This model with logical channel is general, because most message-passing schemes can be represented.

An execution of message-passing program is represented as a finite set of events and the *happened-before* relation [12] defined over those events. If an event a always occurs before another event b in all executions of the program, it satisfies that a happens before b, denoted $a \rightarrow b$. For example, if there exist two events $\{a, b\}$ executed in the same process, $a \rightarrow b \lor b \rightarrow a$ is satisfied. If there exist a send event s and the corresponding receive event r between a pair of processes, $s \rightarrow r$ is satisfied. This binary relation $\rightarrow$ is defined over its irreflexive transitive closure; if there are three events $\{a, b, c\}$ that satisfy $a \rightarrow b \land b \rightarrow c$, it also satisfies $a \rightarrow c$.

Messages may arrive at a process in a nondeterministic order by various causes in the execution environment, such as variations in process scheduling and network latencies. In a large class of message-passing programs [1, 4, 18] that are intended to be deterministic, nondeterministic order of message arrivals

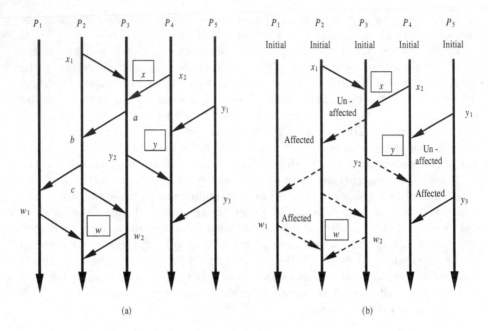

Fig. 1. Message Races

causes unintended nondeterministic executions of a program so that such race conditions of messages should be detected for debugging [7, 9–11, 13, 14, 16]. A *message race* [2, 7, 10, 11, 17] occurs in a receive event, if two or more messages are sent over communication channels on which the receive listens and they are simultaneously in transit without guaranteeing the order of their arrivals. A message race is represented as $\langle r, M \rangle$, where r is a receive event and M is a set of racing messages toward r. Thus, r receives the message delivered first in M, and the send event s which sent a message in M does not satisfy $r \rightarrow s$. We denote a message sent by a send event s as $msg(s)$.

Figure 1.a shows a partial order of events that occurred during an execution of message-passing program. A vertical arc in the figure represents an event stream executed by a process along with time; and a slanting arc between any two vertexes that are optionally labelled with their identifiers represents a delivery of message between a pair of send and receive operations. For instance, two processes, P_2 and P_4, send two messages, $msg(x_1)$ and $msg(x_2)$, to P_3 respectively, and two send events, x_1 and x_2, do not satisfy $x \rightarrow x_1 \land x \rightarrow x_2$ where x is a receive event occurred in P_3. This implies that these two messages race each other toward x. This message race occurred at x in P_3 therefore can be represented as $\langle x, X \rangle$, because it consists of a set of racing messages $X = \{msg(x_1), msg(x_2)\}$ and the first event x to receive one of the racing messages.

Suppose that there exist only two message races $\{\langle m, M \rangle, \langle n, N \rangle\}$ in an execution of a program, and they satisfy $m \rightarrow s \lor m \rightarrow n$ where $msg(s) \in N$. Then $msg(s)$ is an *affected message* by $\langle m, M \rangle$ because $m \rightarrow s$; and $\langle n, N \rangle$

is an *affected race* by $\langle m, M \rangle$. And we say that $\langle m, M \rangle$ is an *unaffected race*, if there does not exist any message $msg(t) \in M$ that satisfies $n \to t$ and there exists no such $\langle n, N \rangle$ that satisfies $n \to m$. For example, Figure 1 shows three races $\{\langle w, W \rangle, \langle x, X \rangle, \langle y, Y \rangle\}$ where $W = \{msg(w_1), msg(w_2)\}$, $X = \{msg(x_1), msg(x_2)\}$, and $Y = \{msg(y_1), msg(y_2), msg(y_3)\}$. The two races $\{\langle w, W \rangle, \langle y, Y \rangle\}$ in the figure are affected by $\langle x, X \rangle$, because $\langle w, W \rangle$ satisfies $x \to a \land a \to b \land b \to w$ followed by $x \to w$ and $msg(y_2) \in Y$ satisfies $x \to y_2$. These affected races may occur depending on the occurrence of $\langle x, X \rangle$, and may disappear when $\langle x, X \rangle$ is eliminated.

A *locally-first race* is the first race to occur in a process. Although a locally-first race is obviously not affected by any other races occurred in the local process, the race is not guaranteed to be unaffected by another race occurred in the other processes. For example, Figure 1 shows that all of the races appeared in the figure are locally-first races, but two locally-first races $\{\langle w, W \rangle, \langle y, Y \rangle\}$ are affected by another locally-first race $\langle x, X \rangle$ occurred in the other process.

The previous techniques [2, 7, 8, 11, 13, 14, 16, 17, 19] to detect efficiently unaffected races do not guarantee that all of the detected races are unaffected. The only efficient technique [16] trying to detect unaffected races in its two monitored executions detects racing messages by halting at the racing receive event of the locally-first race to occur in each process. However, if a process halts at the racing receive, the process cannot send messages thereafter and then does hide chains of affect-relations among those races. This technique therefore does not guarantee that all of the detected races are unaffected, because other processes which did not receive such affected messages may report their affected races as unaffected erroneously.

For example, consider this two-pass technique for the same execution instances shown in Figure 1. In the first execution, each process writes some information into a trace file locating its locally-first race at $w \in P_2$, $x \in P_3$, or $y \in P_4$. In the second execution, it tries to halt the three processes at the locations $\{w, x, y\}$, and then eventually receives the racing messages into their receive buffers except P_2 which stops at non-racing receive b. This results in the three receive buffers of (P_2, P_3, P_4) to contain three sets of messages $(\emptyset, X, \alpha)$ respectively, where $X = \{msg(x_1), msg(x_2)\}$ and $\alpha = \{msg(y_1), msg(y_3)\} \subseteq Y$. Consequently, this two-pass technique reports two races $\{\langle x, X \rangle, \langle y, Y \rangle\}$ as unaffected, but actually $\langle y, Y \rangle$ is affected by $\langle x, X \rangle$ as shown in Figure 1. This kind of erroneous reports is resulted from halting at x of P_3, and then not having delivered $msg(y_2) \in Y$ at P_4.

3   Tracing Race States

To capture affect-relations among processes, our two-pass algorithm traces the state transitions of the locally-first race at each receive event until the executions terminate. In the first pass, it monitors the program execution to determine if each current receive is involved in a race and detects some location information of the locally-first race to occur in the current process. In the second pass, it

0 CheckReceivePass2(*Send*, *recv*, *Msg*)
1 **for all** i **in** *Channels* **do**
2   **if** $(cutoff \rightarrow recv) \wedge (firstChan = i)$
        $\wedge \neg$ *affecting*) **then**
3   *firstRecv* := *recv*;
4   *affecting* := **true**;
5   **endif**
6 **endfor**
7 *affecting* := *affecting* $\vee$ *Msg*[*affecting*];
8 **if** $(firstRecv = \neg$ **null**
        $\wedge firstRecv \nrightarrow Send)$ **then**
9   *racingMsg* := *racingMsg* $\cup$ *Msg*;
10   *racing* := **true**;
11 **endif**
12 *state* := *CheckRace*(*state*, *racing*,
        Msg[*affecting*]);

0 CheckReceivePass1(*Send*, *recv*,
     thisChan)
1 *prevRecv* := *PrevBuf*[*thisChan*];
2 **if** $(prevRecv \nrightarrow Send)$
      $\wedge (Send[me] \rightarrow cutoff)$ **then**
3   *cutoff* := *Send*[*me*];
4   *firstChan* := *thisChan*;
5 **endif**
6 **for all** i **in** *Channels* **do**
7   *PrevBuf*[i] := *recv*
8 **endfor**

Fig. 2. Pass-1 Algorithm   **Fig. 3.** Pass-2 Algorithm

locates the first racing receive with all racing messages toward the receive using the location information obtained in the first pass, and thereafter traces the events and states of the detected race in each process.

Figure 2 shows our pass-1 algorithm to check each receive associated locally with a sequence number *recv* which receives a message delivered over a logical channel *thisChan* from *Send* event of a process. To represent *Send*, a vector timestamp [3, 15] is used for the concurrency information to check the happened-before relation [12] between every two events. First, it examines *prevRecv* in line 1-2 which represents a location of the previous receive with a sequence number loaded from a global data structure *PrevBuf* for *thisChan*. If *prevRecv* did not happen before *Send*, the current receive is involved in a race occurred at *prevRecv*. Secondly, it determines in line 2 if the detected race is the first to occur in the current process *me* by comparing *Send*[*me*] with *cutoff*. *cutoff* represents an approximate location of the current locally-first race. If *Send*[*me*] happened before *cutoff*, it has found a new candidate of the locally-first race to occur and a new *cutoff* which is updated into *Send*[*me*] in the process. Consider $msg(w_2)$ received at P_2 in Figure 1. The new *cutoff* of P_2 becomes a send event c, because it must be *Send*[*me*] which is the most recently happened before w_2 in P_2. Lastly, the current event *recv* is stored into every entry of *PrevBuf* which has as many entries as *Channels*, where *Channels* is a set of logical channels associated with the current receive. The pass-1 algorithm therefore reports two kinds of information to detect the locally-first race: *cutoff* for an approximated location and *firstChan* for its logical channel.

Figure 3 shows our pass-2 algorithm to check each receive *recv* with a delivered message *Msg* using {*cutoff*, *firstChan*} reported by the pass-1. Line 1-2 checks the happened-before relation between *cutoff* and *recv*, examine if *firstChan* is included in *Channels*, and check if the receive has been *affecting*.

The current receive $recv$ must be the first racing receive $firstRecv$ to occur in the process, if $recv$ associated with $firstChan$ is unaffected and occurred first after $cutoff$.

To produce affect-relation information that will be attached to each message to notify other processes of their being affected, line 7 updates the flag as a disjunction of the current $affecting$ and $Msg(affecting)$ which is the value of Msg's $affecting$ field attached by $Send$. It is because messages sent hereafter by me may affect other processes, if either a race occurred or affected messages have been received from other processes. The line 8 checks if the received message is racing toward the first racing receive denoted $firstRecv$ in the process. The $firstRecv$ that is not null means that there exists the first racing receive in the process, and the $firstRecv$ that does not happen before $Send$ means that the received message is racing toward the first racing receive. Therefore, if the condition of the line 8 is satisfied, it stores the received message into $racingMsg$ and sets $racing$ to true at line 9 and 10. Line 12 passes three values to $CheckRace()$ which is a function to trace the race state transitions: $state$ to be updated, $racing$, and $Msg(affecting)$.

$CheckRace()$ uses one enumerated variable $state$ and two logical variables $(racing, affected)$ which correspond to $(racing, Msg[affecting])$ in line 12 of pass-2 algorithm, respectively. In the function, Initial state may be changed to Unaffected or Affected state. If $\neg$ $affected \wedge racing$ in Initial state, the state becomes Unaffected; if $affected$ in Initial state, the state becomes Affected. Unaffected state must be changed to Affected if $affected \wedge racing$ is satisfied; otherwise the state must not be changed.

Figure 1.b illustrates the state transitions of the locally-first races and the affected messages, which are reported when we apply this algorithm to Figure 1.a. In the figure, a dotted line represents an affected message. The state of each process begins from Initial state when the program starts to execute. In case of P_3 and its locally-first race $\langle x, X \rangle$, Initial state becomes Unaffected because $msg(x_1)$ is not affected. In case of P_4 and its locally-first race $\langle y, Y \rangle$, Initial state becomes Unaffected just like P_3, because $msg(y_1)$ is not affected. However, Unaffected state of P_4 becomes Affected after another message $msg(y_2) \in Y$ affected by $\langle x, X \rangle$ has been delivered.

4   Visualizing Affect-Relations

To visualize the affect-relations among the locally-first races in all the processes, we need two types of traces: $event$ $traces$ and $race$ $traces$. These traces are produced per process and combined into one trace file after the second monitored execution terminates. An event trace contains information about each message-passing event occurred in an execution of program: the current process identifier, event type, the communicating process identifier, message identifier, channel identifier, and vector timestamp. A race trace contains a set of events which are involved in any of the locally-first races and represented by the same information as the events included in an event trace.

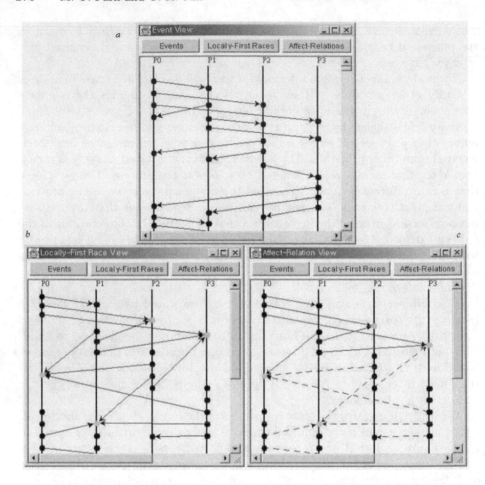

Fig. 4. (a) Event View, (b) Locally-First Race View, and (c) Affect-Relation View

Based on the traces, three kinds of views are generated: *event view, locally-first race view*, and *affect-relation view*. An event view shows visually happened-before relations of all events in the event traces. A locally-first race view shows visually the only events which are involved in the locally-first races. An affect-relation view shows visually affect-relations among the locally-first races. Figure 4 shows three examples of the views respectively.

We tested our technique using a benchmark program *Stress* published in *mpptest* [6] written in C language and MPI [5] library, and modified the benchmark to generate racing messages at will toward a set of receive events with both MPI_ANY_SOURCE and MPI_ANY_TAG in the benchmark. Figure 4.a shows an event view visualizing all of the send and receive events occurred in an execution of the modified *Stress*. In this view, it is hard to discern if there exists any race, even though we can see all of the transmitted messages and all of the events occurred during execution. Figure 4.b shows a locally-first race view visualizing the locally-first races occurred at racing receive in the execution. In the figure, every process has its locally-first race; for example, the locally-first race of P_1 has occurred in the fifth event and three messages race toward the racing receive. Figure 4.c shows an affect-relation view visualizing affect-relations among the locally-first races shown in Figure 4.b. In the figure, the locally-first race of P_1 is an affected race, because its three racing messages are affected. In summary, the locally-first races of P_0, P_1, and P_3 are affected, and then the only unaffected is the locally-first race of P_2.

To support debugging message-passing programs, a toolset [8, 10, 11] for large applications includes a post-mortem visualization scheme of event graphs with horizontal/vertical abstraction to combine a set of functionalities not only for detecting communication errors and race conditions but also analyzing performance bottlenecks in blocking communications. While this visualization scheme effectively guides users to analyze message races and investigate the consequences of the observed nondeterministic choices, unfortunately it does not guarantee that all of the detected races are unaffected.

5   Conclusion

We presented a novel technique that detects and traces the locally-first race to occur in each process and then visualizes effectively the affect-relations of all the locally-first races to detect unaffected races. To detect the locally-first race to occur in each process, we monitored the program execution to determine if each receive event is involved in a race and if the race occurs first in the process. During the other execution, we traced the states of the race and events appeared in each process into two trace files respectively. After the execution, we visualized the locally-first races to occur in all processes and the affect-relations among those races of all processes to detect unaffected races.

Detecting unaffected races helps avoid overwhelming the programmer with too much information when a programmer debugs message-passing programs. Programs with deterministic output may have many intended races. In this case, programmers would have to specify which receive operations can be involved in the intended races and then our technique can ignore them. We showed that our technique is effective in detecting unaffected races visually using a published benchmark program and makes it possible to debug race conditions in message-passing programs effectively. We have been improving the efficiency of our algorithm for the future work to extend the power of message-race detection.

References

1. Cypher, R., and E. Leu, "The Semantics of Blocking and Nonblocking Send and Receive Primitives," *8th Int'l Parallel Proc. Symp.*, pp. 729-735, IEEE, April 1994.
2. Damodaran-Kamal, S. K., and J. M. Francioni, "Testing Races in Parallel Programs with an OtOt Strategy," *Int'l Symp. on Software Testing and Analysis*, pp. 216-227, ACM, August 1994.

3. Fidge, C. J., "Partial Orders for Parallel Debugging," *Sigplan/Sigops Workshop on Parallel and Distributed Debugging*, pp. 183-194, ACM, May 1988.

4. Geist, A., A. Beguelin, J. Dongarra, W. Jiang, R. Manchek, and V. Sunderam. "PVM: Parallel Virtual Machine," *A Users' Guide and Tutorial for Networked Parallel Computing*, Cambridge, MIT Press, 1994.

5. Gropp, W., and E. Lusk, *User's Guide for Mpich, A Portable Implementation of MPI*, TR-ANL-96/6, Argonne National Laboratory, 1996.

6. Gropp, W., and E. Lusk, "Reproducible Measurements of MPI Performance Characteristics," *6th European PVM/MPI Users' Group Conf.*, Lecture Notes in Computer Science, 1697: 11-18, Springer-Verlag, Sept. 1999.

7. Kilgore, R., and C. Chase, "Re-execution of Distributed Programs to Detect Bugs Hidden by Racing Messages," *30th Annual Hawaii Int'l. Conf. on System Sciences*, Vol. 1, pp. 423-432, Jan. 1997.

8. Kranzlmüller, D., S. Grabner, and J. Volkert, "Event Graph Visualization for Debugging Large Applications," *Sigmetrics Symp. on Parallel and Distributed Tools*, pp. 108-117, ACM, Philadelphia, Penn., May 1996.

9. Kranzlmüller, D., and M. Schulz, "Notes on Nondeterminism in Message Passing Programs," *9th European PVM/MPI Users' Group Conf.*, Lecture Notes in Computer Science, 2474: 357-367, Springer-Verlag, Sept. 2002.

10. Kranzlmüller, D., and J. Volkert, "Why Debugging Parallel Programs Needs Visualization," *Workshop on Visual Methods for Parallel and Distributed Programming*, at Symp. on Visual Languages, IEEE, Seattle, Washington, Sept. 2000.

11. Kranzlmüller, D., *Event Graph Analysis for Debugging Massively Parallel Programs*, Ph.D. Dissertation, Joh. Kepler University Linz, Austria, Sept. 2000.

12. Lamport, L., "Time, Clocks, and the Ordering of Events in a Distributed System," *Communications of the ACM*, 21(7): 558-565, ACM, July 1978.

13. Lei, Y., and K. Tai, "Efficient Reachability Testing of Asynchronous Message-Passing Programs," *8th Int'l Conf. on Engineering of Complex Computer Systems* pp. 35-44, IEEE, Dec. 2002.

14. Mittal, N., and V. K. Garg, "Debugging Distributed Programs using Controlled Re-execution," *19th Annual Symp. on Principles of Distributed Computing*, pp. 239-248, ACM, Portland, Oregon, 2000.

15. Mattern, F., "Virtual Time and Global States of Distributed Systems," *Parallel and Distributed Algorithms*, pp. 215-226, Elsevier Science, North Holland, 1989.

16. Netzer, R. H. B., T. W. Brennan, and S. K. Damodaran-Kamal, "Debugging Race Conditions in Message-Passing Programs," *Sigmetrics Symp. on Parallel and Distributed Tools*, pp. 31-40, ACM, May 1996.

17. Netzer, R. H. B., and B. P. Miller, "Optimal Tracing and Replay for Debugging Message-Passing Parallel Programs," *Supercomputing*, pp. 502-511, IEEE/ACM, Nov. 1992.

18. Snir, M., S. Otto, S. Huss-Lederman, D. Walker, and J. Dongarra, *MPI: The Complete Reference*, MIT Press, 1996.

19. Tai, K. C., "Race Analysis of Traces of Asynchronous Message-Passing Programs," *Int'l Conf. Distributed Computing Systems*, pp. 261-268, IEEE, May 1997.

MPI Cluster System Software*

Narayan Desai, Rick Bradshaw, Andrew Lusk, and Ewing Lusk

Mathematics and Computer Science Division
Argonne National Laboratory, Argonne, Illinois 60439

Abstract. We describe the use of MPI for writing system software and tools, an area where it has not been previously applied. By "system software" we mean collections of tools used for system management and operations. We describe the common methodologies used for system software development, together with our experiences in implementing three items of system software with MPI. We demonstrate that MPI can bring significant performance and other benefits to system software.

1 Introduction

In this paper we use the term "cluster system software" to describe the collection of tools used to manage a parallel machine composed of multiple nodes. Such a machine typically includes a local file system and perhaps a shared file system. We are not discussing per node software such as the node OS or compilers. Rather, the tools we refer to are the programs and scripts, either externally provided or locally written, that support the management of the cluster as a parallel machine and the execution of parallel jobs for users.

Research into system software is not by any means a new activity; previous efforts have been discussed in [8] and [11]. These efforts have focused around scaling serial unix services to large cluster scale. The approach described in this paper describes the use of MPI to create scalable, parallel system tools. As clusters have become larger, the lack of scalability in traditional system management tools is becoming a bottleneck. This situation has led to a gap in the amount and quality of parallelism used in system support programs as opposed to application programs.

Parallel *applications* have embraced MPI as a portable, expressive, scalable library for writing parallel programs. MPI is already available for applications on most clusters, but it has seldom been used for writing system programs. (An exception is the collection of Scalable Unix Tools described in [9]. These are MPI versions of common user commands, such as ls, ps, and find. In this paper we focus on a different class of tools, more related to system admainstration and operation.) As we hope to show, MPI is a good choice for systems tools because

---

* This work was supported by the Mathematical, Information, and Computational Sciences Division subprogram of the Office of Advanced Scientific Computing Research, U.S. Department of Energy, SciDAC Program, Office of Science, under Contract W-31-109-ENG-38.

D. Kranzlmüller et al. (Eds.): EuroPVM/MPI 2004, LNCS 3241, pp. 277–286, 2004.

MPI's scalability and flexibility, particularly its collective operations, can provide a new level of efficiency for system software. In this paper we explore the potential of using MPI in in three different system software tasks:

File Staging. Applications need both executable and data files to be made available to the individual nodes before execution, and output data may need to be collected afterward. This process is awkward and can be painfully slow when the cluster, for the sake of scalability, has no globally shared file system.

File Synchronization. The rsync program is a classical Unix tool for ensuring that the content of file systems on two nodes is consistent. A cluster may require many nodes to be synchronized with a "master" node.

Parallel Shell. We have written a parallel shell called MPISH to supervise the execution of parallel jobs for users. It handles many of the same functions that a normal shell such as sh or bash does for serial applications (process startup, environment setup, stdio management, interrupt delivery) but in a parallel and scalable way.

In Section 2 we discuss the shortcomings of current approaches and elaborate on the advantages and disadvantage of an MPI-based approach. In Section 3 we briefly describe the Chiba City [3] cluster at Argonne, where our experiments took place. In Sections 4, 5, and 6 we describe the three example tasks mentioned above. In each case we describe the old and new versions of the tool we have developed for that task, and we assess the general usefulness of our new approach. Section 7 describes plans for applying the MPI approach to other system software.

2   Background

2.1   The Current Situation

The system software community uses various schemes to address scalability concerns. In some cases, large external infrastructures are built to provide resources that can scale to required levels. In other cases, *ad hoc* parallelism is developed on a task-by-task basis. These *ad hoc* systems generally use tools like rsh for process management functionality, and provide crude but functional parallel capabilities.

Such tools suffer from poor performance and a general lack of transparency. Poor performance can be attributed to naive implementations of parallel algorithms and coarse-grained parallelism. Lack of transparency can be attributed to the use of "macroscopic" parallelism: that is, the use of large numbers of independent processes with only exit codes to connect back to the overall computation, leading to difficulty in debugging and profiling the overall task.

2.2   Potential of an MPI Approach

The adoption of *comprehensive* parallelism, such as that provided by MPI, can provide many benefits. The primary benefit is an improved quality of parallelism.

A complete set of parallel functionality is not only available but already optimized; whereas with *ad hoc* parallelism collective operations often are not available. Also, MPI implementations are highly optimized for the high-performance networks available on clusters. While these networks are available for use by serial tools through TCP, its performance, even over high performance networks, tends to lag behind the performance provided by the vendor's MPI implementation.

The use of MPI for system software does have some potential disadvantages. Since an MPI approach is likely to utilize collective operations, the simple mechanisms available for providing fault tolerance to pairs of communicating process may not apply. Techniques for providing fault tolerance in many situations are being developed [7] but are (MPI-)implementation-dependent. In addition, the manner in which MPI programs are launched (e.g. `mpiexec`, `mpirun` in various forms) is not as portable as the MPI library itself. System scripts that invoke MPI-based tools may have to be adapted to specific MPI installations.

3   Experimental Environment

Chiba City [3] is a 256-node cluster at Argonne National Laboratory devoted to scalable software research, although "friendly users" also run applications. It is not, however, dedicated to applications use and hence is available for parallel software development and experimentation, particularly in the area of system software. For the past year, its software stack has consisted largely of components developed in the context of the SciDAC Scalable System Software project [13]. That is, the scheduler, queue manager, process manager, configuration manager, and node monitor are all implemented as communicating peer components [4]. Extending these components to include necessary system utilities and support programs initiated the experiments described in this paper.

In addition to being large enough that scalability of system operation is an issue, Chiba City presents particular challenges for application use because of its lack of a global file system. This was a deliberate choice to force the development of economical, scalable approaches to running user applications in this environment.

Instead, Chiba City is divided into "towns" consisting of 32 nodes and a "mayor" node. The eight mayors are connected via gigabit Ethernet to a "president" node. The mayors mount the individual file systems of each node and are connected to them by Fast Ethernet. The nodes all communicate with each other over both a separate Fast Ethernet connection and Myrinet. The MPI implementation on Chiba City is the latest release of MPICH2 [6]. The experiments reported here were all carried out using TCP over Fast Ethernet as the underlying transport layer for the MPI messages, pending finalization of the Myrinet version of MPICH2.

4   File Staging

Chiba City has no shared file system for user home directories. Over the past five years Chiba City has been in operation, we have used two systems that attempt to provide seamless, on-demand access to home-directory data on nodes. The first, *City Transit*, was developed during Chiba City's initial deployment. The second system, implemented in MPI, has been in use for the past year. We will describe each system's implementation in detail, discuss usage experiences, compare performance, and assess both systems.

4.1   City Transit

City Transit was implemented in Perl in the style of a system administrator tool. Parallel program control is provided by `pdsh`, which executes commands across nodes. The staging process starts with a request, either from an interactive user or from a job specification, identifying the data to be staged and the nodes that need the data. The requested data is archived (uncompressed) into a tar file. This tar file is copied via `NFS` from the home directory file server to the cluster master and then to mayors who manage destination nodes. Finally, the nodes unarchive the tar file, available via an `NFS` mount from the mayors. Once this process has completed, the user has the pertinent portion of his home directory on all assigned compute nodes.

Chiba City's management infrastructure (the president and mayors) is connected using Gigabit Ethernet, while the remainder of the system is connected using Fast Ethernet. This process was optimized to improve performance through knowledge of network topology. By preferentially using the faster Gigiabit Ethernet links, we substantially accelerated the file staging process.

Nevertheless, this process has several shortcomings. First, the tar file of user data is written to disk on the cluster master and some mayors during intermediate staging steps. This process can be wasteful, as each tar file is useful only once. For larger file-staging runs, multiple stages of disk writes can constitute a large fraction of the overall run time. Second, process control is provided by multiple, recursive invocations of `pdsh`. The return-code handling capabilities of `pdsh` require that in each invocation, multiple return codes are compacted into one. While this allows for basic error detection, complex error assessment isn't possible. Third, the coarse-grained parallelism provided by this staged approach effectively places several global barrier synchronizations in the middle of the process. Fourth, several shared resources, including file servers, the cluster master, and a subset of the mayors are heavily used during the staging process. Thus users can seriously impact setup performance of one another's jobs. Fifth, since all data transmissions to compute nodes originate in the system management infrastructure, full network bisection bandwidth isn't available.

4.2   Userstage

The second file staging process in use on Chiba City is implemented in MPI. Users (or the queue manager by proxy) specify a portion of the home file system to be used on compute nodes. This data is archived into a compressed tar file and placed in an HTTP-accessible spool. The nodes start the `stagein` executable, which is an MPI program. Rank 0 of this program downloads the stage tar file from the file server via HTTP. The contents of this file are distributed to all nodes using MPI_Bcast, and then all nodes untar the data to local disk.

This approach addresses many of the shortcomings of the City Transit approach. MPI provides fine-grained parallelism and synchronization and highly optimized collectives that use available networks effectively. Errors can be easily detected and handled properly. Also, only the initial tar file download is dependent on system infrastructure. Hence, users are better isolated from one another during parallel job setup. Moreover, the use of MPI allows us to largely skip the optimization process, which was quite troublesome during the design process of City Transit.

We carried out a set of additional experiments in which the file was broadcast in various sizes of individual chunks rather than all at once, in order to benefit from pipeline parallelism. This approach did indeed improve performance slightly for small numbers of nodes, where the MPICH2 implementation of MPI_Bcast uses a minimal spanning tree algorithm. For more than four nodes, however, MPICH2 uses a more sophisticated scatter/allgather algorithm, and the benefit disappears. This experience confirmed our intuition that one can rely on the MPI library for such optimizations and just code utility programs in the most straightforward way. The only chunking used in the experiment as reported here was to allow for the modest memories on the nodes; this points the way for further optimization of MPI_Bcast in MPICH2.

4.3   Practical Usage Experiences

City Transit was used for four years. During this period, there were frequent problems with transient failures. These problems were difficult to track, as debugging information was split across all systems involved in file staging. The debugging process was improved with the addition of network logging. However, many failure modes could be detected only by manual execution of parts of City Transit, because of poor implementation and pdsh's execution model. More important, the tree design of control used didn't allow communication of errors between peers. In case of errors, non-failing nodes could finish the file staging process. As this process could take upwards for 30 minutes and put a heavy load on the system infrastructure, single node failures would cause a large expenditure of resources for no benefit. Also, even in cases where everything worked properly, file staging operations performed poorly, and users were generally unhappy with the process.

Our experiences over the past year with the MPI-based staging mechanism show a stark contrast with our previous experiences. Performance is substantially improved, as shown in Figure 1. This improvement can be attributed to the highly optimized parallel algorithms implemented in MPICH2. The MPI version performs better overall, even though it does not exploit knowledge of the network topology, and as one expects, the advantages are greater both as the number of nodes increases for fixed file sizes and as the file size increases for a given number of nodes. At 128 nodes for a gigabyte file, the new code is writing an aggregate of 216 Mb/sec compared with only 57 Mb/sec for City Transit.

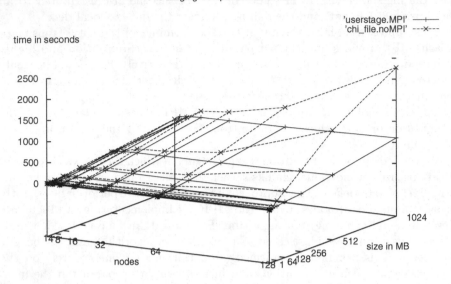

Fig. 1. Performance of old vs. new file staging code.

More important, the reliability of the staging process has been substantially improved. We attribute this to a number of factors. First, the execution model of the new staging tools provides an easier environment for error detection and handling. Because programs aren't called recursively, error codes and standard output are readily available to the executing program. Also, the communication constructs available to MPI programs allow for errors to be easily disseminated to all processors.

As a result of the parallel infrastructure provided by MPI, the volume of code required to implement file staging has been dramatically reduced. *City Transit* consisted of 5,200 lines of Perl code, whereas the MPI-based code is only 84 lines of C. This size reduction makes the code easier to write, debug, and test comprehensively.

5   File Synchronization

File synchronization is an extremely common operation on clusters. On Chiba City, we use this process to distribute software repositories to mayors from the cluster master. This process is executed frequently and can highly tax the resources available on the cluster master. We will describe two schemes for implementing this functionality. The first is based on `rsync`. The second is a file synchronizer written in MPI. We will discuss their implementations and standard usage cases. We will also describe our experiences using each solution and compare relative performance.

5.1   Rsync

Rsync is a popular tool that implements file synchronization functionality. Its use ranges from direct user invocation to system management tools. For example, SystemImager [5], a commonly used system building tool, is implemented on top of rsync.

Rsync is implemented using a client/server architecture. The server inventories the canonical file system and transmits file metadata to the client. The client compares this metadata with the local file system and requests all update files from the server. All communication transactions occur via TCP sockets. This process is efficient for point-to-point synchronizations; however, the workloads we see on parallel machines tend to be parallel, and therefore this process can be wasteful. During a parallel synchronization, the server ends up inventorying the file system several times, and potentially transmitting the same data multiple times. Usually, the files between slaves are already synchronized, so the odds of unnecessary retransmission are high. Many more technical details about rsync are available at [12].

5.2   MPISync

To address several issues with our usage of rsync, we have implemented a parallel file synchronization tool, which performs a one-to-many synchronization. The steps used in mpisync are similar to those used by rsync, with the replacement of serial steps by parallel analogues where appropriate. The first step is a file inventory on the master. The result of this process is broadcast to all nodes, which in turn perform an inventory of the local file system. A list of files that need distribution is produced by a reduction of the set of inventories on all nodes. All processors then iterate through files needing distribution. File contents are then broadcast singly. Through the use of MPI_Comm_Split, only the processors requiring the data participate in these broadcasts.

5.3   Experiences and Assessment

As we mentioned, mpisync was designed to address several shortcomings in the usage for rsync for parallel synchronization. It easily handles several of the shortcomings of the serial approach; that is, local files are read only once on the server, and the MPI library provides much more efficient distribution algorithms than a set of uncoordinated point-to-point links can provide. Considerably improved scalability and performance are shown by the benchmark shown in 2. Standard rsync performance drops off relatively quickly, and 32-way execution runs are the largest ones that properly complete. On the other hand, mpisync runs properly execute concurrently on 128 nodes, and its efficiency is closely related to the underlying MPI implementation. For this reason, we expect that its scalability would remain good at levels larger that this.

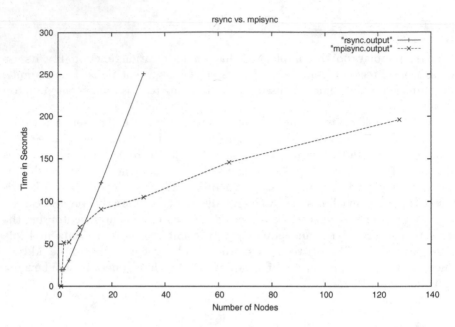

Fig. 2. Performance of old vs. new `rsync` code.

6 A Parallel Execution Environment

Parallel applications are run on clusters in a variety of ways, depending on the resource management system and other tools installed on the cluster. For example, `pdsh` uses a tree to execute the same program on a given set of nodes. MPD [2] is a more flexible process manger, also targeting scalability of parallel job startup, particularly, but not exclusively, for MPI applications. A number of systems (e.g. LAM [1]) require a setup program to be run on one node, which then builds an execution environment for the application on the entire set of allocated nodes. PBS [10] executes user scripts on a single node, which are in turn responsible for initiating MPI jobs via some MPI-implementation-dependent mechanism.

Just as serial jobs typically are run as children of a shell program, which manages process startup, delivery of arguments and environment variables, standard input and output, signal delivery, and collection of return status, so parallel programs need a parallel version of the shell to provide these same services in an efficient, synchronized, and scalable way. Our solution is `MPISH` (MPI Shell), an MPI program that is started by whatever mechanism is available from the MPI implementation, and then manages shell services for the application. An initial script, potentially containing multiple parallel commands, is provided to the MPI process with rank 0, which then interprets it. Its has all the flexibility of MPI, including the collective operations, at its disposal for communicating to the other ranks running on other nodes. It can thus broadcast executables, arguments, and environment variables, monitor application processes started on other nodes, distribute and collect `stdio`, deliver signals, and collect return codes, all in a scalable manner provided by MPI.

`MPISH` implements the PMI interface defined in MPICH2. Thus any MPI implementation whose interface to process management is through PMI (and there are several) can use it to locate processes and establish connections for application communication. Thus `MPISH` both is itself a portable MPI program and supports other MPI programs.

We have used `MPISH` to manage all user jobs on Chiba City for the past year, and found it reliable and scalable. `MPISH` implements a "PBS compatibility mode" in which it accepts PBS scripts. This has greatly eased the transition from the PBS environment we used to run to the component-based system software stack now in use.

7 Summary and Plans

Using MPI for system software has proven itself to be a viable strategy. In all cases, we have found MPI usage to substantially improve the quality, performance, and simplicity of our systems software. Our plans call for work in several areas.

First, we can improve our existing tools in a number of ways. We can incorporate MPI datatype support into applications, specifically `mpisync`, to improve code simplicity. We also plan the development of other standalone system tools. A number of other common tasks on clusters are currently implemented serially, and could greatly benefit from this approach. Parallel approaches could be taken in the cases of several common system management tasks; system monitoring, configuration management, and system build processes could be substantially accelerated if implemented using MPI.

Use of other advanced MPI features for system software is under consideration. I/O intensive systems tasks could easily benefit from the addition of MPI-IO support. The implementation of the MPI-2 process management features in MPICH2 will allow the implementation of persistent systems software components in MPI, using `MPI_Comm_connect` and `MPI_Comm_spawn` to aid in fault tolerance.

A final area for further work is that of parallel execution environments. `MPISH` provides an initial implementation of these concepts. The idea of an execution environment that allows the composition of multiple parallel commands is quite compelling; this appeal only grows as more parallel utilities become available.

References

1. Greg Burns, Raja Daoud, and James Vaigl. LAM: An open cluster environment for MPI. In John W. Ross, editor, *Proceedings of Supercomputing Symposium '94*, pages 379–386. University of Toronto, 1994.
2. R. Butler, N. Desai, A. Lusk, and E. Lusk. The process management component of a scalable system software environment. In *Proceedings of IEEE International Conference on Cluster Computing (CLUSTER03)*, pages 190–198. IEEE Computer Society, 2003.

3. http://www.mcs.anl.gov/chiba.
4. N. Desai, R. Bradshaw, A. Lusk, E. Lusk, and R. Butler. Component-based cluster systems software architecture: A case study. In *Proceedings of IEEE International Conference on Cluster Computing (CLUSTER04)*, 2004.
5. Brian Elliot Finley. VA SystemImager. In USENIX, editor, *Proceedings of the 4th Annual Linux Showcase and Conference, Atlanta, October 10–14, 2000, Atlanta, Georgia, USA*, page 394, Berkeley, CA, USA, 2000. USENIX.
6. William Gropp and Ewing Lusk. MPICH. World Wide Web.
   ftp://info.mcs.anl.gov/pub/mpi.
7. William Gropp and Ewing Lusk. Fault tolerance in mpi programs. *High Performance Computing and Applications*, To Appear.
8. J. P. Navarro, R. Evard, D. Nurmi, and N. Desai. Scalable cluster administration - chiba city i approach and lessons learned. In *Proceedings of IEEE International Conference on Cluster Computing (CLUSTER02)*, pages 215–221, 2002.
9. Emil Ong, Ewing Lusk, and William Gropp. Scalable Unix commands for parallel processors: A high-performance implementation. In Y. Cotronis and J. Dongarra, editors, *Recent Advances in Parallel Virtual Machine and Message Passing Interface*, volume 2131 of *Lecture Notes in Computer Science*, pages 410–418. Springer-Verlag, September 2001. 8th European PVM/MPI Users' Group Meeting.
10. http://www.openpbs.org/.
11. Rolf Riesen, Ron Brightwell, Lee Ann Fisk, Tramm Hudson, Jim Otto, and Arthur B. Maccabe. Cplant. In *Proceedings of the Second Extreme Linux Workshop at the 1999 Usenix Technical Conference*, 1999.
12. http://rsync.samba.org/.
13. http://www.scidac.org/scalablesystems.

A Lightweight Framework
for Executing Task Parallelism on Top of MPI

P.E. Hadjidoukas

IRISA/INRIA, Campus de Beaulieu, 35042 Rennes cedex, France
Panagiotis.Hadjidoukas@irisa.fr

Abstract. This paper presents a directive-based programming and run-time environment that provides a lightweight framework for executing task parallelism on top of MPI. It facilitates the development of message-passing applications that follow the master-slave programming paradigm, supports multiple levels of parallelism and provides transparent load balancing with a combination of static and dynamic scheduling of tasks. A source-to-source translator converts C and Fortran master-slave programs, which express their task (RPC-like) parallelism with a set of OpenMP-like directives, to equivalents programs with calls to a run-time library. The result is a unified programming approach that enables the efficient execution of the same code on both shared and distributed memory multiprocessors. Experimental results on a Linux-cluster demonstrate the successful combination of ease of programming with the performance of MPI.

1 Introduction

Master-slave computing is a fundamental approach for parallel and distributed computing that has been used successfully on a wide class of applications. According to this model, applications consist in a master process that is responsible for distributing tasks to a number of slave processors and receives the results from them. Programming a master-slave application requires knowledge of the primitives provided by the underlying message passing environment and additional programming effort when the application exhibits load imbalance. On the other hand, Remote Procedure Calls (RPC) constitutes a powerful technique for the easier creation of distributed computing applications. A strong similarity exists between RPC and task parallelism of master-slave message passing programs: a procedure executes on a server (slave), processes some data and returns the results to the client (master).

AMWAT [3] and MW [8] are libraries that facilitate the development of such applications on metacomputing environments by providing a higher-level abstraction. They hide from the user the burden of writing code for explicit data transfer and load balancing. OmniRPC [7] is a recent approach that combines OpenMP [6] at the client side with GridRPC, allowing a program to issue multiple asynchronous RPC calls. Although these proposals are applicable to clusters,

D. Kranzlmüller et al. (Eds.): EuroPVM/MPI 2004, LNCS 3241, pp. 287–294, 2004.
© Springer-Verlag Berlin Heidelberg 2004

they target at a higher level than that of cluster computing, excluding available programming libraries and tools.

This paper presents a directive-based programming and runtime environment that provides a convenient and lightweight framework for RPC-like task parallelism on top of MPI [5]. A flexible and portable runtime library implements a two-level thread model on clusters of multiprocessors, trying to exploit the available shared-memory hardware. It provides transparent data movement and load balancing and allows for static and dynamic scheduling of tasks. Furthermore, it supports multiple levels of parallelism and more than one master and enables the same program code to run efficiently on both distributed and shared memory multiprocessors. Finally, a source-to-source translator (DOMP2MP) converts C and Fortran master-slave programs, which express their RPC-like parallelism with a set of OpenMP-like directives, to equivalents programs with calls to this runtime library. This higher-level abstraction not only provides a unified programming approach to master-slave computing but also extends and integrates two different programming models for clusters of multiprocessors: message passing (MPI) and shared memory (OpenMP). Experimental results on a 16-node Linux-cluster demonstrate the successful combination of ease of programming with the performance of MPI.

The rest of this paper is organized as follows: Section 2 reviews the architecture of the runtime library and section 3 describes our OpenMP-like programming environment. Experimental evaluation is reported in Section 4. Finally, we conclude in Section 5.

2   Runtime Support

At the core of our programming environment there is a portable runtime library (DOMP2MP RTL) that implements a two-level thread model on clusters of multiprocessors. This model actually provides a lightweight framework for executing the task parallelism of master-slave message passing applications. Tasks are represented with work descriptors, distributed to the available nodes and eventually executed on top of user-level threads. The same approach has been also followed for the master, which is simply the primary task of the application. Lightweight threads have been also used in Adaptive MPI [2], as an alternative approach to an MPI implementation. In contrast, we use our two-level thread model to support multilevel parallelism and adopt a lazy thread-creation policy, binding a task to a thread only just before its execution.

Architecture. Figure 1 illustrates the general architecture of our runtime environment on a cluster of computing nodes. Each MPI process consists of one or more virtual processors and a special I/O thread, the *Listener*, which is responsible for the dependence and queue management and also supports the transparent and asynchronous movement of data. There are intra- one inter-node ready queues where tasks are submitted for execution. This configuration enables dynamic switching between the fork-join and the SPMD execution model.

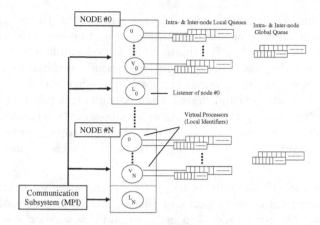

Fig. 1. General Architecture

The insertion (stealing) of a work descriptor in (from) a queue that resides in the same process is performed through hardware shared memory, without any data movement. Otherwise, the operations are performed by sending appropriate messages to the Listener of the remote node. The runtime library also uses this combination of hardware shared-memory with explicit messages to maintain the coherence of the execution model: each work descriptor (task) is associated with an owner-node (home node). If a task finishes on its owner-node, its successor is notified directly through shared memory. Otherwise, a message is sent to the owner-node and this notification is performed by the Listener.

Data Management. When the application is executed on shared memory machines, the runtime library (and accordingly the application) operates exclusively through the available hardware. However, when a task is inserted on a remote node then its data has to be moved explicitly. In this case, we associate each work descriptor with the arguments (data) of its corresponding function, similarly to Remote Procedure Call. For each of the arguments the user specifies its (MPI) data type and number of elements (count), an intent attribute, and optionally a reduction operator. The reduction operations supported by the runtime library can be applied to both scalar variables and arrays that return to the home-node (master) as results. The explicit data movement is performed transparently to the user and the only point that reminds message passing programming is the above description.

As noted before, the master of a message-passing application that follows the master-slave programming paradigm is instantiated with a user-level thread that runs on one of the available virtual processors. Multithreading allows this virtual processor to participate in the computation since the master remains blocked as soon as it has generated and distributed the task-parallelism across the nodes. This virtual processor processes the tasks through hardware shared memory, operating directly on their data, which means that no data copying occurs and arguments are really passed by reference. Thus, the programmer has to allocate the required space for each task in order to ensure that two tasks do not operate on the same data. Fortunately, this issue is unlikely to occur because master-slave message passing applications correspond to embarrassingly parallel algorithms without any data races. Moreover, the efficiency of our hybrid approach for both distributed and shared memory computing relies on this feature. An alternative approach is the execution of the master on a separate process and the use of multiple processes instead of threads. Data copying takes places and the performance relies on the efficiency of the MPI implementation.

Load Balancing. The scheduling loop of a virtual processor is invoked when the current user-level thread finishes or blocks. A virtual processor visits in a hierarchical way the ready queues in order to find a new descriptor to execute. The stealing of a descriptor from a remote queue includes the corresponding data movement, given that it does not return to its owner node. This stealing is performed synchronously: the requesting virtual processor waits for an answer from the Listener of the target node. This answer is either a special descriptor denoting that no work was available at that node, or a ready descriptor, which will be executed directly by the virtual processor. Inter-node work stealing can be enabled or disabled dynamically according to the application needs. When a idle virtual processor tries to find available work on remote nodes, it starts from nodes with consecutive numbers until it has visited the entire set. The number of messages is reduced by maintaining, on each node, information about the status of nodes. A further improvement of this mechanism is that when the Listener of an idle node receives a work stealing request, it includes in the response the identifier of the last non-idle node that is known to its own node.

3 Directive-Based Programming Environment

In order to facilitate the development of master-slave message passing programs, we introduce a higher-level programming abstraction by mapping an extension of the proposed OpenMP workqueuing model [9] to the API exported by the runtime system, a proposal initially presented in [4]. This model is a flexible mechanism that fits very well in master-slave message passing programming. Specifically, our compilation environment exploits the following two constructs of the OpenMP workqueuing model: omp parallel taskq and omp task. In our case, the two OpenMP-like directives have the following format (for C programs):

```
#pragma domp parallel taskq [schedule(processor|node|local))]

#pragma domp task [callback(<procedure_name>)]
{ <task procedure>
 [<callback procedure>] }
```

The first directive (domp parallel taskq) prepares an execution environment for the asynchronous execution of the distributed procedure provided in the structured block defined with the `domp task` directive. The `schedule` clause determines the task distribution scheme that will be followed. Per-node distribution can be useful for tasks that generate intra-node (shared-memory) parallelism, expressed with OpenMP. In this case, a single virtual processor is used per node and, consequently, one task is active at a time. Local distribution corresponds to the traditional approach where all tasks are inserted in the local queue of the master. The second clause (`callback`) allows the user to define a callback routine that is invoked asynchronously on the master (home node) when a task has finished. Furthermore, a description of the procedure arguments has to be provided by the user. In agreement to the previous directives, this is performed as follows:

```
#pragma domp procedure <procedure_name> <number of arguments>
#pragma domp procedure <description of first argument>
...
#pragma domp procedure <description of last argument>
```

where,
<description of argument> → <count>, <datatype>, <intent>[|| <reduction>]
<count> → number of elements
<datatype> → valid MPI datatype
<intent> → `CALL_BY_VAL` | `CALL_BY_PTR` | `CALL_BY_REF` | `CALL_BY_RES`
<reduction> → `OP_ADD` | `OP_SUB` | `OP_MUL` | `OP_DIV`

The intent attributes represent the way procedure arguments are used. Their list can be further expanded in order to include additional forms of data management. A detailed description of them and their correspondence to Fortran 90's `IN`, `OUT`, and `INOUT` is given below:

- `CALL_BY_VAL` (IN): The argument is passed by value. If it is a scalar value then it is stored in the descriptor.
- `CALL_BY_PTR` (IN): As above, but the single value has to be copied from the specified address. This option can be also used for user-defined MPI datatypes (`MPI_Type_struct`).
- `CALL_BY_REF` (INOUT): The argument represents data sent with the descriptor and returned as result in the home node's address space.
- `CALL_BY_RES` (OUT): No data is actually sent but it will be returned as result. It is assumed by the target node that "receives" data initialized to zero.

The presented directives are also available for Fortran programs, introduced however with the `D$OMP` keyword. At the first stage, a source-to-source translator (`DOMP2MP`) parses our directives and generates an equivalent program with calls to the runtime system. This program, which is possible to contain OpenMP directives, is compiled by a native (OpenMP) compiler and the object files are linked with the software components of our runtime environment producing the final executable.

Figure 2 presents two programs that clarify the use of our OpenMP-like extensions. The first application is a recursive version of Fibonacci. It exhibits

```
/* Fibonacci */
#include <domp2mp.h>

void fib(int n, int *res)
{
#pragma domp procedure fib 2
#pragma domp procedure 1,MPI_INT,CALL_BY_VAL
#pragma domp procedure 1,MPI_INT,CALL_BY_REF
    int res1 = 0, res2 = 0;
    if (n < 2) {
        *res = n;
    } else if (n < 38) {
        fib(n-1, &res1);
        fib(n-2, &res2);
        *res = res1+ res2;
    } else {
        #pragma domp parallel taskq schedule(node)
        {
            #pragma domp task
            { fib(n-1, &res1); }
            #pragma domp task
            { fib(n-2, &res2); }
        }
        *res = res1+ res2;
    }
}

int main(void)
{
    int res, n = 48;
    fib(n, &res);

    return 0;
}
```

```
/* Master-Slave Demo Application */
#include <domp2mp.h>

void taskfunc(double in, double *out)
{
    *out = sqrt(in);
}

void cbfunc(double in, double *out)
{
    printf("sqrt(%f)=%f\n", in, *out);
}

int main(vois)
{
    int N = 100, i, count;
    double *result;
    double num;

    result = (double *)malloc(N*sizeof(double));

#pragma domp procedure taskfunc 2
#pragma domp procedure count,MPI_DOUBLE,CALL_BY_VAL
#pragma domp procedure count,MPI_DOUBLE,CALL_BY_RES
    count = 1; /* dynamically */
    #pragma domp parallel taskq
    for (i=0; i<N; i++) {
        num = (double) rand();
        #pragma domp task callback(cbfunc)
        { taskfunc(num, &result[i]);
          cbfunc(num, &result[i]);
        }
    }
    return 0;
}
```

Fig. 2. Examples of our OpenMP-like directives

multilevel parallelism and each task generates two new tasks that are distributed across the nodes. The second example is a typical master-slave (demo) application: each task computes the square root of a given number and returns the result to the master. Tasks are distributed across the processors and whenever a task completes, a callback function (`cbfunc`) is executed asynchronously on the master node, printing the result.

Apart from the above OpenMP-like directives, the runtime library also exports some library functions to assist users in managing their master-slave program:

- `domp_get_num_nodes()`: Returns the number of available nodes (processes).
- `domp_get_node_num()`: Returns the identifier (rank) of the current node (process).
- `domp_broadcast(void *addr, int count, MPI_datatype type)`: This routine broadcasts the data of address `addr`.
- `domp_isomalloc(void **mem, int size)`: Allocates `size` bytes of memory on all nodes (at the same address).

The rationale of the last two functions is that in master-slave programs there can be read-only global data that are initialized by the master and broadcast to the slaves, for example the vector in matrix-vector multiplication. Due to master-slave execution, we must ensure that this broadcast address is valid and available on all nodes. Therefore we also provide a memory allocation routine (`domp_isomalloc`) that allocates memory on all nodes at the same address [1].

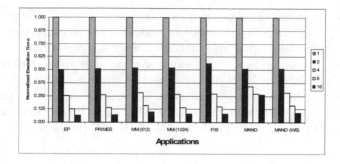

Fig. 3. Normalized Execution Times on the Linux cluster

4 Experimental Evaluation

In this section, we present experimental results that demonstrate the features of our programming environment on the main target platform of this work, i.e. clusters of computing nodes. Preliminary performance results on a small scale shared-memory multiprocessor can be found in [4]. We conducted our experiments on a cluster of 16 Pentium 4 2.4 GHz dual-processor machines, running Debian Linux (2.4.12). Each machine has 1GB of main memory and the cluster is interconnected with Fast Ethernet (100 Mbps). As native compiler we have used GNU gcc (3.2.3), while MPIPro for Linux is the thread-safe MPI implementation.

Figure 3 illustrates the normalized execution times of five embarrassingly parallel master-slave applications on the Linux cluster. The work stealing mechanism is disabled unless otherwise specified. EP is an Embarrassingly Parallel benchmark that generates pairs of Gaussian random deviates. For this application, the number of generated tasks is equal to the number of processors. The second application (PRIMES) computes the prime numbers from 1 to 20M. For achieving load balancing the master generates 128 tasks. MM performs a master-slave matrix multiplication (C=AxB) for N = 512 and 1024. Matrix B is isoallocated and broadcast, without including this operation in our measurements. The support of multiple levels of parallelism is demonstrated with the Fibonacci application, which was presented in Figure 2. Finally, the last two groups of columns demonstrate the effectiveness of our work stealing mechanism to a master-slave application that computes the Mandelbrot set for a 2-d image of 2048x2048 pixels. Each task processes a block of pixels of size 64x64, while a callback routine stores the calculated values in a two-dimensional matrix.

We observe that the first two applications scale very efficiently and manage to exploit the processing power of the cluster, something that has been achieved with minimal programming effort. For matrix multiplication, performance degradation starts to appear on 8 and 16 nodes for matrix dimension N = 512, since the amount of distributed work is not large enough to overwhelm the communication costs. Fibonacci scales satisfactorily, although this benchmark heavily stresses both the runtime system and the MPI library, and despite the low-performance interconnection network of the cluster. Finally, Mandelbrot with work stealing (MAND (WS)) succeeds to scale due to the combination of static and dynamic scheduling that the runtime library supports: tasks are initially distributed to the nodes in a cyclic way and eventually work stealing between them starts, avoiding the bottleneck of a central queue and allowing the master to participate in the parallel computation. The execution times of the applications on a single processor were the following: EP: 62.9, PRIM: 135.8, MM (512): 8.54, MM (1024): 72.56, FIB: 173.5, and MAND: 57.9 (in seconds).

5   Conclusions

This paper presented a directive-based programming and runtime environment for master-slave applications on top of the Message Passing Interface. It integrates features of several parallel programming models, from threads and OpenMP to MPI and RPC. Our future work includes further improvements and in-depth evaluation of the load balancing mechanism, support of additional methods for data movement (MPI one sided), experimentation on heterogeneous platforms, and exploitation of application adaptability on multiprogramming environments.

References

1. Antoniu, G., Bouge, L., Namyst,R., Perez, C.: Compiling Data-parallel Programs to A Distributed Runtime Environment with Thread Isomigration. Parallel Processing Letters, 10(2-3):201–214, June 2000.
2. Huang, C., Lawlor, O., Kale, L. V.: Adaptive MPI. Proc. 16th Intl. Workshop on Languages and Compilers for Parallel Computing, College Station, TX, USA, October 2003.
3. Goux, J-P., Kulkarni, S., Linderoth, J., Yoder, M.: An Enabling Framework for Master-Worker Applications on the Computational Grid. Proc. 9th IEEE Intl. Symposium on High Performance Distributed Computing, Pittsburgh, Pennsylvania, USA, August 2000.
4. Hadjidoukas, P.E., Polychronopoulos, E.D., Papatheodorou, T.S.: OpenMP for Adaptive Master-Slave Message Passing Applications. Proc. Intl. Workshop on OpenMP Expreriences and Implementations, Tokyo, Japan, October 2003.
5. Message Passing Interface Forum: MPI: A message-passing interface standard. Intl. Journal of Supercomputer Applications and High Performance Computing, Volume 8, Number 3/4, 1994.
6. OpenMP Architecture Review Board: OpenMP Specifications. Available at: http://www.openmp.org.
7. Sato, M., Boku, T., Takahashi, D.: OmniRPC:a Grid RPC ystem for Parallel Programming in Cluster and Grid Environment. Proc. 3rd Intl. Symposium on Cluster Computing and the Grid, Tokyo, Japan, May 2003.
8. Shao, G., Berman, F., Wolski, R.: Master/Slave Computing on the Grid. Proc. 9th Heterogeneous Computing Workshop. Cancun, Mexico, May 2000.
9. Su, E., Tian, X., Girkar, M., Grant, H., Shah, S., Peterson, P.: Compiler Support of the Workqueuing Execution Model for Intel SMP Architectures. Proc. 4th European Workshop on OpenMP, Rome, Italy, September 2002.

Easing Message-Passing Parallel Programming Through a Data Balancing Service*

Graciela Román-Alonso[1],
Miguel A. Castro-García[1,2], and Jorge Buenabad-Chávez[2]

[1] Departamento de Ing. Eléctrica, Universidad Autónoma Metropolitana, Izt.
Ap. Postal 55-534, D.F. 09340, México
{grac,mcas}@xanum.uam.mx
[2] Sección de Computación, Centro de Investigación y de Estudios Avanzados del IPN
Ap. Postal 14-740, D.F. 07360, México
jbuenabad@cs.cinvestav.mx

Abstract. The message passing model is now widely used for parallel computing, but is still difficult to use with some applications. The explicit data distribution or the explicit dynamic creation of parallel tasks can require a complex algorithm. In this paper, in order to avoid explicit data distribution, we propose a programming approach based on a data load balancing service for MPI-C. Using a parallel version of the merge sort algorithm, we show how our service avoids explicit data distribution completely, easing parallel programming. Some performance results are presented which compare our approach to a version of merge sort with explicit data distribution.

1 Introduction

Nowadays parallel programming is a powerful way to reduce execution time. In general, programmers can use a shared memory or a message passing model to develop their applications [3]. However, under both models, for some problems it is not easy to partition data among the processors so as to obtain the best performance. In clusters particularly, partitioning data evenly among the nodes may be inadequate because of multiprogramming or heterogeneity of cluster nodes.

In this paper we present an approach to develop parallel applications without specifying data distribution.

In this approach, applications must use a data balancing service, and must manage data as a list. Also, the processing of each list element needs to be independent of other list elements (for the time being). At run time, the data balancing service carries out two tasks. It first, and once only, distributes evenly the data list elements among the processors. It then periodically balances the workload by transferring data list elements from overloaded to underloaded processors in order to reduce execution time.

---

* This work is supported by the CONACyT Project 34230-A: Infraestructura para la construcción de aplicaciones fuertemente distribuidas.

D. Kranzlmüller et al. (Eds.): EuroPVM/MPI 2004, LNCS 3241, pp. 295–302, 2004.

Similar approaches to ease the development of parallel applications have been proposed [1, 9, 11]. From a branch-and-bound sequential code, these approaches derive a parallel version through the use of *skeletons*, software frameworks, which run the sequential code in different nodes and communicate with each other. MALLBA also manages skeletons for Divide and Conquer, Dynamic Programming, Genetic Algorithms and Simulated Annealing applications, among others [1]. Code parallelization is hidden from the user into the skeletons, as is load balancing which, in some cases [9, 11], was added later to the skeleton structure.

In our approach, load balancing is the key aspect to ease parallel programming as described earlier: through an initial data partitioning and then through periodic load balancing to improve performance. From a sequential code that uses data organized into a list, users only need to call our load balancing service within the main loop of their application.

Our service has been implemented for MPI-C. An earlier version was designed to reduce the execution time of a parallel evolutionary algorithm that generates neural networks [6]. In this paper we present a second general version tested on an homogeneous cluster of 8 PCs.

To illustrate the programming advantage of our approach, we compare two versions of a Divide and Conquer problem. In Section 2 we present a typical solution that requires explicit data distribution and in Section 3 we present an implementation using our data list balancing service. The internal architecture of the service is presented in Section 4. In Section 5 we show some performance results of our approach. We then give some conclusions and present our current and future work.

2 Parallel Programming with Explicit Data Distribution into Processors

In message-passing programming it is necessary to distribute data among the processors. For example, the parallel merge sort problem, with a Divide and Conquer behavior [3], starts with a processor P0 having a set of N numbers to be sorted. Successively the set is divided into 2 parts and a half of the numbers is passed to a new processor. This behavior is the same in all the processors that receive a subset of numbers. The division stops until each processor keeps or receives a data set with a minimum size (as shown in Fig. 1). Then each processor sorts its subset. The final solution is built when each processor sends back its data subset sorted. Each processor that receives a partial solution applies the merge sort algorithm and passes its result until P0 has the final result.

The data distribution of parallel merge sort is straightforward. However, the algorithm to determine data distribution of other application could require a lot of design work in order to avoid dead-lock situations between processors. This requirement could be discouraging for beginners who try to develop parallel programs for more complex problems.

In the next section we show how our approach avoids programmers to allocate data to processors explicitly.

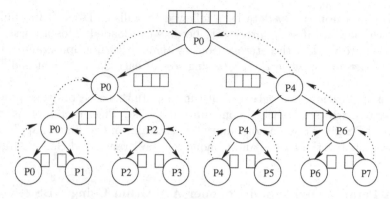

Fig. 1. Distribution of 8 data on 8 processes under a parallel mergesort algorithm

3 Parallel Programming with DBS Data List Balancing Service

The alternative that we propose here to avoid explicit data distribution requires the following:

a) Programming with MPI-C (for the time being), under an SPMD model.

b) Defining the application data as a global data list composed of a set of elements. In this work each element is an integer, but it can be any complex structure.

c) Calling our data balancing service (DBS). The programmer is responsible for using it correctly. Below, we explain the semantics of the DBS primitives and how to use them.

3.1 DBS Interface

The interface of DBS has two procedure calls:

a) MPI_DBS_sendList(L). A process calls this procedure to send its data list to DBS. The process must get the list back as described below to continue execution. It is possible to take one or more elements from the list before calling MPI_DBS_sendList(L). The elements thus taken are to be processed locally. If the list has only one element, DBS doesn't migrate it to other processors.

b) MPI_DBS_recvList(L). The process that calls this function asks DBS to give the list back. If DBS has at least one datum to give back, the process receives a new data list which can have a number of elements greater than 0. Otherwise the process remains blocked either until DBS has some data to give back or until all the application data has been processed by other nodes. This function returns 0 if there are still data to be processed on the system, and returns 1 if all the other processes have finished processing the data (their lists are empty).

Both procedures are called together in an application loop.

Some applications process a fixed amount of data. In this case a good solution could be an initial static allocation of data among the processors (each processor starting with a non empty data list), followed by calls to DBS. DBS would try to balance the data if some processors become overloaded. This aspect can be useful when we work with heterogeneous systems, with multiprogrammed non-dedicated systems or when the processing cost of each data element is different [6].

Other applications generate data at run time. In this case, only one processor in the system will start with a non-empty data list. The other processors will wait until DBS gives them a non-empty list. Example of this kind of applications are those that use the Divide and Conquer or the Branch and Bound strategy.

3.2   A Parallel Divide and Conquer Algorithm Using DBS

In this section we show how the merge sort application described in Section 2 can be built using DBS to avoid explicit data allocation by the programmer. The algorithm starts with only the master processor 0 having a non-empty data list (all other processors wait to receive a list). The list has N elements/numbers to be sorted. Processor 0 first, and once only, divides the initial data list into P lists, where P is the total number of processors, and sends to each processor one of the resulting lists.

Thereafter each processor processes its list. A processor may need to balance its workload/list if it is not processing it as fast as other processors. To this end, it periodically calls MPI_DBS_sendList and MPI_DBS_recvList within the main loop of the merge sort algorithm, as shown in the pseudo-code below.

```
process Parallel_Merge_Sort(List * L, int processid) {
 exit = FALSE;  sorted_list = Empty;  L2 = Empty ;

 while ( NOT(exit) ) {
   if ( size( L ) > 0 ) {
      L2 = Get_First_Elements( &L, MIN_SIZE );
      MPI_DBS_sendList( &L );
      Sort( &L2 );
      Merge( &sorted_list, &L2 ); /* L2 becomes empty */
   }
   exit=MPI_DBS_recvList(&L);
 }
 Coordinator_Gathers_and_Merges_Partial_Results( procid );
 return( sorted_list );
}
```

When a processor receives the exit signal, through calling MPI_DBS_recvList, there is no more data to sort, and gives back the list it has sorted.

This example shows how a programmer can solve a divide and conquer problem without specifying explicit data distribution data, simply using a data list and calling the proposed data balancing service.

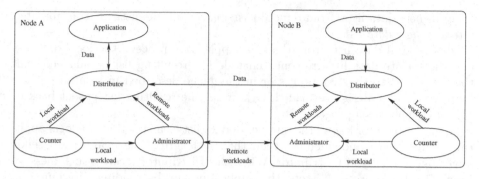

Fig. 2. The data list load balancing service (DBS)

4   The DBS Architecture

Some load balancers work at the operating system level [4] and others at the application level [5, 10]. The first make a general system-wide balance, and the latter usually requires application programmers to modify the code and data structures of the applications to integrate a load balancing algorithm. Other works, like Athapascan [2, 7], manage an implicit assignment of tasks in a parallel system, but it is necessary that the programmer considers the dynamic creation of concurrent tasks in order to process different data spaces. Using our approach the programmer manages data as a list and calls the two DBS procedures as described in the previous section.

DBS consists of 3 processes, the Counter, the Administrator and the Distributor, running along with an application process in each node of a cluster, Fig. 2.

With a given frequency, the Counter measures the local workload of the processor where it is running, and if there is a change from one state to another (described shortly) sends the new state to the Administrator and Distributor. In our implementation we use a threshold value based on CPU usage to determine the load state of a processor. If the workload of a processor is under this threshold the processor state is underloaded (state 0). If the workload is over the threshold the processor state is overloaded (state 1).

The Administrator in each node collects the workload state of *all* other nodes. (This is a global view of the system workload which we chose for its simplicity; however, it is not scalable. We are currently working on managing a partial view, where each processor will only collect the workload of a few processors [8, 11].) If there is a change of the load state of a processor (detected by the Counter), the Administrator sends the new load state to all remote administrators, each of which forwards it to its local Distributor.

The Distributor receives:

a) the local workload state changes, from the local Counter.

b) remote workload state changes, from the local Administrator; workload changes (along with the corresponding processor id.) are organized into an ascending order queue by workload state (0,1).

c) remote data lists, from remote Distributors; all the lists are concatenated into a *remote data list*.

d) the local data list, from the local application process; the local data list and the remote data list are concatenated. The resulting list is balanced, half the list is sent to the first processor in the queue described in b) above, if the local workload state is overloaded. Otherwise, the resulting list is sent back to the local application process.

Each Distributor knows the number of data list elements that the local application process has, and when it detects that the length of the local list is 0, sends a message to a central Distributor. When this Distributor receives such message from all other processors, stops the application run by sending a termination message to all processors.

5   Results

The parallel merge sort algorithm was executed to sort the numbers 1 to 2^{19} in ascending order (previously classified in descending order), under two different scenarios: using an explicit data allocation (EDA), as described in section 2 and using the data list balancing service (DBS), as described in section 3. The runs were carried out on a cluster of 8 homogeneous processors (Pentium III, 450 MHz, RAM 64MB), interconnected through a 100 M-bps Fast Ethernet switch.

We ran both versions, EDA and DBS, both on a dedicated environment and on non-dedicated environment. The frequency to measure the load state of each processor, underloaded or overloaded, was 2 seconds and the threshold value of CPU usage to determine the load state was 60%. Figure 3 shows the execution time of 10 runs of both versions on both environments on 8 processors.

EDA and DBS show about the same run time in the same environment. Obviously in the dedicated environment they show a smaller run time than in the non-dedicated one, where there is another application running (doing some floating point arithmetic operations throughout the run) along either with EDA or with DBS. In each environment, the small difference between EDA and DBS is somewhat surprising. For merge sort, and similar applications, where the amount of data is fixed in size, it is possible to specify an optimal data distribution based on the total amount of data and the number of processors in the system. Hence, EDA should show better performance than DBS, in which the balancing service runs periodically. However, in EDA the partitioning is carried out by several processors which divide the workload among other processors until a minimum size, $2^{19}/8$, is reached in each processor. In DBS, the master processor partitions the workload locally and then gives an equal part to each of the processors.

6   Conclusions and Future Work

We have presented the use of a data balancing service (DBS) to avoid the explicit data distribution by the programmer. This service was implemented with MPI-C and can be invoked by any MPI application that organizes data as a list. The

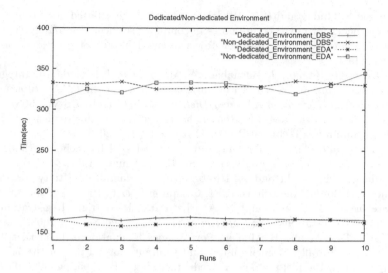

Fig. 3. Execution time for EDA and DBS versions on a dedicated and on a non-dedicated environment, with 8 processors

processing of each list element needs to be independent of other list elements. By calling the two DBS procedures, the data list is balanced among the processors, if need be. An example of a parallel merge sort application using our service was shown and compared to a version with explicit data distribution. Our results show that using our service an application can show as good performance as an application with explicit data distribution.

However, the merge sort algorithm was easily implemented with DBS, through avoiding explicit data distribution, contrary to the version with explicit data allocation. For applications where the programmer does not know the total amount of data to be generated during execution, or when the processing cost of each list element is different, or under multiprogramming, DBS can give better results through balancing the workload in addition to easing the programming.

We are currently evaluating our DBS approach to be used with other types of applications and the use of other data structures instead of a list. We are working on transforming C sequential code that uses data structures other than a list into code that uses a list. We are also investigating automatic, dynamic adjustment of some DBS parameters which determine when to balance data and how much of it.

References

1. E. Alba et al. MALLBA: A Library of Skeletons for Combinatorial Optimisation. Euro-Par 2002: 927-932
2. The Apache Project. http://www-apache.imag.fr/software/
3. M. Allen and B. Wilkinson. *Parallel Programming, Techniques and Applications Using Networked Workstations and Parallel Computers.* Prentice Hall, 1999.

4. A. Barak, S. Guday and R.G. Wheeler. The mosix distributed operating system: load balancing for unix Lecture notes in computer science, Vol. 672, 1993
5. M. Bubak and K. Sowa. Parallel object-oriented library of genetic algorithms. page 12, 1996.
6. M. Castro, G. Román, J. Buenabad, A. Martínez and J. Goddard. Integration of load balancing into a parallel evolutionary algorithm. In *ISADS International Symposium and School on Advanced Distributed Systems*, LNCS Vol. 3061, 2004.
7. T. Gautier, R. Revire and J.L. Roch Athapascan: An API for Asynchronous Parallel Programming. Technical Report INRIA RT-0276, 2003
8. R.M. Keller and C.H. Frank. The gradient model load balancing method. *IEEE Transactions on Software Engineering*, SE-13(1), January 1987.
9. Y. Shinano, M. Higaki and R. Hirabayashi. A Generalized Utility for Parallel Branch and Bound Algorithms. IEEE Computer Society Press, pg. 392-401, 1995.
10. R. Stefanescu, X. Pennec and N. Ayache: Parallel Non-rigid Registration on a Cluster of Workstations. *Proc. of HealthGrid'03* In Sofie Norager 2003.
11. C. Xu, S. Tschöke and B. Monien. Performance Evaluation of Load Distribution Strategies in Parallel Branch and Bound Computations. Proc. of the 7th IEEE Symposium on Parallel and Distributed Processing, SPDP'95, pg. 402-405, 1995.

TEG: A High-Performance, Scalable, Multi-network Point-to-Point Communications Methodology

T.S. Woodall[1], R.L. Graham[1], R.H. Castain[1], D.J. Daniel[1], M.W. Sukalski[2],
G.E. Fagg[3], E. Gabriel[3], G. Bosilca[3], T. Angskun[3], J.J. Dongarra[3], J.M. Squyres[4],
V. Sahay[4], P. Kambadur[4], B. Barrett[4], and A. Lumsdaine[4]

[1] Los Alamos National Lab
[2] Sandia National Laboratories
[3] University of Tennessee
[4] Indiana University

Abstract. TEG is a new component-based methodology for point-to-point messaging. Developed as part of the Open MPI project, TEG provides a configurable fault-tolerant capability for high-performance messaging that utilizes multi-network interfaces where available. Initial performance comparisons with other MPI implementations show comparable ping-pong latencies, but with bandwidths up to 30% higher.

1 Introduction

The high-performance computing arena is currently experiencing two trends that significantly impact the design of message-passing systems: (a) a drive towards petascale class computing systems that encompass thousands of processors, possibly spanning large geographic regions; and (b) a surge in the number of small clusters (typically involving 32 or fewer processors) being deployed in a variety of business and scientific computing environments. Developing a message-passing system capable of providing high performance across this broad size range is a challenging task that requires both an ability to efficiently scale with the number of processors, and dealing with fault scenarios that typically arise on petascale class machines.

A new point-to-point communication methodology (code-named "TEG") has been designed to meet these needs. The methodology has been developed as part of the Open MPI project [3] – an ongoing collaborative effort by the Resilient Technologies Team at Los Alamos National Lab, the Open Systems Laboratory at Indiana University, and the Innovative Computing Laboratory at the University of Tennessee to create a new open-source implementation of the Message Passing Interface (MPI) standard for parallel programming on large-scale distributed systems [4, 7]. This paper describes the concepts underlying TEG, discusses its implementation, and provides some early benchmark results.

A rather large number of MPI implementations currently exist, including LAM/MPI [1], FT-MPI [2], LA-MPI [5], MPICH [6], MPICH2 [8], the Quadrics version of MPICH [10], Sun's MPI [12], and the Virtual Machine Interface (VMI) 2.0 from NCSA [9]. Each of these implementations provides its own methodology for point-to-point

D. Kranzlmüller et al. (Eds.): EuroPVM/MPI 2004, LNCS 3241, pp. 303–310, 2004.

communications, resulting in a wide range of performance characteristics and capabilities:

- LAM/MPI's "Component Architecture" enables the user to select from several different point-to-point methodologies at runtime;
- Sun's and Quadrics' MPIs both will stripe a single message across multiple available similar Network Interface Cards (NICs).
- MPICH2 has a design intended to scale to very large numbers of processors.
- VMI is designed to stripe a single message across heterogeneous NICs, giving it the ability to transparently survive the loss of network resources with out any application modification.
- FT-MPI is designed to allow MPI to recover from processor loss with minimal overhead.
- LA-MPI is designed to stripe a single message across identical NIC's, different messages across different NIC types, recover transparently from loss of network resources, and allows for optional end-to-end data integrity checks.

TEG represents an evolution of the current LA-MPI point-to-point system that incorporates ideas from LAM/MPI and FT-MPI to provide Open MPI with an enhanced set of features that include: (a) fault tolerant message passing in the event of transient network faults – in the form of either dropped or corrupt packets – and NIC failures; (b) concurrent support for multiple network types (e.g. Myrinet, InfiniBand, GigE); (c) single message fragmentation and delivery utilizing multiple NICs, including different NIC types, such as Myrinet and InfiniBand; and (d) heterogeneous platform support within a single job, including different OS types, different addressing modes (32 vs 64 bit mode), and different endianess.

The remainder of this paper describes the design of the TEG point-to-point messaging architecture within Open MPI, the component architecture and frameworks that were established to support it, and the rationale behind various design decisions. Finally, a few results are provided that contrast the performance of TEG with the point-to-point messaging systems found in other production quality implementations. An accompanying paper provides detailed results [13].

2   Design

The Open MPI project's component-based architecture is designed to be independent of any specific hardware and/or software environment, and to make it relatively easy to add support for new environments as they become available in the market. To accomplish this, Open MPI defines an MPI Component Architecture (MCA) framework that provides services separating critical features into individual components, each with clearly delineated functional responsibilities and interfaces. Individual components can then be implemented as plug-in-modules, each representing a particular implementation approach and/or supporting a particular environment. The MCA provides for both static linking and dynamic (runtime) loading of module binaries, thus creating a flexible system that users can configure to meet the needs of both their application and specific computing environment.

Open MPI leverages the MCA framework to map the point-to-point communication system into two distinct components:

1. Point-to-point Management Layer (PML). The PML, the upper layer of the point-to-point communications design, is responsible for accepting messages from the MPI layer, fragmenting and scheduling messages across the available PTL modules, and managing request progression. It is also the layer at which messages are reassembled on the receive side. MPI semantics are implemented at this layer, with the MPI layer providing little functionality beyond optional argument checking. As such, the interface functions exported by the PML to the MPI layer map closely to the MPI-provided point-to-point API.

2. Point-to-point Transport Layer (PTL). At the lower layer, PTL modules handle all aspects of data transport, including making data available to the low level network transport protocols (such as TCP/IP and InfiniBand), receiving data from these transport layers, detection of corrupt and dropped data, ACK/NACK generation, and making decisions as to when a particular data route is no longer available.

At startup, a single instance of the PML component type is selected (either by default or as specified by the user) by the MCA framework and used for the lifetime of the application. All available PTL modules are then loaded and initialized by the MCA framework. The selected PML module discovers and caches the PTLs that are available to reach each destination process. The discovery process allows for the exclusive selection of a single PTL, or provides the capability to select multiple PTLs to reach a given endpoint. For example, on hosts where both the shared-memory and the TCP/IP PTLs are available, the PML may select both PTLs for use, or select the shared-memory PTL exclusively. This allows for a great deal of flexibility in using available system resources in a way that is optimal for a given application.

Another related simplification over prior work includes migrating the scheduling of message fragments into the PML. While LA-MPI supports delivering a message over multiple NICs of the same type, this functionality was implemented at the level corresponding to a PTL, and thus did not extend to multi-network environments. Migrating this into the PML provides the capability to utilize multiple networks of different types to deliver a single large message, resulting in a single level scheduler that can be easily extended or modified to investigate alternate scheduling algorithms.

2.1 PML: TEG

TEG is the first implementation of the Open MPI PML component type. The following sections provide an overview of the TEG component, focusing on aspects of the design that are unique to the Open MPI implementation.

Send Path. TEG utilizes a rendezvous protocol for sending messages longer than a per-PTL threshold. Prior experience has shown that alternative protocols such as eager send can lead to resource exhaustion in large cluster environments and complicate the matching logic when delivering multiple message fragments concurrently over separate paths. The use of a rendezvous protocol – and assigning that functionality into the PML (as opposed to the PTL) – eliminates many of these problems.

To minimize latency incurred by the rendezvous protocol, TEG maintains two lists of PTLs for each destination process: a list of low latency PTLs for sending the first fragment, and a second list of available PTLs for bulk transfer of the remaining fragments. This is done to provide both low latency for short messages as well as high bandwidth for long messages, all in a single implementation.

To initiate a send, the TEG module selects a PTL in round-robin fashion from the low-latency list and makes a down call into the PTL to allocate a send management object, or "send request descriptor". The request descriptor is obtained from the PTL such that, once the request descriptor has been fully initialize for the first time, all subsequent resources required by the PTL to initiate the send can be allocated in a single operation, preferably from a free list. This typically includes both the send request and first fragment descriptors.

Request descriptors are used by the PML to describe the operation and track the overall progression of each request. Fragment descriptors, which describe the portion of data being sent by a specific PTL, specify the offset within the message, the fragment length, and (if the PTL implements a reliable data transfer protocol) a range of sequence numbers that are allocated to the fragment on a per-byte basis. A single sequence number is allocated for a zero length message.

Once the required resources at the PTL level have been allocated for sending a message, Open MPI's data type engine is used to obtain additional information about the data type, including its overall extent, and to prepare the data for transfer. The actual processing is PTL specific, but only the data type engine knows if the data is contiguous or not, and how to split the data to minimize the number of conversions/memory copy operations.

If the overall extent of the message exceeds the PTLs first fragment threshold when the PML initiates the send operation, the PML will limit the first fragment to the threshold and pass a flag to the PTL indicating that an acknowledgment is required. The PML will then defer scheduling the remaining data until an acknowledgment is received from the peer. Since the first fragment must be buffered at the receiver if a match is not made upon receipt, the size of this fragment is a compromise between the memory requirements to buffer the fragment and the desire to hide the latency of sending the acknowledgment when a match is made. For this reason, this threshold may be adjusted at runtime on a per-PTL basis.

The data-type engine can begin processing a message at any arbitrary offset into the data description and can advance by steps of a specified length. If the data is not contiguous, additional buffers are required to do a pack operation – these will be allocated in concert with the PTL layer. Once the buffer is available, the data-type engine will do all necessary operations (conversion or just memory copies), including the computation of the optional CRC if required by the PTL layer. Finally, the PTL initiates sending the data. Note that in the case of contiguous data, the PTL may choose to send the data directly from the users buffer without copying.

Upon receipt of an acknowledgment of the first fragment, the PML schedules the remaining message fragments across the PTLs that are available for bulk transfer. The PML scheduler, operating within the constraints imposed by each PTL on minimum and maximum fragment size, assigns each PTL a fraction of the message (based on the weighting assigned to the PTL during startup) in a round-robin fashion.

As an example, assume that two processors can communicate with each other using an OS-bypass protocol over 4X InfiniBand (IB) and TCP/IP over two Gigabit-Ethernet (GigE) NICs, and that a run-time decision is made to use all of these paths. In this case, the initial fragment will be sent over the low latency IB route, and – once the acknowledgment arrives – the remaining part of the message will be scheduled over some combination of both the IB and GigE NICs.

Receive Path. Matching of received fragments to posted receives in Open MPI is event-based and occurs either on receipt of an envelope corresponding to the first fragment of a message, or as receive requests are posted by the application. On receipt of an envelope requiring a match, the PTL will attempt to match the fragment header against posted receives, potentially prior to receiving the remainder of the data associated with the fragment. If the match is made and resources are available, an acknowledgment can be sent back immediately, thereby allowing for some overlap with receiving the fragment's data. In the case where multiple threads are involved, a mutex is used to ensure that only one thread at a time will try and match a posted receive or incoming fragment.

If the receive is not matched, a receive fragment descriptor is queued in either a per-peer unexpected message or out-of-order message list, depending upon the sequence number associated with the message. As additional receives requests are posted, the PML layer checks the unexpected list for a match, and will make a down call into the PTL that received the fragment when a match is made. This allows the PTL to process the fragment and send an acknowledgment if required.

Since messages may be received out of order over multiple PTLs, an out-of-order fragment list is employed to maintain correct MPI semantics. When a match is made from the unexpected list, the out-of-order list is consulted for additional possible matches based on the next expected sequence number. To minimize the search time when wild-card receives are posted, an additional list is maintained of pointers to lists that contain pending fragments.

When acknowledgments are sent after a match is made, a pointer to the receive descriptor is passed back in the acknowledgment. This pointer is passed back to the destination in the remaining fragments, thereby eliminating the overhead of the matching logic for subsequent fragments.

Request Progression. TEG allows for either asynchronous thread-based progression of requests (used to minimize the CPU cycles used by the library), a polling mode implementation (if threading is not supported or desired), or a hybrid of these. When thread support is enabled, no down calls are made into the PTL to progress outstanding requests. Each PTL is responsible for implementing asynchronous progression. If threading is not enabled, the PML will poll each of the available PTL modules to progress outstanding requests during MPI test/wait and other MPI entry points.

As fragments complete at the PTL layer, an upcall into the PML is required to update the overall status of the request and allow the user thread to complete any pending test/wait operations depending upon completion of the request.

2.2   PTL: TCP/IP

The first implementation of an Open MPI PTL module is based on TCP/IP to provide a general network transport with wide availability. The TCP PTL is also fairly simple, as the TCP/IP stack provides for end-to-end reliability with data validation in main memory by the host CPU, and as such does not require the PTL layer to guarantee reliability itself. It also provides in-order delivery along a single connection.

Initialization. During initialization, the TCP module queries the list of kernel interfaces that are available, and creates a TCP PTL instance for each exported interface. The user may choose to restrict this to a subset of the available interfaces via optional run-time parameters. Interface discovery includes determining the bandwidth available for each interface, and exporting this from the PTL module, so that an accurate weighting can be established for each PTL.

During module initialization, a TCP listen socket is created at each process, and the set of endpoints {address:port} supported by each peer are published to all other peers via the MCA component framework, which utilizes Open MPI's out-of-band (OOB) messaging component to exchange this information, when MPI_COMM_WORLD is created.

Connection Establishment. Connections to peers are deferred until the PML attempts to utilize the PTL to deliver a fragment. When this occurs, the TCP PTL queues the fragment descriptor, and initiates a connection. If the connection completes successfully, any queued fragment descriptors are progressed as the connection becomes available. If the connection fails, an upcall into the PML indicates that the fragment could not be delivered, and should be sent via an alternate PTL. When this occurs, the PTL with the failure will be removed from the list of PTLs used to reach this destination, after several attempts to reconnect fail.

Event Based Progress. The TCP PTL utilizes non-blocking I/O for all socket connect/send/receive operations and an event based model for progressing pending operations. An event dispatching library, libevent [], is utilized to receive callbacks when registered file descriptors become available for send or receive. The library additionally provides an abstraction layer over multiple O/S facilities for I/O multiplexing, and by default enables the most efficient facility available from /dev/epoll (Linux), BSD kernel queues, Real-Time signals, poll(), or select().

When the PML/PTLs are operating in an asynchronous mode, a separate thread is created to dispatch events, which blocks on the appropriate system call until there is activity to dispatch. As the set of file descriptors of interest may change while this dispatch thread is asleep in the kernel, for some facilities (e.g. select/poll) it is necessary to wake up the dispatch thread on changes to the descriptor set. To resolve this issue without polling, a pipe is created and monitored by the event library such that the file descriptor associated w/ the pipe can be signaled where there are changes.

2.3   Results

This section presents a brief set of results, including latency (using a ping-pong test code and measuring half round-trip time of zero byte message) and single NIC bandwidth results using NetPIPE v3.6 [11]. We compare initial performance from Open MPI using the TEG module with data from FT-MPI v1.0.2, LA-MPI v1.5.1, LAM/MPI v7.0.4, and MPICH2 v0.96p2. Experiments were performed on a two processor system based on 2.0GHz Xeon processors sharing 512kB of cache and 2GB of RAM. The system utilized a 64-bit, 100MHz, PCI-X bus with two Intel Pro/1000 NICs (based on the Super P4Dp6, Intel E7500 chipset), and one Myricom PCI64C NIC running LANai 9.2 on a 66MHz PCI interface. A second PCI bus (64-bit, 133MHz PCI-X) hosted a second, identical Myricom NIC. The processors were running the Red Hat 9.0 Linux operating system based on the 2.4.20-6smp kernel.

Table 1 compares the resulting measured latencies and peak bandwidths. As the table indicates, the initial TCP/IP latency of Open MPI/TEG over Myrinet is comparable to that of the other MPI's. The event based progression, which does not consume CPU cycles while waiting for data, gives slightly better results than the polling mode, but this is within the range of measurement noise. We expect to spend some more effort on optimizing TEG's latency.

Table 1. Open MPI/TEG Single NIC latency and peak bandwidth measurements compared to other MPI implementations (non-blocking MPI semantics).

Implementation	Myrinet Latency (μs)	Peak Bandwidth (Mbps)
Open MPI/TEG (Polled)	51.5	1855.92
Open MPI/TEG (Threaded)	51.2	1853.00
LAM7	51.5	1326.13
MPICH2	51.5	1372.49
LA-MPI	51.6	1422.40
FT-MPI	51.4	1476.04
TCP	–	1857.95

The third column of the table contrasts the peak bandwidth of raw TCP, Open MPI (utilizing the TEG module), and other MPI implementations over a single NIC using Myrinet. With Open MPI/TEG, we measured the bandwidth using both single-threaded polling for message passing progress and a separate thread for asynchronous message passing progress. These yield almost identical performance profiles, peaking out just above 1800 Mb/sec.

The peak bandwidths of Open MPI/TEG are almost identical to TCP, indicating that very little overhead has been added to TCP. However, the other four implementations show noticeable performance degradation for large message sizes, with Open MPI/TEG peaking out at least 30% above them.

3   Summary

TEG is a new MPI point-to-point communications methodology implemented in Open MPI that extends prior efforts to provide a design more suitable for heterogeneous network and OS environments. TEG also addresses fault tolerance for both data and process, issues that are expected to arise in the context of petascale computing. Initial performance data shows ping-pong latencies comparable to existing high performance

MPI implementations, but with bandwidths up to 30% higher. Future work will continue to investigate the wide range of runtime capabilities enabled with this new design.

Acknowledgments

This work was supported by a grant from the Lilly Endowment, National Science Foundation grants 0116050, EIA-0202048, EIA-9972889, and ANI-0330620, and Department of Energy Contract DE-FG02-02ER25536. Los Alamos National Laboratory is operated by the University of California for the National Nuclear Security Administration of the United States Department of Energy under contract W-7405-ENG-36. Project support was provided through ASCI/PSE and the Los Alamos Computer Science Institute, and the Center for Information Technology Research (CITR) of the University of Tennessee.

References

1. G. Burns, R. Daoud, and J. Vaigl. LAM: An Open Cluster Environment for MPI. In *Proceedings of Supercomputing Symposium*, pages 379–386, 1994.
2. Graham E. Fagg, Edgar Gabriel, Zizhong Chen, Thara Angskun, George Bosilca, Antonin Bukovski, and Jack J. Dongarra. Fault tolerant communication library and applications for high perofrmance. In *Los Alamos Computer Science Institute Symposium*, Santa Fee, NM, October 27-29 2003.
3. E. Garbriel, G.E. Fagg, G. Bosilica, T. Angskun, J. J. Dongarra J.M. Squyres, V. Sahay, P. Kambadur, B. Barrett, A. Lumsdaine, R.H. Castain, D.J. Daniel, R.L. Graham, and T.S. Woodall. Open mpi: Goals, concept, and design of a next generation mpi implementation. In *Proceedings, 11th European PVM/MPI Users' Group Meeting*, 2004.
4. A. Geist, W. Gropp, S. Huss-Lederman, A. Lumsdaine, E. Lusk, W. Saphir, T. Skjellum, and M. Snir. MPI-2: Extending the Message-Passing Interface. In *Euro-Par '96 Parallel Processing*, pages 128–135. Springer Verlag, 1996.
5. R. L. Graham, S.-E. Choi, D. J. Daniel, N. N. Desai, R. G. Minnich, C. E. Rasmussen, L. D. Risinger, and M. W. Sukalksi. A network-failure-tolerant message-passing system for terascale clusters. *International Journal of Parallel Programming*, 31(4), August 2003.
6. W. Gropp, E. Lusk, N. Doss, and A. Skjellum. A high-performance, portable implementation of the MPI message passing interface standard. *Parallel Computing*, 22(6):789–828, September 1996.
7. Message Passing Interface Forum. MPI: A Message Passing Interface. In *Proc. of Supercomputing '93*, pages 878–883. IEEE Computer Society Press, November 1993.
8. Mpich2, argonne. http://www-unix.mcs.anl.gov/mpi/mpich2/.
9. S. Pakin and A. Pant. . In *Proceedings of The 8th International Symposium on High Performance Computer Architecture (HPCA-8)*, Cambridge, MA, February 2002.
10. Quadrics, llc web page. http://www.quadrics.com/.
11. Q.O. Snell, A.R. Mikler, and J.L. Gustafson. In *IASTED International Conference on Intelligent Information Management and Systems*, June 1996.
12. Sun, llc web page. http://www.sun.com/.
13. T.S. Woodall, R.L. Graham, R.H. Castain, D.J. Daniel, M.W. Sukalsi, G.E. Fagg, E. Garbriel, G. Bosilica, T. Angskun, J. J. Dongarra, J.M. Squyres, V. Sahay, P. Kambadur, B. Barrett, and A. Lumsdaine. Open mpi's teg point-to-point communications methodology : Comparison to existing implementations. In *Proceedings, 11th European PVM/MPI Users' Group Meeting*, 2004.

Efficient Execution on Long-Distance Geographically Distributed Dedicated Clusters

E. Argollo[1], J.R. de Souza[2], D. Rexachs[1], and E. Luque[1]

[1] Computer Architecture and Operating System Group
Universidad Autónoma de Barcelona. 08193 Barcelona, Spain
eduardo.argollo@aomail.uab.es,
{Dolores.Rexachs,Emilio.Luque}@uab.es
[2] Curso de Informática, Universidade Católica do Salvador, Bahia, Brasil
josemar@ucsal.br

Abstract. Joining, through Internet, geographically distributed dedicated heterogeneous clusters of workstations can be an inexpensive approach to achieve data-intensive computation. This paper describes a master/worker based system architecture and the strategies used to obtain effective collaboration in such a collection of clusters. Based on this architecture an analytical model was built to predict and tune applications' execution performance and behaviour over time This architecture and model were employed for the matrix multiplication algorithm over two heterogeneous clusters, one in Brazil and the other in Spain. Our approach results show that the model reaches 94% prediction, achieving 91% of the clusters' total performance.

1 Introduction

Joining available commodity-off-the-shelf workstations with standard software and libraries into dedicated Heterogeneous Networks of Workstations (HNOW) could be an easy and economic solution for achieving data intensive computation. With the spread of Internet use and its increasing bandwidth and reliability enhancement, the possibility of geographically interconnecting scattered groups of low-cost parallel machines has become a reality. However, it is no trivial matter to efficiently achieve the collaboration on application execution over a collection of these machines (CoHNOW) [1].

Not only is the obtaining of cluster collaboration challenged by in-cluster machines heterogeneity. We also need to face the throughput, latency and reliability differences between the intra-cluster local area network (LAN) and the inter-clusters long-distance network (LDN). These differences are magnified when Internet is used.

Our study is based on the master/worker (M/W) paradigm: thus, the framework and policies proposed and implemented can be easily applied to a wide range of applications. The M/W conception also allows dynamic re-arrangement and change of strategies depending on variations in environment conditions. Although its scalability limitations are known, this can be reduced by a hierarchical approach.

This paper describes a system architecture and presents an analytical model that helps to predict and tune applications' execution performance and behaviour over time on dedicated long-distance geographically distributed CoHNOW. Two geo-

D. Kranzlmüller et al. (Eds.): EuroPVM/MPI 2004, LNCS 3241, pp. 311–318, 2004.
© Springer-Verlag Berlin Heidelberg 2004

graphically separate HNOW compose our testbed system: one is located in Brazil and the other is located in Spain.

MPI [2] is used for intra-cluster communication. To optimize inter-cluster communication performance, guarantee transparency and hide the undesirable effects of the LDN connection (high latency and unpredictable throughput), a logic element was added to the architecture [3] [4]: the Communication Manager (CM). The use of communication elements for different levels of networks is not new and can also be seen on PACX-MPI [5]. Although, new problems are presented when facing standard long-distance Internet like the lack of reliability on the transport layer protocol (TCP) on which most of MPI solutions are based.

The mentioned communication problems lead us to the implementation of a MPI-like communication library. This library adds transport layer fault tolerance through reconnections and resending of lost data.

The accomplished analytical model allows prediction of the execution behaviour, through examination of algorithm and environment characteristics. It also helps in finding the best strategies to tune this in order to reach efficient execution performance.

The Matrix Multiplication (MM) algorithm is the benchmark application. This is a highly scalable algorithm and the amount of work can be modified with no great effort. According to Beaumont et. al.[6], the MM problem with different-speed processors turns out to be surprisingly difficult. In fact its NP-completeness was proved over heterogeneous platforms.

The following sections present our study in further detail. Section 2 describes the system architecture. The analytical model for the system and benchmark application is explained in Section 3. In Section 4, experiment preparation and results are shown, and finally, conclusions and further work are presented in Section 5.

2 System Architecture

The first step on the way to achieving collaboration between HNOWs was defining an architecture for the system. The master/worker paradigm was chosen not only because it permits framework portability to a wide range of applications but also because the existence of a central element allows dynamic control on the policies and strategies.

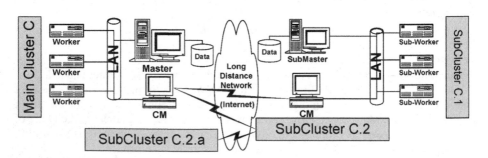

Fig. 1. System architecture as a Hierarchical Master Worker CoHNOW.

The architecture of CoHNOW can be seen as a M/W hierarchically organized collection of HNOW(Fig. 1). The cluster that contains the master, on which all data resides and from which the application execution starts, is considered the local cluster or main cluster. All other remote clusters are also called sub-clusters, their masters, sub-masters and their workers, sub-workers.

Communication Managers proved to be a necessary resource to guarantee transparency and handle properly the LDN high latency and unpredictable throughput. The employment of workstations to centralize communication between parallel machines is not new. It was also applied on the PACX project [5] that keeps two stations in each cluster for this job. Although PACX, as most MPI solutions, uses the Internet transport layer protocol., this protocol is not reliable on the LDN link between the clusters.

To solve this lack of reliability an MPI-like API library was built. This library also uses the Internet transport layer but it divides the message into smaller packets in a circular numbered buffer. When a disconnections occurs, all data lost is resent. This buffer strategy also allows the best exploit of bandwidth peaks.

CM architecture is organized around four threads (Fig. 2). The threads permit the independent and concurrent transmission and reception of data. A mutex semaphore is necessary to avoid two simultaneous MPI calls, preventing failures.

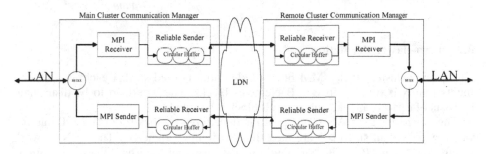

Fig. 2. Communication Managers architecture.

3   System Model

In such a heterogeneous system, to predict the execution behaviour and determine the possible levels of collaboration it is a key factor. To reach these goals an analytical model was developed. This model inputs are some of the CoHNOW and application characteristics. The model can also be used in the design and tune of the application to reach the best possible performance over the available computational resources.

The Matrix Multiplication is a simple but important Linear Algebra Kernel [7] and was selected as a benchmark application. Efficient and scalable solutions like Scala-PACK [8] already exists for this problem. Although, in our approach the MM problem is just a well defined and known application, a test for a generic framework. So, for this paper implementation, no use has been made of any compiler enhancements or pre-built functions. In order to permit the system requirement of granularity changes a block optimization technique is used. Blocked algorithms operate on sub-matrices or blocks, so that data loaded into faster levels of the memory hierarchy are reused [9].

On the next sections we explain a general and simple communication-computation ratio analysis. Then we emply this analysis to the intra-cluster and to the inter-cluster collaboration, applying the equations to the MM application.

3.1 Communication – Computation Analysis

The work distribution is essential to the communication-computation ratio analysis. Inside the cluster the master element distributes packets of data to be processed by the workers. For each result data answer, provided by the worker, the master sends a new workload. This simple approach permits the dynamic system balance. The workers idle time is avoided by a pipeline strategy: workers communication and communication threads run simultaneously.

Computation time (*CptTime*) can be defined as the ratio between operations (*Oper*) and performance (*Perf*). Communication time (*CommTime*) is the ratio between communication (*Comm*) and throughput (*TPut*). There will be no workers idle time whenever communication time is smaller or equal to computation time.. We can then conclude that the system performance will be limited as show on (1).

$$CommunicationTime \leq ComputationTime \Rightarrow Perf \leq \frac{Oper*TPut}{Comm} \tag{1}$$

3.2 Intra-cluster Prediction

At the intra-cluster level a MM blocked algorithm is used so that each M x M elements matrix is divided in B x B elements blocks. The workload to be distributed consists of a pair of blocks, and a result block is returned.

The *Comm* for the LAN (*LanComm*) is three times a block size (2). M, B and the floating point data size (α) are then the algorithms parameters. The LAN network throughput (*LanTPut*) is an environment parameter. The block operations (*BlockOper*) for the MM is known (3).

$$LanComm = \partial * BlockSize \,; \;\; BlockSize = \alpha * B^2 \,; \partial = 3; \tag{2}$$

$$BlockOper = (2B-1)*B^2 = 2B^3 - B^2 \tag{3}$$

Applying those equations to the equation 1, we can infer the local cluster performance limit (*ClusterPerfLimit*) equation (4).

$$ClusterPerfLimit \leq \frac{(2*B-1)*LanTPut}{\partial*\alpha} \tag{4}$$

To verify our model we first made experiments to determine each single workstation execution performance as a function of the block. The total addition of these values is the maximum possible reachable performance (*LocalClusterPerf*). The real Cluster Expected Performance (*ClusterExpectedPerf*) for this cluster must be the minimum value between the *LocalClusterPerf* and *ClusterPerfLimit*(5).

$$ClusterExpectedPerf(B) = min(LocalClusterPerf(B), ClusterPerfLimit) \tag{5}$$

Applying the purposed architecture to the MM application we obtained 94% prediction precision of the real *Execution Performance*, as can be seen at Fig. 3.

Fig. 3 also demonstrate that 400 x 400 block is adequate to obtain the maximum real clusters performance.

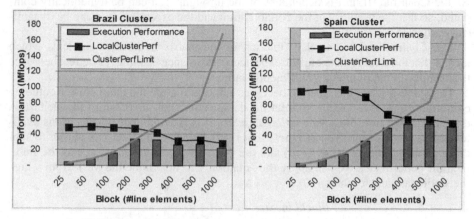

Fig. 3. Spain and Brazil cluster performance graphics per block.

3.3 Inter-cluster Prediction

The next target is to consider a cluster as a single element and analyse communication and workload distribution between clusters. The goal is to determine the amount of the remote cluster performance that can be attained to the CoHNOW.

To avoid or minimize idle time, the sub-master should receive enough data to feed all the workers. To do this, despite low LDN throughput, the data must be distributed to the sub-cluster in a way that the data locality can be exploit: old sent data should be reused when joined with that which is newly arrived.

At MM application, this can be obtained through the distribution of complete operands' sets of rows/columns. Each new row/column (r/c) pair that reaches the sub-master can be computed with the previously received columns/rows. When a new r/c P is received by the sub-master, a total of P^2*N block operations are possible. N is the amount of blocks per line (N=M/B).

The newly possible operations for the received r/c P is the subtraction of the new total for possible operations P^2 from the already available $(P-1)^2$ operations. It can be concluded that, for a new r/c P sent, $(2*P-1)*N$ new block operations are possible.

Just one block is returned for the whole row per column multiplication (*Comm = BlockSize*). This block represents the answer for equation 6 LDN operations (*LdnResOper*). Applying those values at equation 1, considering a remote LDN average throughput value (*LdnResTPut*), it is possible to infer the cluster contribution limit (ContribPerfLimit (7)).

$$LdnResOper = 2 * M * B^2 - B^2 \qquad (6)$$

$$ContribPerfLimit \leq \frac{(2*M-1)*LdnResTPut}{\alpha} \qquad (7)$$

This approach denotes a pipeline strategy. The more r/c arrives the closer the cluster contribution will be until it stabilizes on the minimum value between the cluster capacity (*ClusterExpectedPerf*) and its possible contribution (*ContribPerfLimit*). This stabilization value is the expected contribution performance (*ExpectedContribPerf*). As the ContribPerfLimit is dependent of the throughput, it can be dynamically computed, so that the usability of the collaboration and the execution time can be dynamically adjusted.

The *ContribPerfLimit* is also function of the matrix number of elements which means that the total remote cluster expected performance can always be reached once the computational problem is sufficiently large.

To establish when the pipeline becomes full, we analyse the main cluster outgoing communication. It is possible to infer the data to communicate (Comm) for each r/c as double the size of a block line (*BlockLineSize* (8)). The number of operations available for a new r/c pair (*NewRCOper*) sent is (9). Considering the LDN cluster to subcluster throughput (*LdnOperTPut*) and the *ClusterExpectedPerformance* as the stabilization performance we can apply the equation 1 to determine the r/c P from which the stabilization is reached (10).

$$BlockLineSize = \alpha * M * B \tag{8}$$

$$NewRCOper = (2 * P - 1) * (2 * M * B^2 - B^2) \tag{9}$$

$$P \geq \frac{\alpha * M * ExpectedContribPerf}{(2 * M * B - B) * LdnOperTPut} + \frac{1}{2} \tag{10}$$

It is important to observe that, although throughput between the clusters is unpredictable, its average can easily experimentally obtained. By attaining this average, the availability of the collaboration can be established and equations can be dynamically calculated through these parameters variations, so that workload and flow actions can be taken in order to maintain or achieve new levels of collaboration.

4 Experiments

In order to ascertain the accuracy of the developed model, two large tests were executed on the testbed. One was to illustrate the capability of prediction and the other to evaluate the possibilities of this approach for future design methodology through the adaptation of those parameters targeting obtention of the best possible performance on CoHNOW.

At Table 1 we can see, for both tests, the matrix parameters M and B, Brazil and Spain prediction expected performances, the expected stabilization column and time to send those columns (stabilization time), considering an average link of 10Kbytes/sec. The table still contains the predicted CoHNOW performance, the experimental obtained real CoHNOW performance and their ratio representing the precision of the model.

For the first test, a matrix of 10,000 x 10,000 and a block of 100 x 100 elements was selected. In this case both clusters will have their own performance limited by the LAN. The prediction formulations target the stabilization point at 35 sent line/col-

Table 1. Execution prediction and real execution comparison to different workloads.

M	B	Brazil Cluster Expected Perf (MFlops)	Spain Cluster Expected Perf (MFlops)	Expected Contrib Perf (MFlops)	Stabilization Column	Stabilization time (min)	Predicted CoHNOW Perf (MFlops)	Real CoHNOW Perf	Real / Predicted ratio
10.000	100	17,39	17,39	17,39	34	448,75	34,78	32,72	94%
20.000	400	28,97	58,10	58,10	29	3.007,35	87,07	79,02	91%

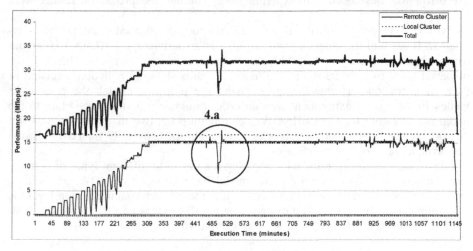

Fig. 4. Two cluster execution behaviour.

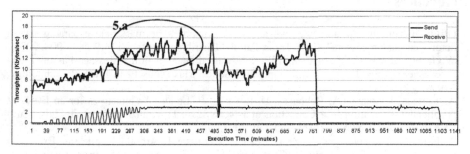

Fig. 5. Throughput on the CoHNOW execution.

umns, after 450 minutes. This test was of 19 hours' duration and its performance execution behaviour over time, for each cluster on CoHNOW, is shown in Fig. 4; experiment throughput evolution is indicated in Fig. 5.

The system was first stabilized at minute 316 due to the throughput improvement between minute 206 and 400 (5.a). A gap on the remote collaboration is seen at minute 498 as a consequence of a sudden decrease in the links throughput (4.a). The second stabilization point is reached at 512 when column 35 is sent. The predicted stabilized performance was reached with 94% precision.

The second experiment was as an attempt to approximate maximum possible CoHNOW performance; thus, the 400 x 400 block size was chosen with a 20,000 x 20,000 matrix. The experiment lasted for 89 hours and the obtained stabilized performance was 91% of the maximum CoHNOW performance.

5 Conclusion

In order to attain effective collaboration between long-distance geographically distributed CoHNOW interconnected through high latency and unpredictable throughput LDN, a system architecture was defined, a model to predict and tune the application execution was developed and experiments to validate the prediction results were made.

Two HNOW geographically distributed compose our testbed system. One is located in Brazil and the other in Spain. In the first experiment, a 10,000 x 10,000 matrix multiplication with 100x100 blocks was executed over 19 hours. This test was prepared so as to validate the model, and its results show a prediction precision of 94% of execution-performance value. Using the proposed model, the parameters values in the second experiment were in order to obtain a performance close to the maximum possible for the CoHNOW. For this purpose, two matrixes of 20,000 x 20,000 elements with 400x400 blocks were multiplied, over 89 hours. Prediction performance precision was 91%.

Future lines of research includes applying the model to different types of applications, extending the methodology and model for n-cluster CoHNOW, and adding intra-cluster workstation fault-tolerance. At the inter-cluster communication level the communication library could be extended as an MPI extension for long-distance allowing the whole CoHNOW to be seen as a single virtual machine.

References

1. Olivier Beaumont, Arnaud Legrand, and Yves Robert. "The master-slave paradigm with heterogeneous processors". IEEE Trans. Parallel Distributed Systems, 14(9):897-908, 2003.
2. W.Gropp, E.Lusk, R.Thakur,Using MPI-2:Advanced Features of the Message-Passing Interface, Scientific and Engineering Computation Series, Massachusetts Institute of Technology, 1999.
3. A. Furtado, J. Souza, A. Rebouças, D. Rexachs, E. Luque, Architectures for an Efficient Application Execution in a Collection of HNOWS. In: D. Kranzlmüller et al. (Eds.):Euro PVM/MPI 2002, LNCS 2474, pp.450-460, 2002.
4. A. Furtado, A. Rebouças, J. Souza, D. Rexachs, E. Luque, E. Argollo, Application Execution Over a CoHNOWS, International Conference on Computer Science, Software Engineering, Information Technology, e-Business, and Applications. ISBN: 0-9742059-0-7, 2003.
5. E. Gabriel, M. Resch, T. Beisel, and R. Keller. Distributed Computing in a Heterogeneous Computing Environment. In Proc. 5th European PVM/MPI Users' Group Meeting, number 1497 in LNCS, pages 180-- 187, Liverpool, UK, 1998.
6. O. Beaumont, F. Rastello and Y. Robert. Matrix Multiplication on Heterogeneous Platforms, IEEE Trans. On Parallel and Distributed Systems, vol. 12, No. 10, October 2001.
7. Dongarra J., D. Walker, Libraries for Linear Algebra, in Sabot G. W. (Ed.), High Performance Computing: Problem Solving with Parallel and Vector Architectures, Addison-Wesley Publishing Company, Inc., pp. 93-134, 1995.
8. J. Choi, J. Demmel, I. Dhillon, J. Dongarra et al., ScaLAPACK: a portable linear algebra library for distribution memory computers - design issues and performance, LAPACK Working Note 95, University of Tennessee, 1995.
9. M. S. Lam, E. Rothberg, M. E. Wolf, "The Cache Performance and Optimizations of Blocked Algorithms", Fourth Intern. Conference on Architectural Support for Programming Languages and Operating Systems, Palo Alto CA, April 1999.

Identifying Logical Homogeneous Clusters for Efficient Wide-Area Communications

Luiz Angelo Barchet-Estefanel* and Grégory Mounié

Laboratoire ID - IMAG, Project APACHE**
51, Avenue Jean Kuntzmann, F-38330 Montbonnot St. Martin, France
{Luiz-Angelo.Estefanel,Gregory.Mounie}@imag.fr

Abstract. Recently, many works focus on the implementation of collective communication operations adapted to wide area computational systems, like computational Grids or global-computing. Due to the inherently heterogeneity of such environments, most works separate "clusters" in different hierarchy levels. to better model the communication. However, in our opinion, such works do not give enough attention to the delimitation of such clusters, as they normally use the locality or the IP subnet from the machines to delimit a cluster without verifying the "homogeneity" of such clusters. In this paper, we describe a strategy to gather network information from different local-area networks and to construct "logical homogeneous clusters", better suited to the performance modelling.

1 Introduction

In recent years, many works focus on the implementation of collective communications adapted to large-scale systems, like Grids. While the initial efforts to optimise such communications just simplified the models to assume equal point to point latencies between any two processes, it becomes obvious that any tentative to model practical systems should take in account the inherently heterogeneity of such systems. This heterogeneity represents a great challenge to the prediction of communication performance, as it may come from the distribution of processors (as for example, in a cluster of SMP machines), from the distance between machines and clusters (specially in the case of a computational Grid) and even from variations in the machines performance (network cards, disks, age of the material, etc.). It is also a true concern for users that run parallel applications over their LANs, where there can be combined different machines and network supports.

As the inherent heterogeneity and the growth of computational Grids make too complex the creation of full-customised collective operations, as proposed in the past by [1,14], a solution followed by many authors is to subdivide the network in communication layers. Most systems only separate inter and intra-cluster communications, optimising communication across wide-area networks,

---

* Supported by grant BEX 1364/00-6 from CAPES - Brazil.
** This project is supported by CNRS, INPG, INRIA and UJF.

D. Kranzlmüller et al. (Eds.): EuroPVM/MPI 2004, LNCS 3241, pp. 319–326, 2004.

which are usually slower than communication inside LANs. Some examples of this "two-layered" approach include [7, 9, 11, 12], where ECO [7, 11] and MagPIe [7] apply this concept for wide-area networks, and LAM-MPI 7 [9] applies it to SMP clusters. Even though, there is no real restriction on the number of layer and, indeed, the performance of collective communications can still be improved by the use of multi-level communication layers, as observed by [4, 5].

While the separation of the network in different levels can improve the communication performance, it still needs to be well tuned to achieve optimal performance levels. To avoid too much complexity, the optimisation of two-layer communication or the composition of multiple layers relies on a good communication modelling of the network. While in this work we use pLogP [8], the main concern for the accuracy of a network model relies on the homogeneous behaviour of each cluster. If there are some nodes that behave differently from what was modelled, they will interfere with the undergoing operation. It is worth to note, however, that most of the works on network-aware collective communication seem to ignore this problem, and define clusters according to simple "locality" parameters, as for example, the IP subnet of the nodes.

While there are many network monitoring tools that could help on the identification of such heterogeneities like, for example, NWS [15], REMOS [12] or TopoMon [2], they still do not provide information about machines that hold multiple application processes, like SMP machines. Further, these tools are unable to identify heterogeneities due to the application environment, as for example, the use of an IMPI [3] server to interconnect different MPI distributions, or an SSH tunnel among different clusters protected by a firewall.

In this paper, we describe a framework to allow the gathering of independent network information from different clusters and the identification of "logical clusters". Our proposal combines the detection of "homogeneity islands" inside each cluster with the detection of SMP processors, allowing the stratification of the network view, from the node layer (specially in the case of SMP machines) to the wide-area network.

Section 2 presents our proposal for automatic topology discovery. The framework is divided in two phases. The first one, presented on Section 3, explains how connectivity data collected by different clusters can be put together. Section 4 presents the second phase, which explains how "logical clusters" are defined from the collected data, and how SMP nodes can be identified. Section 5 presents the results from a practical experiment, and some considerations on the benefits from the use of our framework. Finally, Section 6 presents our conclusions and perspective for future works.

2   What We Propose

We propose a method to automatically discover network topology in order to allow the construction of optimised multilevel collective operations. We prefer automatic topology discovery instead of a predefined topology because if there are hidden heterogeneities inside a cluster, they may interfere with the communication and induce a non negligible imprecision in the communication models.

The automatic discovery we propose should be done in two phases: the first phase collects reachability data from different networks. The second phase, executed at the application startup, identifies SMP nodes (or processes in the same machine), subdivides the networks in homogeneous clusters and acquires pLogP parameters to model collective communications.

As the first step is independent from the application, it can use information from different monitoring services, which are used to construct a distance matrix. This distance matrix does not need to be complete, in the sense that a cluster does not need to monitor its interconnection with other clusters, and several connectivity parameters can be used to classify the links and the nodes as, for example, latency and throughput.

When the network is subdivided in homogeneous subnets, we can acquire pLogP parameters, necessary to model the collective communications and to determine the best communication schedule or hierarchy. Due to the homogeneity inside each subnet, pLogP parameters can be obtained in an efficient way, which reflects in a small impact on the application initialisation time.

At the end of this process we have logical clusters of homogeneous machines and accurate interconnection parameters, that can be used to construct an interconnection tree (communicators and sub-communicators) that optimises both inter and intra-cluster communication.

3   First Phase: Gathering Network Information

While there are many works that focus on the optimisation of collective communications in Grid environments, they consider for simplicity that a cluster is defined by its locality or IP subnet, and that all machines inside a cluster behave similarly. Unfortunately, this "locality" assumption is not adequate to real systems, which may contain machines that behave differently both in performance and in communication. In fact, even in clusters with similar material, machines can behave differently (we believe that it is nearly impossible to have homogeneity in a cluster with hundreds of machines). Hence, to better optimise collective communications in a Grid environment, the choice of the topologies must be based on operational aspects that reflect the real performance level of each machine or network.

3.1   Obtaining Network Metrics

There are many tools specialised on network monitoring. These tools can obtain interconnectivity data from direct probing, like for example NWS [15], from SNMP queries to network equipments, like REMOS [12], or even combine both approaches, like TopoMon [2]. For simplicity, this work obtains data at the application level, with operations built according to NWS definition. We chose NWS as it is a *de facto* standard in the Grid community, and can be configured to provide information like communication latency, throughput, CPU load and available memory. To our interest, we can use communication latency and throughput, obtained from NWS, to identify sets of machines with similar communication parameters.

However, contrarily to some tools like TopoMon, our method does not require total interconnection among all nodes in all clusters. Indeed, the objective of the first step of our topology discovery is to identify heterogeneity inside each cluster, and by this reason, each cluster may use its own monitoring tool, without being aware of other clusters. This strategy allows the use of regular monitoring data from each cluster, while does not create useless traffic between different clusters. Hence, the data obtained from different clusters is collected and used to construct a distance matrix, which will guide the elaboration of the cluster hierarchy for our collective operations.

As clusters are not aware of each other, the "missing interconnections" clearly delimit their boundaries, which reduces the cost of the clustering process. Moreover, we are not strongly concerned with the problem of shared links, like [2] or [10], because the reduction of the number of messages exchanged among different clusters is part of the collective communication optimisation.

4 Second Phase: Application-Level Clustering

One reason for our emphasis on the construction of logical clusters is that machines may behave differently, and the easiest way to optimise collective communications is to group machines with similar performances. In the following section we describe how to separate machines in different logical clusters according to the interconnection data we obtained in the First Phase, how to identify processes that are in the same machine (SMP or not), and how this topology knowledge may be used to obtain pLogP parameters in an efficient way.

4.1 Clustering

From the interconnection data from each cluster acquired on the previous phase, we can separate the nodes in different "logical cluster". To execute this classification, we can use an algorithm similar to the Algorithm 1, presented by ECO[12].

This algorithm analyses each interconnection on the distance matrix, grouping nodes for wich their incident edges respect a latency bound (20%, by default) inside that subnet. As this algorithm does not generate a complete hierarchy, just a list of subnets, it does not impose any hierarchical structure that would "freeze" the topology, forbidding the construction of dynamic inter-cluster trees adapted to each collective communication operation and its parameters (message size, segments size, etc.).

4.2 SMP Nodes and Group Communicators

While NWS-like tools provide enough information to identify logical clusters, they cannot provide information about processes in SMP nodes, as they are created by the application. Actually, as the processes distribution depends on

Algorithm 1 ECO[12] algorithm for partitioning the network in subnets

```
initialize subnets to empty
for all nodes
  node.min_edge = minimum cost edge incident on node
sort edges by nondecreasing cost
for all edges (a,b)
  if a and b are in the same subnet
    continue
  if edge.weight>1.20 * node(a).min_edge or edge.weight>1.20 * node(b).min_edge
    continue
  if node (a) in a subnet
    if (edge.weight>1.20 * node(a).subnet_min_edge)
      continue
  if node (b) in a subnet
    if (edge.weight>1.20 * node(b).subnet_min_edge)
      continue
  merge node(a).subnet and node(b).subnet

  set subnet_min_edge to min(edge,node(a).subnet_min_edge, node(b).subnet_min_edge)
```

the application parameters (and environment initialisation), the identification of processes in SMP nodes shall be done during the application startup.

However, the implementation of an SMP-aware MPI is not easy, because the definition of MPI does not provides any procedure to map process ranks into real machine names. To exemplify this difficulty, we take as example the recent version 7 from LAM/MPI [9]. Their SMP aware collective communications, based on MagPIe [7], rely on the identification of processes that are started in the same machine, but they use proprietary structures to identify the location of each process. To avoid be dependent on a single MPI distribution, we adopted a more general solution, where each process, during its initialisation, call *gethostname()*, and sends this data to a "root" process that will centralise the analysis. If perhaps this approach is not as efficient as the one used by LAM, it still allows the identification of processes in the same machine (what can be assumed as an SMP machine).

As the data received by the root contains both the machine name and the process rank, it can translate the logical clusters into communicators and sub-communicators, adapted to the MPI environment.

5 Practical Results

5.1 Clustering

To validate our proposal, we looked for a network environment that could raise some interesting scenarios to our work. Hence, we decided to run our tests on our experimental cluster, IDPOT[1]. IDPOT can be considered as a "distributed cluster", as its nodes are all spread through our laboratory, while connected with a dedicated Gigabit Ethernet (two switches). All machines are Bi-Xeon 2.5 GHz, with Debian Linux 2.4.26, but they have network card from two different manufacturers, and the distribution of the machines in the building may play

---

[1] http://idpot.imag.fr

an important role in the interconnection distance between them and the Gigabit switches.

Applying the methods described in this paper over a group of 20 machines from IDPOT gives the following result, depicted on Fig. 1. This figure presents the resulting subnets, as well as the interconnection times between each subnet and among nodes in the same subnet. It is interesting to note how the relative latency among each cluster would affect a collective communication that was not aware of such information. For example, in the case of a two-level model, subnets C and E would affect the expected performance, as their interconnections are twice or even three times slower than others. This would also reflect in the case of a multi-layer model, where an unaware algorithm could prefer to connect directly subnets C and E, while it is more interesting to forward communications through subnet D.

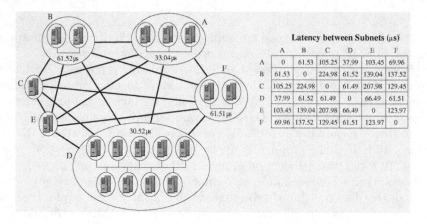

Fig. 1. IDPOT network partition, with latency among nodes and subnets

We identified as the main factor for such differences the presence of network cards from one manufacturer on subnets A and D, while subnets B, C, E and F have onboard cards from other manufacturer. As second factor, we can list the location of the nodes. While it played a less important role, the location was the main cause for separation between subnet A and subnet D. Actually, the distance between those machines, which are under different switches, affected the latency just enough to force ECO's algorithm to separate them in two different subnets. A correct tuning on the parameters from ECO's algorithm may allow subnets A and D to be merged in a single one, a more interesting configuration for collective communications.

5.2   Efficient Acquisition of pLogP Parameters

While the logical clusters generated by our framework allow a better understanding of the network effective structure, we are still unable to model communications with precision. This first reason is that interconnection data may be incomplete. As said in Section 3.1, the monitoring tools act locally to each LAN, and by this reason, they do not provide data from the inter-cluster connections.

Besides this, the data acquired by the monitoring tools is not the same as the data used in our models. For example, the latency, which originally should have the same meaning to the monitoring tool and the application, is obtained differently by NWS and pLogP. In NWS, the latency is obtained directly from the round-trip time, while pLogP separates the round-trip time in latency and gap, as depicted by Figure 2, with differences that may interfere on the communication model. In addition, the information collected by the monitoring tools is not obtained by the application itself, and thus, is not submitted to the same constraints that the application will find at runtime, as for example, the use of an Interoperable MPI (IMPI) server to interconnect the clusters.

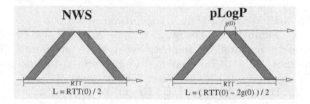

Fig. 2. Differences between NWS and pLogP "latency"

Hence, to model the communication in our network, we need to obtain parameters specifically for pLogP. Hopefully, there is no need to execute $n(n-1)$ pLogP measures, one for each possible interconnection. The first reason is that processes belonging to the same machine were already identified as SMP processes and grouped in specific sub-communicators. And second, the subnets are relatively homogeneous, and thus, we can get pLogP parameters in an efficient way by considering a single measure inside each subnet as a sample from the pLogP parameters common to the entire cluster. As one single measure may represents the entire subnet, the total number of pLogP measures is fairly reduced. If we sum up the measures to obtain the parameters for the inter-clusters connections, we shall execute at most $C(C+1)$ experiments, where C means the number of subnets. Further, if we suppose symmetrical links, we can reduce this number of measures by half, as $a \rightarrow b = b \rightarrow a$. By consequence, the acquisition of pLogP parameters for our experimental 20-machines cluster would need at most $(6 * (6 + 1))/2 = 21$ measures.

6   Conclusions

This paper proposes a simple and efficient strategy to identify communication homogeneities inside computational clusters. The presence of communication heterogeneities reduces the accuracy from the communication models used to optimise collective communications in wide-area networks. We propose a low cost method that gathers connectivity information from independent clusters and groups nodes with similar characteristics. Using real experiments on one of our clusters, we show that even minor differences may have a direct impact on the communication performance. Our framework allowed us to identify such differences, classifying nodes accordingly to their effective performance. Using such classification, we can ensure a better accuracy for the communication models, allowing the improvement of collective communication performances, specially those structured on multiple layers.

References

1. Bhat, P., Raharendra, C., Prasanna, V.: Efficient Collective Communication in Distributed Heterogeneous Systems. Journal of Parallel and Distributed Computing, No. 63, Elsevier Science. (2003) 251-263
2. Burger, M., Kielmann, T., Bal, H.: TopoMon: a monitoring tool for Grid network topology. Intl. Conference on Computational Science'02. Springer-Verlag, LNCS Vol. 2330 (2002) pp. 558-567.
3. Interoperable MPI Web page. *http://impi.nist.gov*
4. Karonis, N. T., Supinski, B., Foster, I., Gropp, W., Lusk, E., Bresnahan, J.: Exploiting Hierarchy in Parallel Computer Networks to Optimize Collective Operation Performance. In: 14th International Conference on Parallel and Distributed Processing Symposium. IEEE Computer Society (2000) 377-384.
5. Karonis, N. T., Foster, I., Supinski, B., Gropp, W., Lusk, E., Lacour, S.: A Multilevel Approach to Topology-Aware Collective Operations in Computational Grids. Technical report ANL/MCS-P948-0402, Mathematics and Computer Science Division, Argonne National Laboratory (2002).
6. Karonis, N. T., Toonen, B., Foster, I.: MPICH-G2: A Grid-enabled implementation of the Message Passing Interface. Journal of Parallel and Distributed Computing, No. 63, Elsevier Science. (2003) 551-563
7. Kielmann, T., Hofman, R., Bal, H., Plaat, A., Bhoedjang, R.: MagPIe: MPI's Collective Communication Operations for Clustered Wide Area Systems. In: 7th ACM SIGPLAN Symposium on Principles and Practice of Parallel Programming, ACM Press. (1999) 131-140
8. Kielman, T., Bal, E., Gorlatch, S., Verstoep, K, Hofman, R.: Network Performance-aware Collective Communication for Clustered Wide Area Systems. Parallel Computing, Vol. 27, No. 11, Elsevier Science. (2001) 1431-1456
9. LAM-MPI Team, LAM/MPI Version 7, http://www.lam-mpi.org/ (2004)
10. Legrand, A., Quinson, M.: Automatic deployment of the Network Weather Service using the Effective Network View. In: High-Performance Grid Computing Workshop (associated to IPDPS'04), IEEE Computer Society (2004)
11. Lowekamp, B., Beguelin, A.: ECO: Efficient Collective Operations for communication on heterogeneous networks. In: 10th International Parallel Processing Symposium. (1996) 399-405
12. Lowekamp, B.: Discovery and Application of Network Information. PhD Thesis, Carnegie Mellon University. (2000)
13. Thakur, R., Gropp, W.: Improving the Performance of Collective Operations in MPICH. In: Euro PVM/MPI 2003, Springer-Verlag, LNCS Vol. 2840 (2003) 257-267.
14. Vadhiyar, S., Fagg, G., Dongarra, J.: Automatically Tuned Collective Communications. In: Supercomputing 2000, Dallas TX. IEEE Computer Society (2000)
15. Wolski, R., Spring, N., Peterson, C.: Implementing a Performance Forecasting System for Metacomputing: The Network Weather Service. In: Supercomputing 1997 (1997)

Coscheduling and Multiprogramming Level in a Non-dedicated Cluster*

Mauricio Hanzich[2], Francesc Giné[1], Porfidio Hernández[2],
Francesc Solsona[1], and Emilio Luque[2]

[1] Departamento de Informática e Ingeniería Industrial
Universitat de Lleida, Spain
{sisco,francesc}@eps.udl.es
[2] Departamento de Informática
Universitat Autònoma de Barcelona, Spain
{porfidio.hernandez,emilio.luque}@uab.es, mauricio@aows10.uab.es

Abstract. Our interest is oriented towards keeping both local and parallel jobs together in a time-sharing non-dedicated cluster. In such systems,
dynamic coscheduling techniques, without memory restriction, that consider the MultiProgramming Level for parallel applications (MPL), is
a main goal in current cluster research. In this paper, a new technique
called Cooperating Coscheduling (CCS), that combines a dynamic coscheduling system and a resource balancing schema, is applied.
The main aim of CCS is to allow the efficient execution of parallel tasks
from the system and parallel user points of view without disturbing the
local jobs. Its feasibility is shown experimentally in a PVM-Linux cluster.

1 Introduction

The studies in [4] indicate that the workstations in a cluster are underloaded.
Time-slicing scheduling techniques exploit these idle cycles by running both parallel and local jobs together. Coscheduling problems arise in this kind of environment. Coscheduling ensures that no parallel process will wait for a non-scheduled
process for synchronization /communication and will minimize the waiting time
at the synchronization points.

One such form of coscheduling is explicit coscheduling [10]. This technique
schedules and de-schedules all tasks in a job together using global context
switches. The centralized nature of explicit coscheduling limits its efficient development in a cluster. Alternatively, this coscheduling can be achieved if every node
identifies the need for coscheduling by gathering and analyzing implicit runtime
information, basically communication events, leading to *dynamic coscheduling*
techniques [2].

Over the past few years, the rapid improvement in the computational capabilities of cluster systems, together with the high rate of idle resources, has driven

* This work was supported by the MCyT under contract TIC 2001-2592 and partially
supported by the Generalitat de Catalunya -Grup de Recerca Consolidat 2001SGR-
00218.

D. Kranzlmüller et al. (Eds.): EuroPVM/MPI 2004, LNCS 3241, pp. 327–336, 2004.

researchers to increase the MultiProgramming Level of parallel jobs (MPL). Thus, the complexity of the coscheduling problem is increased due to: a) the CPU and memory resource assignation throughout the cluster may be balanced; b) the performance of local jobs must be guaranteed; c) the MPL should be adapted according to the CPU and memory requirements.

Previous work on scheduling has focused on coordinating the allocation of CPU and memory, using ad-hoc methods for generating schedules. In [1], memory and CPU requirements were used as a lower bound for constraining the number of CPUs assigned to a job. This approach, used for space slicing by some researchers, could be used to solve some of the three problems presented.

Our contribution is to provide new algorithms for solving the coscheduling problem without hard restrictions. Specifically, our algorithm, called Cooperating CoScheduling (CCS), extends the dynamic coscheduling approach, including the reservation of a percentage of CPU and memory resources in order to assure the progress of parallel jobs without disturbing the local users. Besides, our CCS algorithm uses status information from the cooperating nodes to re-balance the resources throughout the cluster when necessary. The main aim of this article is to show that, without disturbing the local user, it is possible to increase the MPL of the parallel tasks and still obtain positive speedups (i.e.: > 1) from the system and parallel user points of view.

The remainder of this paper is outlined as follows: in section 2 the related work is explained. In section 3, the CCS algorithm is presented. The efficiency measurements of CCS are performed in Section 3. Finally, the main conclusions and future work are explained in Section 4.

2   Related Work

Dynamic coscheduling techniques in non-dedicated NOWs [9, 2] have shown that they can perform well when many local users compete with a single parallel job. Unfortunately, the single analysis of communication events is not enough to guarantee the performance of local and parallel jobs when the MPL is increased. Basically, this is due to: a) the priority assignation throughout the cluster to the same job may be variable [5]; b) when the competing parallel jobs have similar communication behavior it is difficult to achieve coscheduling [7]; c) the level of paging due to the parallel MPL > 1 may slow the entire system down [3]. These facts have driven researchers to limit the MPL to one in non-dedicated clusters. As a consequence, the efficiency of using idle resources is very low.

In [6], a new coscheduling technique, named Cooperating CoScheduling (CCS), was presented. Unlike *traditional dynamic techniques*, under CCS, each node takes its scheduling decisions from the occurrence of local events, basically communication, memory and CPU, together with events received from remote events. This way, CCS preserves the performance of local jobs while guaranteeing the performance of local users. Thus, CCS allows the limits of the MPL of parallel jobs in non-dedicated clusters to be explored. With this aim, these limits are evaluated in this paper according to different parallel and local workloads.

3   CCS: Cooperating CoScheduling

This section explains our proposal for the coscheduling of non-dedicated clusters. The next subsection gives some of the notations and assumptions considered.

3.1   Assumptions and Notation

Given that our aim is the implementation of CCS in the Linux operating system, some of the assumptions are made taking its properties into account.

The scheduler is assumed to be time-sharing with process preemption based on ranking processes according to their priority. The CPU time is divided into *epochs* where each process (*task*) is assigned a specified quantum, which it is allowed to run. When the running process has expired, its quantum is either blocked waiting for an event (for example a communication event or a page fault), or another process is selected to run from the Ready Queue (RQ). The epoch ends when all the processes in the RQ have exhausted their quantum.

We assume that each node maintains the following information for each task:

- $task.q$: a *static quantum* which is set by default to $DEF\_QUANTUM$[1]. This is the priority of the process and the quantum assigned at each epoch.
- $task.rss$: resident set size of *task*.
- $task.uid$:user identification of *task*. For simplicity, we assume that all the parallel jobs belong to the same user, which is identified as $PARAL$.
- $task.jid$: job identification of *task*. We assume that all the tasks belonging to the same parallel job *(cooperating tasks)*, share it.
- $task.nr\_locals$: counts the number of remote $LOCAL$ and NO_LOCAL events received by the *task*. Section 3.2 explains how this is increased/decreased.

In addition, each node maintains the following information about its state:

- $node_k.M$: main memory size of $node_k$.
- $node_k.mem$: the total used memory into $node_k$.
- $node_k.mem\_par$: the sum of memory used by parallel tasks in $node_k$.
- $node_k.LOCAL\_USER$: set to TRUE when there is a local user in $node_k$.
- $node_k.cooperating(task)$: Each node manages a list, denoted as *cooperating*, which contains the addresses of all the nodes where the cooperating tasks of *task* are being executed. This is provided by the PVM daemon.

3.2   CCS Algorithm

The three main aims of CCS can be summarized in the following points:

- **The performance of local jobs** must be preserved by CCS so that local users allow the execution of the parallel tasks in their own machines.

---

[1] This is *210ms* in the Linux Kernel v.2.2.

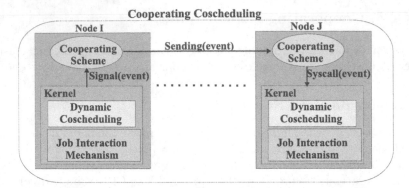

Fig. 1. CCS architecture.

– The *coscheduling* **of communicating parallel jobs** has to be achieved in order to minimize the communication waiting time of the parallel tasks. For doing this, a dynamic coscheduling strategy is applied by CCS.
– The *uniform allocation of CPU and memory resources* **assigned to each parallel job throughout the cluster.** Due to an imbalance produced by a dynamic local user activity, a distributed algorithm is applied in order to improve this situation.

Figure 1 shows the CCS architecture and the modules that achieves each of the objectives set out above. How each module works is detailed below.

Dynamic Coscheduling. Coscheduling strategies enable parallel applications to dynamically share the machines in a NOW with interactive, CPU and IO-bound local jobs. To do so and in the same way as some previous studies [1], we have used *dynamic coscheduling with immediate blocking* [2]. This means that a receiving process is then immediately blocked and awoken by the o.s. when the message eventually arrives. The priority of the receiving process is boosted according to the number of packets in the receive socket queue and the currently running process is preempted if necessary. In this way, coscheduling between cooperating tasks is achieved.

Job Interaction Mechanism. In order to assure the progress of parallel jobs without excessively disturbing local users, we propose to apply a social contract. This means that a percentage of CPU and memory resources (L) is guaranteed for parallel tasks. The best value for L is examined in subsection 4.1.

The main memory (M), is divided into two pools: one for the parallel tasks ($M^D = M * L$) and one for the local tasks ($M^L = M - M^D$). If the parallel tasks require more memory than M^D then they will be able to use the M^L portion whenever local tasks are not using it. However, when swapping is activated and the memory used by the parallel tasks exceeds $M^D (mem\_par > M * L)$, the one with the most mapped pages in memory will be stopped (*stoptask*). Thus, the

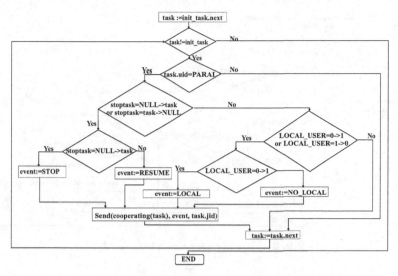

Fig. 2. Algorithm for sending events.

resident size of the remaining tasks will be increased and the swapping activity reduced [3]. Therefore, the MPL of parallel jobs is adapted dynamically to the memory requirements of the local and parallel users.

At the end of each epoch, the scheduler reassigns the quantum of all active tasks ($task.q$). When there is some local user activity, the time slice will be $DEF\_QUANTUM * L$ for every parallel task, or 0 if the task is stopped. In all the other cases, the quantum will be $DEF\_QUANTUM$.

Cooperating Scheme. In order to balance resources throughout the cluster, the cooperating nodes should interchange status information. However the excess of information could slow down the performance of the system and may limit its scalability. For this reason, only events that provoke a modification of the resources assigned to parallel processes are sent. These are the following:

1. $LOCAL$: whenever the local user activity begins in a node. It means that the $LOCAL\_USER$ variable changes from 0 to 1 ($LOCAL\_USER=0->1$).
2. NO_LOCAL: whenever the local activity finishes ($LOCAL\_USER=1->0$).
3. $STOP$: whenever one parallel task has been stopped as a consequence of the job interaction mechanism. This means that the $stoptask$ variable makes the transition from NULL to $task$ ($stoptask=NULL->task$).
4. $RESUME$: whenever the $stoptask$ restarts execution ($stoptask=task->$ NULL).

At the beginning of each epoch, CCS checks the value of the $LOCAL\_USER$ and $stoptask$ variable according to the algorithm shown in fig. 2, which is used for sending the notifications. Note that while a notification of $LOCAL$ (NO_LOCAL)

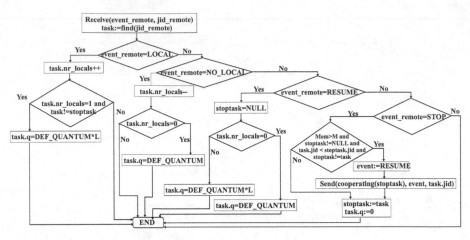

Fig. 3. Algorithm for receiving events.

events is sent by each active parallel task, the notification of the *STOP (RE-SUME)* events is only sent to the nodes which have cooperating tasks of the stopped (resumed) task.

When a cooperating node receives one of the above events, it will reassign the resources according to the algorithm in fig. 3. In function of the received event (*event_remote*), the algorithm performs as follows:

1. LOCAL: The scheduler will assign a quantum equal to $DEF\_QUANTUM *$ L to the parallel task notified by the event, whenever it is not *stopped*.
2. NO_LOCAL: The scheduler will assign $DEF\_QUANTUM$ to the *task* whenever the number of *LOCAL* received events coincides with the *NO_LOCAL* received events (task.nr_locals=0). In this way, *CCS* ensures that the parallel job is not slowed down by any local user in the cluster.
3. RESUME: CCS will assign $DEF\_QUANTUM$ or $DEF\_QUANTUM * L$, according to the local user activity in other nodes for *task*.
4. STOP: If a node with overloaded memory receives the *STOP* event for a task different from s*toptask*, CCS will not stop the notified task if $task.jid >$ $stoptask.jid$. In any other case, *task* will be stopped *and stoptask* resumed.

4   Experimentation

The performance of CCS was evaluated in a cluster composed of eight Pentium III with 256MB of memory each, interconnected through a fast Ethernet network.

The evaluation was carried out by running several workloads composed of a set of NAS PVM applications merged in such a way that it was possible to characterize the system, bounding it by computation and communication parameters (see fig. 4.left). Each workload was created exercising each class several times, choosing 1 to 4 instances of parallel applications in a round-robin manner (e.g.: Workload i: "class A - MPL 4" = SP.A, BT.A, SP.A, BT.A. Workload j:

"class D - MPL 2" = IS.B, FT.A). Each of the workloads was run in *route-direct* PVM mode with an eight task (node) size for every job. The cluster execution of these workloads was done in two ways: in *parallel* (i.e.: all the applications in the workload were executed concurrently, MPL > 1), or *sequentially* (MPL=1). For both modes of executing, we measured the *speedup* earnings of the workloads (*Wrk*) and applications (*Appl*) according to the following equation:

$$Speedup(Wrk/Appl) = \frac{Texec_{node}(Wrk/Appl)}{Texec_{cluster}(Wrk/Appl)},$$

where $Texec_{cluster}(Wrk/Appl)$ and $Texec_{node}(Wrk/Appl)$ are the execution time of the workload (also called makespan), or a single application (appl) obtained in the cluster and on a single node, respectively.

In order to establish a non-dedicated cluster, we used a synthetic workload which models the local user activity (CPU load = 0.15, Memory load = 85MB and Network traffic = 3KB/s) in four of the eight nodes. These parameters were settled by monitoring the resources used by the users in our laboratory. At the end of its execution, the *local* benchmark returns the system call latency (responsiveness) and wall-clock execution time.

Our experimentation is divided into two sections. First, the best percentage of resources (*L*) that could be given to parallel tasks is evaluated. Next, the performance of parallel benchmarks is studied in relation to the MPL using the *L* value found. With this we want to show that, without disturbing the local user at all, it is possible to obtain considerable profit from executing the parallel workloads in a multiprogrammed manner.

Class	Benchmarks	% Comp.	% Comm.	Mem (MB)
A	SP.A	88	12	23
	BT.A	93	7	7
B	MG.B	82	18	63
	LU.A	84	16	25
C	SP.A	88	12	23
	CG.A	40	60	55
D	IS.B	58	42	117
	FT.A	63	37	61

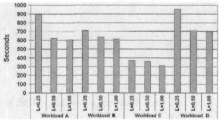

Fig. 4. Class definitions (left). Exec. time for different *L* values (right).

4.1 Effect of the *L* Factor

In this subsection, an analysis of the effect of the resource distribution (*L*), on the performance of the parallel and local workloads is detailed. This is evaluated by running the workloads for a MPL=4 and three values for *L*: 0.25, 0.5 and 1.

Fig. 4.right shows that the earnings of using *L*=0,5 are greater than those when using *L*=0,25. This improvement is due to the following:

1. The time slice is not big enough to exploit the locality of the parallel application data into the cache[5] with a L=0,25.
2. With L=0,5, a more relaxed quantum improves the coscheduling probability.
3. In the cases where paging is activated (classes B and D), the likelihood of stopping one parallel task increases when the value of L decreases.

On the other hand, the impact of the increment of L to a value of 1 does not produce a significant benefit with respect to the parallel workload at the time that the impact on the local workload is notable, as can be seen from fig. 5.

Fig. 5 shows the system call response time and bounded slowdown for the local workload running jointly with the A (computation) and D (communication) classes. The system call response time measures the overhead introduced into the interactivity perception while the bounded slowdown reflects the overhead introduced into the computation phase of the local user.

From fig. 5.left it is possible to see that the response time for L=0,5 never exceeds significantly the 400ms stated [8] as the acceptable limit in disturbing the local user responsiveness. Note as well that the chosen limit of MPL=4 is due to the fact that with an L value of 0,5, in the worst situation, the local task response time will be $DEF\_QUANTUM * L * MPL = 400ms$ for a $DEF\_QUANTUM = 200$ ms (the default value in a Linux environment).

This figure also shows that L=1 means an intrusion that should not be tolerated by the local user, especially when the *response time* > 800ms. Note that the bounded slowdown increases with the value of L but always below the MPL.

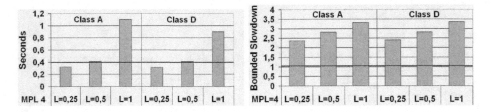

Fig. 5. System call response time (left) and Bounded Slowdown (right) for A and D class workloads.

In conclusion, L=0,5 is the best compromise choice given that the parallel workloads do not benefit as much as the local ones lose for L=1, while for L=0,25 the parallel workload decreases its performance without giving any significant benefit to the local user.

4.2   The Effect of MPL on the Parallel Workload Performance

This subsection evaluates the speedup of the parallel workloads for the *sequential* and *parallel* mode for several values of MPL (1 to 4).

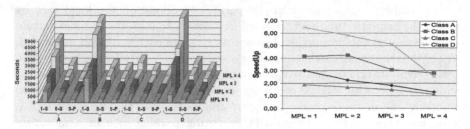

Fig. 6. Workloads exec. times (left). Average NAS app. speedup for each class (right).

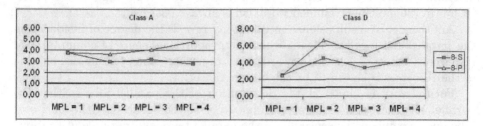

Fig. 7. Parallel (8-P) and Sequential (8-S) workloads speedup.

Fig. 6.left shows the workload execution times for each of the classes in three different sceneries that can be called: 1-S (monoprocessor), 8-S (*sequential* case) and 8-P (*parallel* case).

As can be seen from the figure, in almost all the cases the parallel version of the workload (i.e. 8-P) has an improved execution time with regard to 8-S and this is better than 1-S. This shows the feasibility of increasing the MPL in a non-dedicated NOW even when the local users are totally preserved.

The speedup curves for the *parallel*, and *sequential* workloads can be observed in fig. 7. As the figures show, the 8-P workload has an enhanced speedup for every exercised case. It is important to emphasize that these earnings rely in part on the CCS capacity to diminish the paging and also augment the coscheduling.

Although useful, the previous figure say nothing about the "parallel user" perception of the convenience of the system. Therefore the fig. 6.right shows how the parallel MPL affects the performance of each parallel workload job.

It can be seen from this figure that the speedup for the parallel tasks is positive (i.e. Speedup > 1) in every case and therefore useful for the parallel user.

It is worth pointing out that the workload speedup profit is almost always greater than the reduction in the speedup of each parallel application instance. Therefore, each parallel application loses less than the whole workload gains.

5   Conclusions and Future Work

This paper studies how the parallel MPL affects a PVM workload, as well as the kind of earnings that can be expected from the system (workload speedup) and user point of view (job speedup), in a non-dedicated Linux NOW. Also, the study embraces the impact of the parallel workload on the local user by means of the response time and bounded slowdown. To keep them under certain acceptable values, the proportional resources given to each workload are varied.

All of the trials were carried out using a wide range of parallel workloads from intensive computation to intensive communication in a coscheduling environment provided by CCS. Our scheme combines: balancing of resources, job interaction mechanisms and dynamic coscheduling techniques to allow the use of MPL > 1.

Our results show that the gain in executing parallel jobs in a multiprogrammed manner in a non-dedicated environment is not only acceptable but desirable based on the achieved speedup. Moreover, the speedup for each parallel job in the multiprogrammed workload is always positive (i.e.: > 1).

Another set of results shown in this study reveals the degree of incidence on the parallel and local workloads when the proportional share of computational resources (L factor) is varied towards one or another. The even-handed sharing of the resources is shown to be the better choice.

Future work is aimed at the distribution of the parallel jobs so that the parallel MPL across the cluster need not to be the same, keeping the local user jobs and maximizing the speedup of the parallel tasks and resource utilization.

References

1. C. Anglano. "A Comparative Evaluation of Implicit Coscheduling Strategies for Networks of Workstations". *9th HPDC*, 2000.
2. P.G. Sobalvarro, S. Pakin, W.E. Weihl and A.A. Chien. "Dynamic Coscheduling on Workstation Clusters". *JSSPP'98, LNCS*, vol. 1459, pp. 231-256, 1998.
3. F. Giné, F. Solsona, P. Hernández and E. Luque. "Dealing with memory constraints in a non-dedicated Linux CLuster". *International Journal of High Performance Computing Applications, vol.17*, pp. 39-48, 2003.
4. R.H. Arpaci, A.C. Dusseau, A.M. Vahdat, L.T. Liu, T.E. Anderson and D.A. Patterson. "The Interaction of Parallel and Sequential Workloads on a Network of Workstations". *ACM SIGMETRICS'95*, pp.267-277, 1995.
5. F. Giné, F. Solsona, P. Hernández and E. Luque. "Adjusting time slices to apply coscheduling techniques in a non-dedicated NOW". *Euro-Par'2002, LNCS*, vol. 2400, pp. 234-239, 2002.
6. F. Giné, F. Solsona, P. Hernández and E. Luque. "Cooperating Coscheduling in a non-dedicated cluster". *Euro-Par'2003*, LNCS, vol. 2790, pp.212-218, 2003.
7. F. Giné, M. Hanzich, F. Solsona, P. Hernández and E. Luque. "Multiprogramming Level of PVM Jobs in a on-dedicated Linux NOW". *EuroPVM/MPI'2003*, LNCS, vol. 2840, pp.577-586, 2003.
8. R. Miller. "Response Time in Man-Computer Conversational Transactions". In AFIPS Fall Joint Computer Conference Proceedings, Vol. 33, pp. 267-277, 1968.
9. C. Mc Cann and J. Zahorjan. Scheduling memory constrained jobs on distributed memory computers. *ACM SIGMETRICS'95*, pp. 208-219, 1996.
10. D. Feitelson and L. Rudolph. "Gang Scheduling Performance Benefits for Fine-grain Synchronization". *J. Parallel and Distributed Computing*, vol. 164, pp. 306-318, 1992.

Heterogeneous Parallel Computing
Across Multidomain Clusters

Peter Hwang, Dawid Kurzyniec, and Vaidy Sunderam

Emory University, Atlanta, GA 30322, USA
{vss,dawidk}@mathcs.emory.edu

Abstract. We propose lightweight middleware solutions that facilitate
and simplify the execution of MPI programs across multidomain clusters.
The system described in this paper leverages H2O, a distributed metacomputing framework, to route MPI message passing across heterogeneous aggregates located in different administrative or network domains.
MPI programs instantiate a specially written H2O pluglet; messages that
are destined for remote sites are intercepted and transparently forwarded
to their final destinations. The software was written and tested in a simulated environment, with a focus on clusters behind firewalls. Qualitatively
it was demonstrated that the proposed technique is indeed effective in
enabling communication across firewalls by MPI programs. In addition,
tests showed only a small drop in performance, acceptable considering
the substantial added functionality of sharing new resources across different administrative domains.

1   Introduction

Cooperative resource sharing typically takes place across multiple networks that
are geographically and administratively separate, and heterogeneous in terms
of architecture, software, and capacities. In such "multidomain" systems, some
parts of such a collection may also be behind firewalls, or on a non-routable
network, leading to difficulties in access to and from a counterpart entity in
another portion of the system. This situation is exacerbated in large scale parallel programs, such as those based on the message-passing paradigm, e.g. using
MPI. In such programming models, individual processes may send and receive
messages to and from any other process, irrespective of its physical location. It
is easy to see that this model does not translate very effectively to multidomain
systems in which communication from one domain to another is not straightforward. Moreover, other issues relating to security, permissions to start processes
or stage executables at remote locations, and dynamic monitoring also arise and
must be resolved.

In the context of the H2O project, we are devising lightweight schemes to
address the above. In this paper, we present a prototype that focuses on cross-
firewall operation; the full system will provide comprehensive multidomain support. Our approach involves the instantiation of customizable agents at selected
locations; by leveraging the security and reconfigurability features of H2O, such

D. Kranzlmüller et al. (Eds.): EuroPVM/MPI 2004, LNCS 3241, pp. 337–344, 2004.

"pluglets" serve as proxies that relay messages between individual domains as appropriate, transparently performing address and other translations that may be necessary. We describe the overall architecture and design of this system, and provide details of using this scheme to enable MPI programs to operate across firewalls, including some preliminary performance results.

2   Background and Related Work

Several other projects with similar goals have adopted comparable approaches and are outlined in this section. The design presented in this paper leverages the component oriented and provider-centric architecture of the H2O framework to accomplish a "proxy agent" based solution.

2.1   The H2O Metasystem Framework

Among general purpose software infrastructures considered most promising today are "grid" [11, 6] and "metacomputing" frameworks. These infrastructures primarily aim to aggregate or cross-access varied resources, often across different administrative domains, networks, and institutions [5]. Some software toolkits implementing a virtual organization that have been developed and deployed include Globus [5] and Globe. Our ongoing work (upon which the research described in this paper is based) has pursued alternative approaches to metasystem middleware; its mainstay is the H2O substrate [10] for lightweight and flexible cooperative computing. In the H2O system, a software backplane architecture supporting component-based services hosts pluggable components that provide composable services. Such components may be uploaded by resource providers but also by clients or third-parties who have appropriate permissions. Resource owners retain complete and fine-grained control over sharing policies; yet, authorized clients have much flexibility in configuring and securely using compute, data, and application services. By utilizing a model in which service providers are central and independent entities, global distributed state is significantly reduced at the lower levels of the system, thereby resulting in increased failure resilience and dynamic adaptability. Further, since providers themselves can be clients, a distributed computing environment can be created that operates in true peer-to-peer mode, but effectively subsumes the farm computing, private (metacomputing) virtual machine, and grid paradigms for resource sharing. H2O assumes that each individual resource may be represented by a software component that provides services through well defined remote interfaces. Providers supply a runtime environment in the form of a component container. The containers proposed in H2O are similar to those in other environments (e.g. J2EE), but are more flexible and capable of of hosting dynamic components that are supplied and deployed by (authorized) external entities, thereby facilitating (re)configuration of services according to client needs. A comprehensive depiction of the H2O architecture is shown in Figure 1. Components, called pluglets, follow a standardized paradigm and implement composable functionality; details may be found in the H2O programming guide and other papers [8].

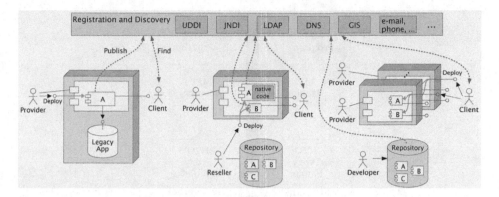

Fig. 1. H2O component model illustrating pluglets deployable by clients, providers, or third-party resellers, and different modes of operation, including client-side aggregation.

2.2   Grid-Enabled MPI

The popularity of MPI and of grid software toolkits has motivated a number of efforts to marry the two. One of the earliest attempts was the MPICH-G project [4] that successfully demonstrated the deployment of MPI programs on multiple clusters interconnected by TCP/IP links. Based on the MPICH implementation of MPI, this project leveraged the Globus toolkit to spawn MPI jobs across different clusters. More recently, a newer version called MPICH-G2 [9] has evolved to be compatible with MPICH2. These approaches mainly concentrated on resource allocation and process spawning issues, although heterogeneity and multinetwork systems were catered for. Another project that complements this approach by focusing on the message passing substrate is PACX-MPI [1], that augments the MPI library with multidomain facilities. A more basic approach is adopted by Madeleine-III [2] whose goal is to provide a true multi-protocol implementation of MPI on top of a generic and multi-protocol communication layer called Madeleine. StaMPI [12] is yet another endeavor along similar lines, and is now being extended to include MPI-IO. These efforts have been quite successful and have addressed a variety of issues concerning multidomain systems. Our ongoing project seeks to build upon these experiences and (1) comprehensively support machine, interconnection network, and operating system heterogeneity; non-routable and non-IP networks; operation across firewalls; and failure resilience by supporting FT-MPI [3], and (2) leverage the component architecture of H2O to provide value added features like dynamic staging, updating of proxy modules, and selective, streamlined functionality as appropriate to the given situation. We describe our prototype focusing on firewalls in this paper.

3   The H2O Proxy Pluglet

The H2O proxy pluglet essentially serves as a forwarding and demultiplexing engine at the edge of each resource in a multidomain system. In this section, we

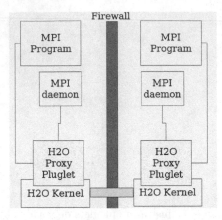

Fig. 2. Simplified H2O MPI Firewall Architecture.

describe the design and implementation of a version of this pluglet that handles MPICH2 communication across firewalls. A simplified operational diagram is shown in Figure 2.

From an operational viewpoint, H2O proxy pluglets are loaded onto H2O kernels by kernel owners, end-users, or other authorized third-parties, by exploiting the hot-deployment feature of H2O. A startup program is provided in the distribution to help load H2O proxy pluglets into the H2O kernels. The proxy pluglets leverage H2O communication facilities to forward messages across clusters. Due to the fact that H2O uses well-known port numbers and is capable of tunnelling communication via HTTP, it is possible to configure firewalls appropriately to allow the H2O forwarding. The startup program takes two arguments: a filename of kernel references corresponding to the H2O kernels where the proxy pluglets will be loaded, and a codebase URL where the proxy pluglet binaries may be found. The startup program (1) reads the kernel references from the argument file; (2) logs in to each of the H2O kernels referenced in the file; (3) loads the proxy pluglets onto the kernels; (4) obtains a handle to each proxy pluglet; and (5) distributes handle data to each proxy pluglet, as shown in the sequence diagram of Figure 3.

Our prototype implementation extends the generic MPICH2 implementation. Using a modified connect method within the daemons, the system reroutes communication between different hosts hosts through the H2O proxy pluglet. MPI programs are run using modified libraries (that will eventually include fault tolerant functionality using the FTMPI [3] library to connect to the H2O proxy pluglet instead of directly to other MPI programs. Steps in the interaction between MPI and the H2O proxy pluglet are listed below, and are shown diagrammatically in Figure 4. The proxy pluglet enables outgoing connections using a locally published port (steps 1, 2, 3); spawns a connection handler thread to service the connection request (4); establishes a mapping to the destination pluglet (5,6,7); connects to the destination user process (8); and sets up direct channels via helper threads (9,10,11).

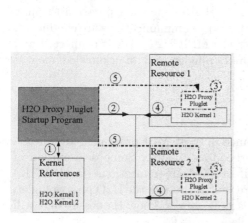

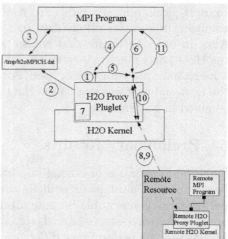

Fig. 3. Startup event sequence.   **Fig. 4.** Message forwarding actions.

4   MPICH2 Library Enhancement

In order to attain greatest possible portability and applicability, this project is
built upon one of the most widespread machine-independent implementations
of MPI, viz. MPICH2. At the same time, it is important to isolate the changes
needed to the MPICH2 library so that evolutionary changes may be made easily. The core modifications to the library involve the re-routing of all messages
through the H2O proxy pluglet rather than directly to the remote MPI process
– an action that would not work across firewalls and non-routable networks.

The goal of minimizing the number and extent of changes was achieved by
locating the lowest level socket connections, which were modified to connect to
the H2O proxy pluglet instead of their previous destination. Only two socket
functions were modified to reroute all MPI program communications through
the H2O Proxy: *sock_post_connect(host, port)* – which asynchronously connects
to a host and port, and *sock_handle_connect()* – which handles newly formed
connections. After MPI programs were set up to successfully communicate over
H2O channels, tests have shown that MPI was still unable to completely communicate through firewalls. This was due to the fact that MPICH2 daemons had
unresolved connections that were blocked by the firewall. To solve this problem
the MPICH2 daemon Python code was also modified to reroute its communication in an analogous manner by modifying the lowest level socket function
mpd_get_inet_socket_and_connect(host, port).

A few other issues are worthy of mention. First, due to the use of lazy connections, overheads are confined to only those communication channels that are
required, at the relatively low expense of first-time delay. The second issue involves connecting to the proxy pluglet. Since rerouting all communications to the
H2O proxy pluglet requires changing the host and port parameters of the MPI

connect() call, a locally published connection endpoint scheme was used, again resulting in isolating all changes to *the sock_post_connect(host, port)* function. Also, in order to ensure that connection to the remote MPI program through the H2O proxy pluglets is established before any communications are conducted, a handshake protocol was devised. After the MPICH2 socket is connected and the host and port are sent to the H2O proxy pluglet, and an acknowledgment is received in confirmation.

5   Preliminary Results

Since performance is of crucial importance in most MPI programs, any proxy-based solution must incur as little overhead as possible. Some degradation of performance is inevitable because all direct TCP/IP connections (in MPICH2) now become three-stage channels that are routed through the H2O proxy pluglet. Measuring this performance difference is necessary to know the losses that will be incurred when access through firewalls is desired.

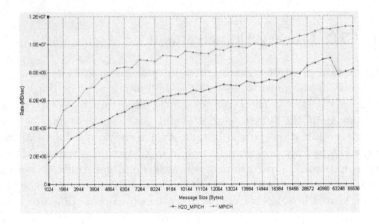

Fig. 5. Results of *mpptest* from MPICH2 package.

The MPICH2 release is supplied with a suite of performance tests to provide reproducible and reliable benchmarks for any MPI implementation. *Mpptest* provides both point-to-point and collective operations tests; and *goptest* is used to study the scalability of collective routines as a function of number of processors [7]. For space reasons, only representative tests on MPICH2 and MPICH2 over H2O on Intel PCs running Linux on 100MBit Ethernet are included in the paper. Mpptest results are shown in Figure 5. MPICH throughput plateaus at 11.2 MB/sec (near the theoretical bandwidth limit) and MPICH over H2O plateaus at 8.8 MB/sec showing a 26.8% decrease. The anomalous drop in performance around message sizes of 50 KB for MPICH over H2O is sometimes seen in parallel computing performance evaluations, and is likely due to a benchmark dependent maximum buffer size.

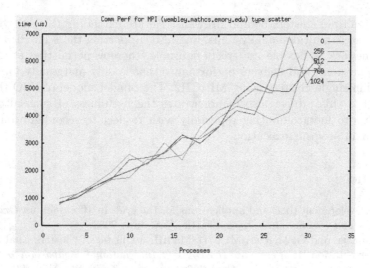

Fig. 6. Results of scatter *goptest* from MPICH2 suite.

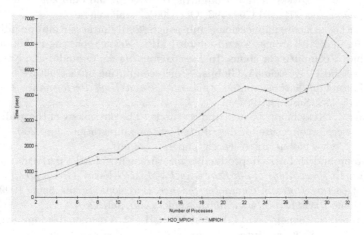

Fig. 7. Scalability comparison between MPICH2 and MPICH2-over-H2O.

Goptest shows the scalability of calling collective routines such as scatter and broadcast with respect to the number of processes. MPICH2 and MPICH2 over H2O scalability with different message sizes are shown in Figure 6. A scalability comparison between MPICH2 and MPICH2 over H2O of the 256-byte message size is shown in Figure 7. Very little difference between MPICH2 and MPICH2 over H2O exists because the added communication is outweighed by the synchronization time of collective routines.

6 Discussion and Future Work

In this paper, we have described a prototype system that supports the operation of the standard MPI model over multidomain systems, exemplified by aggregates of multiple clusters behind firewalls. We are extending this subsystem to comprehensive multidomain systems, and also supporting the FTMPI interface. The framework is flexible and reconfigurable, thereby permitting straightforward deployment, and delivers performance that is only marginally lower than that within single clusters using MPICH2. The pluglet model of H2O that was leveraged in this project also demonstrates the usefulness of such alternative approaches to metacomputing especially with respect to reconfigurability and multidomain program execution.

References

1. MPI development tools and applications for the grid. In *Workshop on Grid Applications and Programming Tools*, Seattle, WA, June 2003.
2. O. Aumage and G. Mercier. MPICH/MadIII: a Cluster of Clusters Enabled MPI Implementation. In *Proc. 3rd IEEE/ACM International Symposium on Cluster Computing and the Grid (CCGrid 2003)*, pages 26–35, Tokyo, May 2003. IEEE.
3. G. Fagg, A. Bukovsky, and J. Dongarra. HARNESS and fault tolerant MPI. *Parallel Computing*, 27(11):1479–1496, Oct. 2001. Available at http://icl.cs.utk.edu/publications/pub-papers/2001/harness-ftmpi-pc.pdf.
4. I. Foster and N. Karonis. A grid-enabled MPI: Message passing in heterogeneous distributed computing systems. In *Supercomputing 98*, Orlando, FL, Nov. 1998.
5. I. Foster and C. Kesselman. Globus: A metacomputing infrastructure toolkit. *The Intl Journal of Supercomputer Applications and High Performance Computing*, 11(2):115–128, Summer 1997.
6. I. Foster, C. Kesselman, J. Nick, and S. Tuecke. The physiology of the grid: An open grid services architecture for distributed systems integration, Jan. 2002. Available at http://www.globus.org/research/papers/ogsa.pdf.
7. W. Gropp and E. Lusk. Reproducible measurements of MPI performance characteristics. In *Proceedings of 6th European PVM/MPI Users' Group Meeting*, volume 1697 of *Lecture Notes in Computer Science*, Barcelona, Spain, Sept. 1999.
8. H2O Home Page. http://www.mathcs.emory.edu/dcl/h2o/.
9. N. Karonis, B. Toonen, , and I. Foster. MPICH-G2: A grid-enabled implementation of the Message Passing Interface. *Journal of Parallel and Distributed Computing (JPDC)*, 63(5):551–563, May 2003. Available at ftp://ftp.cs.niu.edu/pub/karonis/papers/JPDC_G2/JPDC_G2.ps.gz.
10. D. Kurzyniec, T. Wrzosek, D. Drzewiecki, and V. Sunderam. Towards self-organizing distributed computing frameworks: The H2O approach. *Parallel Processing Letters*, 13(2):273–290, 2003.
11. Z. Nemeth and V. Sunderam. A comparison of conventional distributed computing environments and computational grids. In *International Conference on Computational Science (ICCS)*, Amsterdam, Apr. 2002. Available at http://www.mathcs.emory.edu/harness/pub/general/zsolt1.ps.gz.
12. Y. Tsujita, T. Imamura, H. Takemiya, and N. Yamagishi. Stampi-I/O: A flexible parallel-I/O library for heterogeneous computing environment. In *Recent Advances in Parallel Virtual Machine and Message Passing Interface*, volume 2474 of *Lecture Notes in Computer Science*. Springer-Verlag, 2002. Available at http://link.springer.de/link/service/series/0558/bibs/2474/24740288.htm.

Performance Evaluation and Monitoring of Interactive Grid Applications*

Bartosz Baliś[1], Marian Bubak[1,2], Włodzimierz Funika[1], Roland Wismüller[3],
Marcin Radecki[2], Tomasz Szepieniec[2], Tomasz Arodź[2], and Marcin Kurdziel[2]

[1] Institute of Computer Science, AGH, al. Mickiewicza 30, 30-059 Kraków, Poland
Phone: (+48 12) 617 39 64, Fax: (+48 12) 633 80 54
{balis,bubak,funika}@uci.agh.edu.pl
[2] Academic Computer Centre – CYFRONET, Nawojki 11, 30-950 Kraków, Poland
{M.Radecki,T.Szepieniec,T.Arodz,M.Kurdziel}@cyf-kr.edu.pl
[3] LRR-TUM – Technische Universität München, D-85747 Garching, Germany
Phone: (+49 89) 289 17676
wismuell@in.tum.de

Abstract. This paper presents how the OCM-G monitoring system together with the G-PM performance evaluation tool is used for observing and improving the performance of interactive grid applications. The OCM-G is an on-line grid-enabled monitoring system while the G-PM is an advanced graphical tool which enables to evaluate and present the results of performance monitoring, to support optimization of the application execution. The G-PM communicates with the OCM-G via OMIS. We show how how the application developer can analyze the performance of a real application using G-PM visualization.

1 Introduction

The focus of the EU CrossGrid project [5] is on interactive applications, and a part of the software created for this purpose is a monitoring environment for applications which is composed of the OCM-G monitoring system and the G-PM performance analysis tool. Our experience has a background [4] going back to 1995 when the specification of OMIS – a standardized interface for interaction between tools and monitoring systems [12] – was defined. In the following years the first monitoring system implementing OMIS – the OCM – was developed, supporting cluster environments and message passing PVM and MPI applications, and, in addition, several tools were implemented on top of OMIS/OCM. The development of the OCM-G, a monitoring system for grid applications, and the G-PM, a grid-enabled performance analysis tool, is a direct continuation of the previous work [2,3,1].

To supply reliable and meaningful performance data on interactive applications, a monitoring system should meet the following requirements. The monitoring system should provide transparency to the location of the objects monitored

---

* This work was partly funded by the EU project IST-2001-32243, CrossGrid.

D. Kranzlmüller et al. (Eds.): EuroPVM/MPI 2004, LNCS 3241, pp. 345–352, 2004.

and the concurrency of performance analysis tools, as monitoring data comes from different locations, the monitoring system should provide a synchronization of event time-stamps. To provide on-line measurements, the performance tool in the interactive applications needs to get monitoring data on-line. Analysis of interactive applications should enable in-depth investigation, so the monitoring system should enable user-defined metrics, monitoring system should provide sufficiently fast data acquisition to allow for fine-grained measurements. The tool must be enabled to program the monitoring system to perform actions for a particular tool activity.

There are several efforts to enable monitoring of grid applications [7] such as Autopilot [16] in the GRaDS project [8], GRM/R-GMA [14, 15] in the DataGrid project [6], and GRM/Mercury [11, 13] in the Gridlab project [9].

In the following secions, we describe the architecture of both the OCM-G and G-PM, and then we show how to find performance flaws. The results presented were obtained with the Davef application [10] which simulates the flow of a river.

2   OCM-G – Grid Application Monitoring System

The architecture of the OCM-G (OMIS-Compliant Monitoring system for the Grid) is shown in Fig. 1. An instance of the OCM-G is running per each user. The system is composed of per-host Local Monitors (LMs) and per-site Service Managers (SMs). There is also one additional SM, called Main Service Manager (MainSM). Application Module (AM) is part of the OCM-G that is linked against application processes (AP).

The system is started-up in the following way. The user manually starts the MainSM on a user interface (UI) machine; a connection string is returned. Then, the user submits the application in the usual way passing two additional command-line parameters: the connections string obtained in step 1 and an arbitrarily chosen application name. The application processes are started on some worker nodes (WN) and attempt to register in Local Monitors; if the LMs are not running, they are forked off first. Next, one of the LMs starts up the SM on local computing element (CE), unless one is already running. All LMs register to their respective SMs passing the connection string to the MainSM, and, finally SMs connect to the MainSM using the connection string. All connections are established in the "bottom-up" manner.

All OCM-G components and tools use Globus GSI security model and the user certificates for authentication. If the MainSM is running and the application has been submitted, the tool (e.g. G-PM) can be started.

3   G-PM – Grid-Oriented Performance Evaluation Tool

The G-PM tool (Grid-oriented Performance Measurement tool) (see Fig. 1) consists of the User Interface and Visualization Component (UIVC), Performance Measurement Component (PMC) allowing for measurements based on standard

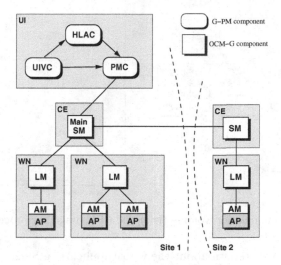

Fig. 1. Architecture of the G-PM and OCM-G

metrics, like "data volume", "time spent in" etc., and High Level Analysis Component (HLAC) which provides a powerful facility for the user to define own metrics and realize measurements, meaningful in the context of the application investigated. The tool cooperates with the OCM-G via a well defined interface deriving from OMIS.

The tool supports a wide set of built-in metrics. These metrics fall into two categories, i.e. *function-based* and *sampled* metrics. The function-based metrics are associated with instrumented versions of specific functions. They are suitable for monitoring the behavior of applications with respect to various libraries they use. The sampled metrics, on the other hand, are not related to any functions or events. They enable monitoring of such quantities as CPU load on the node or memory used by the process. The G-PM can be customized by the user or application developer by means of a dedicated Performance Measurement Specification Language (PMSL) [17]. The language allows for processing and combining the existing metrics into a form more suitable for a particular evaluation purpose. Furthermore, in PMSL a new category of metrics is available i.e. *probe-based* metrics. These metrics are associated with a call to a function artificially inserted into the application's code by the developer i.e. the *probe*.

4   Performance Evaluation

From the programmer's point of view probably the most useful information are those about CPU usage and communication volume. Both should be given on a *per process* basis, otherwise they are of little use since the level of details is too low to evaluate whether the application works in the way it has been designed. In the G-PM the user can choose between showing details regarding particular

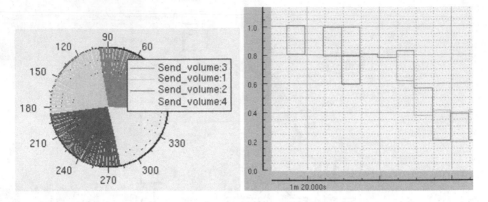

Fig. 2. Communication volume (left) and CPU usage per process (right)

process and aggregated data about the whole application. Such an approach has the advantage that instrumentation (i.e. monitoring data collection) is activated only for processes which are actually involved in a performance evaluation scope.

Fig. 2 depicts the communication volume of four worker-processes along with CPU usage over a period of time. The master process is not included since it does different work. The worker processes send comparable amounts of data: they implement the same algorithm and synchronize after each loop interaction. The diagram on the left illustrate a per-process communication volume realized within MPI_Send() calls since the start of the computation. To measure the communication volume, the G-PM instructs the OCM-G to create a counter which keeps the relevant value. All MPI library functions are instrumented, so each call to MPI_Send() function performed within a specified process first reports an event to the OCM-G and then calls the original MPI_Send. The event report contains some event-specific information, which in our case is the amount of data being sent in the related call. The event triggers some actions, namely, the counter is increased by the message size. Depending on update interval the counter value is read and left unmodified by the G-PM. The picture on the right (Fig. 2) illustrates a per process CPU usage.

The data displayed on a diagram is gathered during the program execution. Once the user has defined a display for the measurement, the G-PM starts to send requests to the OCM-G. In case of the CPU usage it sends a request for information about a particular process to the SM. Then the SM distributes the request to the appropriate LMs and the latter obtains the needed information from the /proc filesystem. Next, the monitoring information passes the data back to the G-PM. In the right-side graph (Fig. 2) we can observe that for some reason the CPU usage for each process decreases.

5   Finding the Reasons of Performance Flaw

To determine the source of performance problems we measured communication delays. The time spent in waiting in MPI_Send() and MPI_Recv() is shown in

Figure 3 on the left and right, accordingly. The value of delay expresses how much of its time the process spends in MPI_Send() on a percentage basis. In the left figure we can see that process 2 has increased the time spent in MPI_Send() from 1% to about 3% while the right picture indicates that process 3 significantly (to value of 40%) increased the receive time. The increase means here that the execution conditions got worse. Having these results in mind we can deduce that the load on the execution machine of process 2 has increased. So, both computation and communication done by process 2 was performed slower and as a result process 3 waited for the completion of computation and for the results from the partner process. The other processes encounter no changes in communication latencies. Knowing that the application synchronizes the processes after each loop iteration and that there is no load balancing implemented, we can explain why the whole application suffers if one process is not well performing. The performance of all processes is limited by the slowest one.

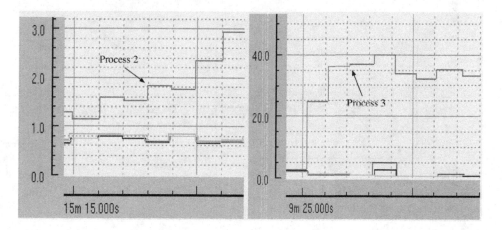

Fig. 3. Delay due to MPI_Send (left) and MPI_Recv (right) per process

6   Performance Analysis with User Metrics and Probes

G-PM enables the measurement and visualization of data more closely related to the application's structure and execution behavior. G-PM supports this via user-defined metrics, i.e. metrics (measurable quantities), which can be defined by the user *at runtime* by providing a specification how to compute the value of this metrics from a set of other metrics. However, the definition of a user-defined metrics can also take into account the occurrences of certain events in the application program. These events usually mark interesting points in the application's execution, like transitions between execution phases or iterations of a program's main loop. Since these points are application-specific, the programmer has to mark them in the application's source code by inserting *probes*, which are just calls to (empty) functions. A probe must receive an integer parameter, the *virtual time* that indicates which events (e.g. on different processes) belong together. In addition, other application-specific data can be passed to a probe. Once probes have been inserted into an application, they can be monitored by the OCM-G in the same way as it monitors routines in programming libraries, like e.g. MPI.

In the DaveF application we inserted a probe at the end of the time loop. The probe receives the simulated time as an additional parameter. User defined metrics in G-PM also allow more sophisticated uses. For example, it is possible to determine the number of communications for each time step by defining a proper metrics that computes the difference of the total number of communications at the end of one loop iteration and the number at the end of the previous iteration. In the same way, other metrics, like the amount of data sent, can be computed (and displayed) for each loop iteration. The result of such measurements is shown in Fig. 4. It shows that the DaveF application needs only very few communications.

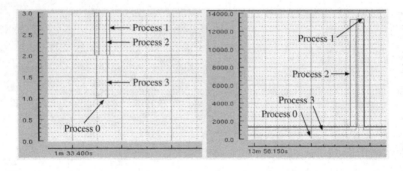

Fig. 4. Measurements of the number of messages sent (left) and message size (right) in each iteration of the DaveF time loop

Metrics like these are translated to the following activities by G-PM: First, counters are defined in the OCM-G to measure the base metrics (i.e. total amount of messages send, total size of these messages). Then, the OCM-G is instructed to read the proper counter when the probe event is detected. On the one hand, this value is now buffered in a FIFO, on the other hand, the previous value in this FIFO is subtracted from the current one, giving the result for the current loop iteration. This result is again buffered in a FIFO, which is read whenever G-PM updates its display.

The user defined metrics are also useful for assessing performance. We are using two metrics: one of them returns the percentage of time spent waiting for messages in each iteration of the time loop, whle the second metrics computes the maximum over all measured processes, and then computes the mean value of all loop iterations executed in the measured time interval. The result for DaveF is the lower curve in Fig. 5, which shows that the application performs nearly perfect: In the mean, no process waits more than 2% of its time for messages. By a simple variation of the second metrics, namely computing the *maximum* value

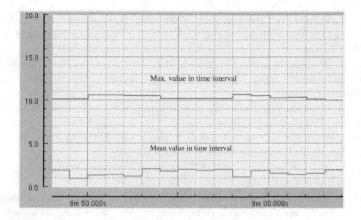

Fig. 5. Communication delay in the DaveF time loop

in the measured time interval, we can see that there are some loop iterations
with a high waiting time, since the value now is much higher (upper curve in
Fig. 5): In the worst case, a process spends up to 10% of the iteration's time
waiting for messages. This is caused by background load on the compute cluster
and the network.

7 Summary

The OCM-G acts as a middleware exposing monitoring services which can be
used by the G-PM or other tools to perform various activities on the monitored
application. The G-PM tool is developed for performance evaluation of grid applications, based on the monitoring functionality of the OCM-G. Being targeted
at interactive grid applications, G-PM provides performance evaluation based on
the basic performance metrics applicable to grid applications. The user defined
metrics of G-PM provide a great flexibility during performance analysis, since
new, adequate metrics can be defined based on the results of the measurements
performed so far, e.g. for checking hypotheses, or for getting more detailed data.
In combination with probes, user defined metrics allow to link performance data
to computational phases of an application, thus supporting a more in-depth
analysis. The metrics are specified at high level, an efficient implementation of
measurements is generated automatically [17]. To our knowledge, G-PM is the
first *on-line* performance analysis tool that provides this feature.

References

1. B. Baliś, M. Bubak, W. Funika, T. Szepieniec, R. Wismüller, and M. Radecki.
   Monitoring Grid Applications with Grid-enabled OMIS Monitor. In F. Fernández
   River, M. Bubak, A. Gómez, R. Doallo: Proc. First European Across Grids Conference, Santiago de Compostela, Spain, February 2003, Springer LNCS 2970, pp.
   230-239.

2. Bubak, M., Funika, W., Balis, B., and Wismüller, R. Concept For Grid Application Monitoring. In: Proceedings of the PPAM 2001 Conference, Spinger LNCS 2328, pp. 307-314, Naleczow, Poland, September 2001. Springer LNCS.
3. Bubak, M., Funika, W., and Wismüller, R.: The CrossGrid Performance Analysis Tool for Interactive Grid Applications. Proc. EuroPVM/MPI 2002, Linz, Sept. 2002, Springer LNCS.
4. Bubak, M., Funika W., Baliś B., and Wismüller R.. On-line OCM-based Tool Support for Parallel Applications. In: Y. Ch. Kwong, editor, *Annual Review of Scalable Computing*, volume 3, chapter 2, pages 32–62. World Scientific Publishing Co. and Singapore University Press, Singapore, 2001.
5. The CrossGrid Project (IST-2001-32243): http://www.eu-crossgrid.org
6. The DataGrid Project: http://www.eu-datagrid.org
7. Gerndt, M. (Ed.): Performance Tools for the Grid: State of the Art and Future. APART White Paper. Research Report Series. Vol. 30. LRR, Technische Universität München. Shaker Verlag, 2004
8. The GrADS Project: http://hipersoft.cs.rice.edu/grads
9. The GridLab Project: http://www.gridlab.org
10. Hluchy L., Tran V.D., Froehlich D., Castaings W.: Methods and Experiences for Parallelizing flood Models. In: Recent Advances in Parallel Virtual Machine and Message Passing Interface, 10th European PVM/MPI Users' Group Meeting 2003, LNCS 2840, Springer-Verlag, 2003, pp. 677-680, ISSN 0302-9743, ISBN 3-540-20149-1. September/October 2003, Venice, Italy.
11. Kacsuk, P.: Parallel Program Development and Execution in the Grid. Proc. PAR-ELEC 2002, International Conference on Parallel Computing in Electrical Engineering. pp. 131-138, Warsaw, 2002.
12. Ludwig, T., Wismüller, R., Sunderam, V., and Bode, A.: OMIS – On-line Monitoring Interface Specification (Version 2.0). Shaker Verlag, Aachen, vol. 9, LRR-TUM Research Report Series, 1997.
http://wwwbode.in.tum.de/~omis/
13. N. Podhorszki, Z. Balaton, G. Gombas: Monitoring Message-Passing Parallel Applications in the Grid with GRM and Mercury Monitor. In: Proc. 2nd European Across Grids Conference, Nicosia, CY, 28-30 Jan. 2004. To appear in Lecture Notes in Computer Science, Springer Verlag.
14. N. Podhorszki and P. Kacsuk Monitoring Message Passing Applications in the Grid with GRM and R-GMA Proceedings of EuroPVM/MPI'2003, Venice, Italy, 2003. Springer LNCS, 2003.
15. R-GMA: A Grid Information and Monitoring System.
http://www.gridpp.ac.uk/abstracts/AllHands_RGMA.pdf
16. Vetter, J.S., and Reed, D.A.: Real-time Monitoring, Adaptive Control and Interactive Steering of Computational Grids. The International Journal of High Performance Computing Applications, 14 357-366, 2000.
17. Wismüller, R., Bubak, M., Funika, W., Arodź, T., and Kurdziel, M. Support for User-Defined Metrics in the Online Performance Analysis Tool G-PM. In: Proc. 2nd European Across Grids Conference, Nicosia, CY, 28-30 Jan. 2004. To appear in Lecture Notes in Computer Science, Springer Verlag.

A Domain Decomposition Strategy for GRID Environments

Beatriz Otero, José M. Cela, Rosa M. Badia, and Jesús Labarta

Departament d'Arquitectura de Computadors
Universitat Politécnica de Catalunya
Campus Nord, Jordi Girona, 1-3, Módulo D6
08034, Barcelona, Spain
{botero,cela,rosab,jesus}@ac.upc.es

Abstract. In this paper, we evaluate the performance of domain decomposition applications with message-passing in GRID environments. We compare two domain decomposition strategies, the balanced and the unbalanced one. The balanced strategy is the normal strategy used in homogenous computing environment. This strategy present some problems related with the heterogeneous communications in a grid environment. We propose an unbalanced domain decomposition strategy in order to overlap the slow communications with computation. The influence of the communication patterns on execution times is analysed. DIMEMAS simulator is used to predict the performance of distributed applications in GRID environments.

1 Introduction

Previous works make reference to the relationship between architecture and domain decomposition algorithms [13]. There are studies on the latency, bandwidth and optimum workload to take full advantage of the available resources [11] [12]. There are also references in the literature to MPI application behaviour in GRID environments [2] [3]. In all these cases, the workload is the same for all the processors.

We consider distributed applications that carry out matrix-vector product operation. These applications solve problems that arise from the discretization of partial differential equations on meshes using explicit solvers, i. e. sheet stamping or car crash simulations. The objective of this study is to improve the execution times of the distributed applications in GRID environments by overlapping remote communications and computation. We propose a new distribution pattern for the data in which the workload is different depending on the processor. We used DIMEMAS tool [1] to simulate the behaviour of the distributed applications in GRID environments.

This work is organised as follows. Section 2 describes the tool used to simulate and define the GRID environment. Section 3 addresses the workload assignment patterns. Section 4 shows the results obtained for box meshes. The conclusions of the work are presented in Section 5.

D. Kranzlmüller et al. (Eds.): EuroPVM/MPI 2004, LNCS 3241, pp. 353–361, 2004.
© Springer-Verlag Berlin Heidelberg 2004

2 Grid Environment

We use a performance predictor simulator called DIMEMAS. DIMEMAS is a tool developed by CEPBA[1] for simulating parallel environments [1] [2] [3]. DIMEMAS requires as input parameters a trace file and a configuration file. In order to obtain the trace file, it is necessary execute the parallel application with an instrumented version of MPI [4]. The configuration file contains details of the simulated architecture such as number of nodes, latency and bandwidth between nodes, etc. DIMEMAS generates an output file that contains the execution times of the application for the parameters specified in the configuration. Moreover, it is possible to obtain a graphical representation of the parallel execution.

We model a GRID environment as a set of full connected hosts. We define a host as a set of highly coupled processors. The hosts are connected through dedicated connections. Each host has a direct full-duplex connection with any other host.

DIMEMAS simulator considers a simple model for point to point communications [2] [3]. This model decomposes the communication time in five components. Figure 1 shows these components.

1. Latency time ($T_{Latency}$) is a fix time to start the communication.
2. Resource contention time ($T_{Resource}$) is dependent of the global load in the local host.
3. The transfer time (T_{Send}) is dependent of the message size. We model this time with a bandwidth parameter.
4. The WAN contention time (T_{WAN}) is dependent of the global traffic in the WAN.
5. The flight time (T_{Flight}) is the time to transfer the message in a WAN. It depends on the distance between hosts.

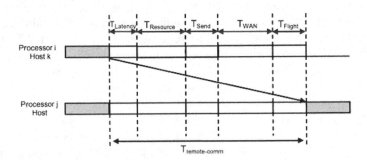

Fig. 1. Point to point communication model

We consider an ideal environment where $T_{Resource}$ is negligible and T_{WAN} depends on a constant traffic model in the WAN. Then, we model the communications with just three parameters: latency, bandwidth and flight time. These parameters are set

---

[1] European Center for Parallelism of Barcelona.

according to what is commonly found in present networks [16]. Other works [17] [18] [19] use different values for these parameters, which have been also considered in this work. Table 1 shows the values of these parameters for the host internal and external communications.

Table 1. Latency, bandwidth and flight time values

Parameters	Internal	External
Latency	20 µs	From 1 ms to 100 ms
Bandwidth	100 MB/s	10MB/s
Flight-time	-	From 200 µs to 100 ms

Because the size of the messages are small (some Kbytes) the time spent in the communication is mainly due to the latency and the flight time. Then, we take constant values for the bandwidth.

Figure 2 shows the general topology of the host connections. We consider two parameters: the number of hosts (n) and the number of processors per host (m). We analyse two cases, $n>>m$ and $n<<m$.

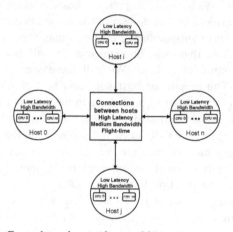

Fig. 2. General topology: **n** hosts with **m** processors per host

3 Data Distribution

Our target problems arise from the discretization of partial differential equations, when explicit methods are used. These algorithms are parallelized using domain decomposition data distribution. One parallel process is associated with one domain.

A matrix-vector product operation is carried out in each iteration of the solver. The matrix-vector product is performed using a domain decomposition algorithm, i.e. as a set of independent computations and a final set of communications. The communications in a domain decomposition algorithm are associated with the boundaries of the domains. Each process must exchange the boundary values with all its neighbours. Then each process has as many communication exchanges as neighbour domains [9]

[10]. For each communication exchange the size of the message is the length of the common boundary between the two domains.

We use METIS to perform the domain decomposition of the initial mesh. The algorithms implemented in METIS are based on multi-level graph decomposition schemes and are described in [5] [6] [7] [8].

In this paper we consider two domain decomposition strategies: the balanced and the unbalanced distributions.

Balanced Distribution. We make as many domains as processors in the grid. The computational load is perfectly balanced between domains. This is the usual strategy of a domain decomposition algorithm. This balanced strategy is suitable in homogeneous parallel computing environments, where all the communications have the same cost.

Unbalanced Distribution. The main idea is to create some domains with a negligible computational load. Those domains are devoted only to manage the slow communications. To perform this we make the domain decomposition in two phases. Firstly, a balanced domain decomposition with the number of hosts is done. This guarantees that the computational load is balanced between hosts. Secondly, an unbalanced domain decomposition inside a host is done. For the second decomposition we split the boundary nodes of the host subgraph. We create as many special domains as remote communications are. Note that these domains contain only boundary nodes, so they have a negligible computational load. We call these special domains *B-domains* (boundary domains). The remainder host subgraph is decomposed in *nproc-b* domains, where *nproc* is the number of processors in the host and *b* stands for the number of *B-domains*. We call these domains *C-domains* (computational domains). As a first approximation we assign 1 CPU to each domain. Then the CPUs assigned to *B-domains* remain inactive most of the time. We take this policy in order to obtain the worst case for our decomposition algorithm. Clearly this inefficiency could be solved assigning all the *B-domains* in a host to the same CPU.

For example, consider a mesh of 16x16 nodes and a grid with 4 hosts and 8 processors per host. Figure 3 shows the balanced domain decomposition and figure 4 shows the unbalanced decomposition. We must remark that the communication pattern of the balance and the unbalanced domain decomposition are different, because the number of neighbours of each domain is different.

Note that the number of external communications in the unbalanced decomposition is reducing to just one communication with any neighbour host. However, in the balanced decomposition there are several external communications with the same neighbour host.

Figure 5 illustrates the communication pattern of the unbalanced/balanced distribution of this example. The arrows in the diagram determine the processors that interchange data. The origin of the arrow identifies the sender. The end of the arrow identifies the receiver. A short arrow represents local communications inside a host, whereas a long arrow represents remote communications between hosts. In (5.a) all the processors are busy and the remote communications are done at the end. In (5.b) the remote communication takes place overlapped with the computation.

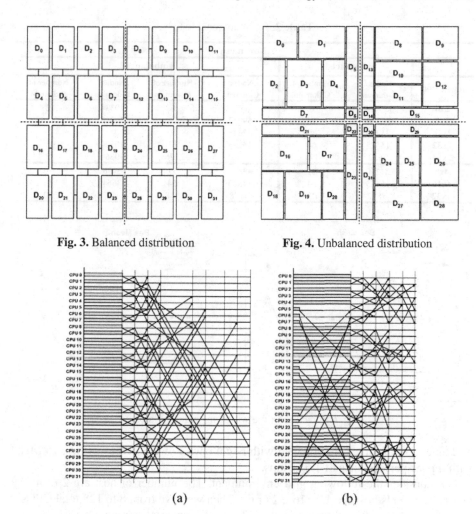

Fig. 3. Balanced distribution   **Fig. 4.** Unbalanced distribution

Fig. 5. Communication pattern: (a) Balanced distribution. (b) Unbalanced distribution

The optimum value of the execution time is obtained when the overlap of the slow communications and the computation is perfect. The question is what will be the overlap between communications and computations with the usual values of communication parameters and usual mesh sizes.

4 Results

In this section we show the results obtained using DIMEMAS. As data set, we consider a finite element mesh with 1,000,000 dofs. This size is usual for car crash or sheet stamping models. The dimensions of the mesh are $10^2 \times 10^2 \times 10^2$ nodes.

Table 2. Computational load

Host x CPU	Balanced		Unbalanced			
	Nodes / C-domain	Number of C-domain	Nodes/ C-domain	Number of C-domain	Nodes/ B-domain	Number of B-domain
			Box mesh			
2x4	125000	8	163222	6	10334	2
2x8	62500	16	69952	14	10336	2
2x16	31250	32	32645	30	10325	2
2x32	15625	64	15796	62	10324	2
2x64	7812	128	7772	126	10364	2
4x8	31250	32	43870	22	3486	10
4x16	15625	64	17840	54	3664	10
4x32	7812	128	8178	118	3499	10

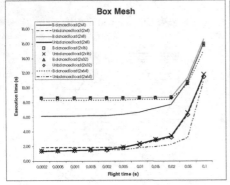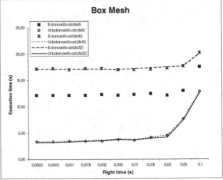

Fig. 6. Execution time with 1 ms external latency

The execution times are obtained with Nighthawk Power3 processors at 375 MHz, with a peak performance of 1.5 Gflops.

We consider the following grid environment. The number of host are 2 or 4, the number of CPUs/host are 4, 8, 16, 32 or 64. Then we have from 8 to 128 total CPUs.

Table 2 shows the computational load and the external communication requirements in the balanced and unbalanced decompositions.

Figure 6 shows the execution time as function of the flight time with a latency time equal to 1 ms. Figure 7 shows the same function for a latency of 100 ms. We can observe that the unbalanced decomposition reduces the execution time of the balanced distribution in all the cases.

The benefit of the unbalanced decomposition goes from 42% to 75% of time reduction. Table 3 shows the reduction time percentages for each grid environment. We want to remark that this is a pessimistic bound, because we waste as many CPUs as *B-domains*.

It is also important to look the MPI implementation. The ability to overlap communications and computations depends on this implementation. A multithread MPI implementation could overlap communication and computation, but problems with context switching between threads and interferences between processes appear. In a single thread MPI implementation we can use non-blocking send/receive with a

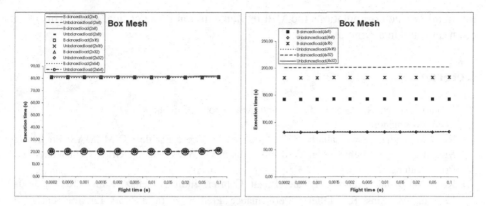

Fig. 7. Execution time with 100 ms external latency

Table 3. Reduction time

Host x CPU	Latency 1 ms	Latency 100 ms
2x4	62 %	75 %
2x8	70 %	74 %
2x16	69 %	74 %
2x32	70 %	74%
2x64	74 %	74 %
4x8	64 %	42 %
4x16	73 %	55 %
4x32	73 %	59 %

`wait_all` routine. However, we have observed some problems with this approach. The problems are associated with the internal order in the non-blocking MPI routines of send and receive actions [14] [15]. In our experiment this could be solved programming explicitly the proper order of the communications. But the problem remains in a general case. We conclude that is very important to have non-blocking MPI primitives that really exploit the full duplex channel capability [16].

5 Conclusions

In this work we show how an unbalanced domain decomposition strategy for grid environments. Our strategy is based in a two level decomposition. In the first step a balanced decomposition in the number of host is done. Then, in every host special boundary domains are created to gather all the host external communications. This strategy reduces the number of external communications and it allows overlapping the external communications with the computation. This strategy reduces the execution times of explicit finite element solvers by a mean factor of 59%. For a practical use of this strategy a simple modification of the usual mesh partitioner is need. To increase the efficiency of this method it is recommended to assign all the boundary domains in a host to the same CPU. To obtain the best overlap between remote communications and computations the MPI implementation is the critical point. This is an open question in a general case.

References

1. Dimemas, Internet. October 2002. http://www.cepba.upc.es/dimemas/ (Tool documentation).
2. Rosa M. Badia, Jesús Labarta, Judit Giménez, Francesc Escale. "DIMEMAS: Predicting MPI applications behavior in Grid environments". Workshop on Grid Applications and Programming Tools (GGF8), 2003.
3. Rosa M. Badia, Francesc Escale, Edgar Gabriel, Judit Gimenez, Rainer Keller, Jesús Labarta, Matthias S. Müller. "Performance Prediction in a Grid Environment". 1st European Across Grid Conference, Santiago de Compostela, 2003.
4. Paraver, Internet. October 2002. http://www.cepba.upc.es/paraver/ (Instrumentation packages).
5. Karypis George and Kumar Vipin, "Multilevel Algorithms for Multi-Constraint Graph Partitioning". University of Minnesota, Department of Computer Science/Army HPC Research Center. Minneapolis. MN 55455. Technical Report #98-019, 1998.
6. Karypis George and Kumar Vipin, "Multilevel k-way Partitioning Schemefor Irregular Graphs". University of Minnesota, Department of Computer Science/Army HPC Research Center. Minneapolis. MN 55455. Technical Report #95-064, 1998. To appear in SIAM Journal of Parallel and Distributed Computing.
7. Karypis George and Kumar Vipin, "A fast and High Quality Multilevel Scheme for Partitioning Irregular Graphs". University of Minnesota, Department of Computer Science/Army HPC Research Center. Minneapolis. MN 55455. Technical Report #95-035, 1998. To appear in SIAM on Scientific Computing.
8. Metis, Internet. http://www.cs.umn.edu/~metis
9. D. E. Keyes. "Domain Decomposition Methods in the Mainstream of Computational Science". Proceedings of the 14th International Conference on Domain Decomposition Methods, UNAM Press, Mexico City, pp. 79-93, 2003.
10. X. C. Cai. "Some Domain Decomposition Algorithms for Nonselfadjoint Elliptic and Parabolic Partial Differential Equations". Technical Report 461, Courant Institute, New York, 1989.
11. D. K. Kaushik, D. E. Keyes and B. F. Smith. "On the Interaction of Architecture and Algorithm in the Domain-based Parallelization of an Unstructured Grid Incompressible Flow Code". In J. Mandel et al., editor. Proceedings of the 10th International Conference on Domain Decomposition Methods, pp. 311-319. Wiley, 1997.
12. W. D. Gropp, D. K. Kaushik, D. E. Keyes and B. F. Smith. "Latency, bandwidth, and concurrent issue limitations in high-performance CFD". Proceedings of the First M.I.T. Conference on Computational Fluid and Solid Mechanics, pp. 839-841. Cambridge, MA, June, 2001.
13. W. D. Gropp and, D. E. Keyes. "Complexity of Parallel Implementation of Domain Decomposition Techniques for Elliptic Partial Differential Equations". SIAM Journal on Scientific and Statistical Computing, Vol. 9, n° 2, pp. 312-326, March, 1988.
14. F. García, A. Calderón, J. Carretero. "MiMPI: A Multithread-Safe Implementation of MPI". In Proceedings of PVM/MPI 99, Recent Advances in Parallel Virtual Machine and Message Passing Interface, 6th European PVM/MPI Users' Group Meeting, Vol. 1697 of Lectures Notes on Computer Science, pp. 207-214, Springer Verlag, Barcelona, 1999.

15. N. Karonis, B. Toonen, and I. Foster. "Mpich-g2: A Grid-enabled implementation of the message passing interface". Accepted for publication in Journal of Parallel and Distributed Computing, 2003.
16. R. Keller, E. Gabriel, B. Krammer, M. S. Müller and M. M. Resch. "Towards efficient execution of MPI applications on the Grid: Porting and Optimization issues". Journal of Grid Computing 1(2): 133-149, 2003 Kluwer Academic Publishers.
17. K. Ranganathan and I. Foster. "Decoupling Computation and Data Scheduling in Distributed Data-Intensive Applications". Proceedings of the 11th International Symposium for High Performance Distributed Computing (HPDC-11), Edinburgh, July 2002.
18. E. Deelman, J. Blythe, Y. Gil, C. Kesselman, G. Mehta, K. Vahi, K. Blackburn, A. Lazzarini, A. Arbree, R. Cavanaugh and S. Koranda. "Mapping Abstract Complex Workflows onto Grid Environments". Journal of Grid Computing. Vol. 1. No. 1, pp. 25-39, 2003.
19. H. Lamenhamedi, B. Szymanski and Z. Shentu. "Data Replication Strategies in Grid Environments". Fifth International Conference on Algorithms and Architectures for Parallel Processing (ICA3PP'02). October 23-25, 2002, Beijing, China.

A PVM Extension to Exploit Cluster Grids

Franco Frattolillo

Research Centre on Software Technology
Department of Engineering, University of Sannio, Italy
frattolillo@unisannio.it

Abstract. Workstations belonging to networked clusters cannot often be exploited since they are not provided with valid IP addresses and are hidden from the Internet by IP addressable front-end machines. The paper presents a PVM extension that enables such clustered machines hidden from the Internet to take part in a PVM computation as hosts. This way, programmers can exploit low-cost, networked computing resources widely available within departmental organizations in a well-known programming environment, such as PVM, without having to adapt their large-scale applications to new grid programming environments.

1 Introduction and Motivations

In the past, large investments have been made by the scientific community in developing large-scale PVM [1] applications, and today PVM is still considered an ideal system for programmers and scientists who are not well experienced in the high performance parallel programming, such as chemists, biologists, etc. However, PVM appears to be inflexible in many respects that can be constraining when the main goal is to exploit the computing resources available within departmental or enterprise networks. In fact, PVM requires that all the computing nodes making up a virtual machine are IP addressable, and this appears as a strong limitation, since workstations are often grouped in networked clusters, each provided with only one IP visible front-end machine that hides from the Internet all the other internal machines of the cluster. Consequently, PVM running on hosts outside a cluster cannot exploit the cluster's internal nodes.

To overcome this limitation, the BEOLIN port [2] has been included in the version 3.4.3 of PVM. The port sees a cluster of workstation (COW) as a single system image. This means that, if a COW has an IP addressable front-end node, PVM can start the *pvmd* daemon on it. Then, an environment variable set by user specifies which clustered nodes are available for running PVM tasks. Each subsequent request to spawn tasks on the COW front-end node causes tasks to be allocated to the clustered nodes specified as available, with each task going onto a separate node. Consequently, no two tasks can share the same node, and if the user attempts to spawn more tasks than there are nodes, an error is returned.

Although BEOLIN hides all the low-level details of a COW and allows users to re-use their PVM applications, it prevents expert programmers from taking full advantage of the computing power made available by a COW. In particular,

D. Kranzlmüller et al. (Eds.): EuroPVM/MPI 2004, LNCS 3241, pp. 362–369, 2004.

a programmer is not allowed to explicitly control task allocation on clustered nodes, and this prevents him from both allocating tasks on the basis of the node computational power and implementing load balancing strategies within the cluster. This can be considered a severe drawback when programmers want to achieve high performance in executing their parallel applications.

In this paper an extension of the PVM system, called ePVM, is presented. The main goal of ePVM is to enable PVM programs to exploit workstations and computers, not provided with valid IP addresses but connected to the Internet through IP visible front-end machines, as hosts in a unique virtual machine. ePVM overcomes the limitations affecting PVM and BEOLIN. In fact, unlike BEOLIN, ePVM does not see a COW as a single system image, since it allows to explicitly control task allocation within a cluster and enables any clustered node to be shared among multiple tasks. In particular, ePVM allows programmers to build and dynamically reconfigure a PVM compliant parallel virtual machine, which can however be extended to comprise collections of COWs provided that these are provided with IP addressable front-end nodes. As a consequence, even if all nodes are actually arranged according to a hierarchical physical network topology consisting of two levels (the level of the IP addressable nodes and the level of the not IP addressable nodes belonging to COWs), they virtually appear as arranged according to a flat network topology.

2   The ePVM Architecture

ePVM extends PVM by introducing the abstraction of the *cluster*. A *cluster* is a set of interconnected computing nodes provided with not valid IP addresses and hidden behind an IP addressable front-end computing node (see Figure 1). During computation, it is managed as a normal PVM virtual machine where a master *pvmd* daemon is started on the front-end node, while slaves are started on all the other nodes of the cluster. However, the front-end node is also provided with an ePVM daemon, called *epvmd*, which allows the cluster to interact with all other clusters. In fact, *epvmd* is a multi-process daemon that serves as a message router and worker able to provide a point of contact among distinct PVM virtual machines, each managing a cluster. Thus, ePVM enables the building of "extended virtual machines" (EVMs). Such a machine is made up by: (1) normal IP visible nodes, that are grouped in a first PVM virtual machine, called the "main" machine; (2) further PVM virtual machines, one for each cluster provided with an IP addressable front-end node (see Figure 1). In fact, due to ePVM, both IP visible nodes and those ones belonging to a PVM virtual machine managing a cluster can be directly referred to as hosts. This means that, even though distinct PVM virtual machines can coexist in an EVM, all the computing nodes can run PVM tasks as in a single, flat distributed computing platform.

At application startup, a "master" *epvmd* daemon is created onto the host from which the application is started. This daemon manages the "main" PVM virtual machine, but it can also manage another PVM virtual machine if it is allocated onto a host that is the front-end node of a grabbed cluster (see Figure 1).

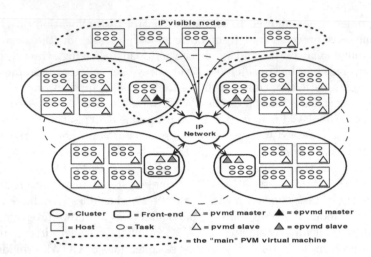

Fig. 1. The architecture of an "extended virtual machine".

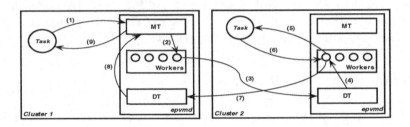

Fig. 2. The *epvmd* architecture and the interactions with application tasks.

Then, a "slave" *epvmd* is started on each front-end node of the other grabbed clusters. All *epvmd*s perform the same actions, but only the master can manage the configuration of the EVM. Such a configuration comprises information about all the clusters and hosts taking part in the EVM.

*epvmd*s act in conjunction with an extension of the PVM library, called *libepvm*, which enables tasks within a cluster to interface with *epvmd* in order to exchange messages with all the other tasks in the EVM, and to perform system service requests. Moreover, to support a flat task network, the task identification mechanism has been modified. In fact, tasks in an EVM are identified by a 32 bit integer, called "virtual task identifier" (VTID), unique across the entire EVM. However, unlike original TID, which contains only two main fields (hostID and taskID), a VTID also contains a third field, called "cluster identifier" (CID), that specifies the virtual machine which the pair (hostID, taskID) belongs to.

When a task allocated onto a host belonging to a cluster of an EVM invokes a *libepvm* routine, the software library decides whether the actions to be performed can be confined to the cluster or involve other clusters of the EVM. In the former case, the routine is served as in a PVM virtual machine, that is, the task invoking the routine contacts the *pvmd* running on its host, which takes charge of completing the routine service. In the latter case, the invoked routine contacts the *epvmd* managing its cluster, which serves the routine within the EVM. To this end, *epvmd* has been internally structured as a multi-process daemon composed of a "manager task" (MT), a "dispatcher task" (DT), and a pool of "workers" (Ws) (see Figure 2). All these tasks are implemented as normal PVM tasks, and this enables them and application tasks to communicate by exploiting the standard routines made available by PVM, such as pvm_send and pvm_receive.

Whenever a task within a cluster wants to communicate with a task running onto a host belonging to a different cluster or requires a system service involving actions to be performed onto a host belonging to a different cluster, it contacts MT, which is the local front-end section of *epvmd*. MT is the sole point of contact among application tasks and *epvmd*, and so it is not allowed to block while waiting for a *libepvm* routine involving inter-cluster operations to be performed. To assure this, *epvmd* exploits Ws, each of which takes charge of performing the operations needed to serve a *libepvm* routine. In particular, a W can serve a *libepvm* routine by contacting the *epvmd* managing the target cluster. In this case, the W interacts with the DT of the remote *epvmd*, which has the task of receiving the service requests coming from remote *epvmd*s and dispatching them to local Ws, which have the task of serving *libepvm* routines (see Figure 2). Once a W has served a remote service request, it can also return a result back to the *epvmd* originating the request. In this case, the W contacts the DT of this *epvmd*, which, once it has received the result, dispatches it to the MT, which takes charge of delivering the result to the task that invoked the routine.

3   The Library Extension

This section focuses on the ePVM routines that are new or implement a semantics different from the corresponding PVM routines (see Figure 3).

To run an application, a user typically starts a first task by hand from the "root" node, which can subsequently build the EVM and start other tasks. Tasks can be also started manually through the PVM "console", which is inherited in ePVM and allows the user to interactively start, query, and modify the EVM.

To set up an EVM, it is necessary to add clusters and hosts. Clusters can be added by invoking the pvm_addcluster. It accepts the host name of a cluster front-end node and returns the CID identifying the cluster. Consequently, all the nodes belonging to the cluster and indicated in the *hostfile* specified on the cluster front-end node are automatically added to the EVM as hosts. The *hostfile* specifies the cluster configuration in a format accepted by the PVM console.

Hosts are added by invoking the pvm_addhosts, which preserves its original PVM semantics, being able to add only IP addressable nodes to the EVM.

Once the EVM is configured, tasks can be created by invoking two routines: pvm_spawn and pvm_cspawn. The former can start up ntask copies of the executable task file on the EVM according to a semantics controlled by the flag argument. In particular, when flag assumes the values 0, 1, 2, 4, 16 and 32, pvm_spawn

```
int info = pvm_addcluster(char *hostname)
int info = pvm_addhosts(char **hosts, int nhost, int *infos)
int nt = pvm_spawn(char *task, char **argv, int flag, char *where, int ntask, int *vtids)
int nt = pvm_cspawn(int cid, char *task, char **argv, int flag,
                    char *where, int ntask, int *vtids)
char *where = make_spawn_param(int cid, char *hostname)
int info = pvm_cconfig(int cid, int *nhost, int *narch, struct pvmhostinfo **hostp)
int info = pvm_globalconfig(struct pvmclusterinfo **cp, int *size)
int info = pvm_getclusters(int **cids, int *ncids)
int cid = pvm_mycid(void)
int info = pvm_notify(int what, int msgtag, int cnt, int *tids)
int info = pvm_removecluster(int cid)
```

Fig. 3. The main ePVM routines in C.

behaves as documented in PVM, thus being able to start up tasks only on the hosts belonging to the cluster within which the routine has been invoked. If flag assumes the new value 64, multiple tasks can be started up according to a new semantics controlled by the where argument. This argument can be specified as the result of the new ePVM routine make_spawn_param. It accepts two parameters, a cluster identifier and a host name, by which it is possible to control the task allocation carried out by the pvm_spawn. Different combinations are allowed: (1) when the cid and the hostname are both specified, tasks are explicitly allocated; (2) if only hostname is specified, tasks are spawned on the host named hostname belonging to the local cluster; (3) if only cid is specified, tasks are spawned on hosts automatically selected by ePVM within the specified cluster; (4) otherwise, tasks are spawned on hosts selected by ePVM within the EVM.

The pvm_cspawn behaves as the PVM pvm_spawn, but within the cluster specified by cid. Consequently, the value 64 for the flag argument is no more allowed.

Information about the EVM can be obtained by using pvm_cconfig, pvm_globalconfig and pvm_getclusters. The pvm_cconfig returns information about the cluster identified by the cid argument. The information includes the number of hosts nhost, the number of different architectures narch, and an array of structures hostp in which each structure contains the PVM information about each host of the cluster. The pvm_globalconfig returns information about the clusters composing the EVM. In particular, the argument cp refers to an array of structures, each containing information about a single cluster. In fact, a structure pvmclusterinfo contains as many structures pvmhostinfo as the hosts belonging to the cluster, as well as other information, such as the cid. Moreover, size reports the number of clusters in the EVM. The pvm_getclusters returns an array containing the cids of all the ncids clusters in the EVM. Furthermore, pvm_mycid specifies the cid of the cluster containing the host on which the task invoking the routine is running.

The pvm_notify allows the caller to be notified on events that may happen in the EVM. Besides the common PVM events, the routine can also manage events regarding clusters, such as the adding or the deletion of a cluster. This is obtained by using two new constants for the what argument.

The pvm_removecluster deletes the cluster identified by the cid in the EVM.

4   Experimental Results

The tests have been conducted on two PC clusters interconnected by an Ethernet network. The first cluster is composed of 16 PCs, each equipped with a Pentium II 350 MHz and 128 MB of RAM. The PCs are interconnected by a Fast Ethernet hub and managed by a further PC that interfaces the cluster to the Internet. The second cluster is composed of 8 PCs, each equipped with an AMD Athlon 1800+ and 256 MB of RAM. The PCs are interconnected by a Fast Ethernet switch. All the PCs run Red Hat Linux rel. 9.

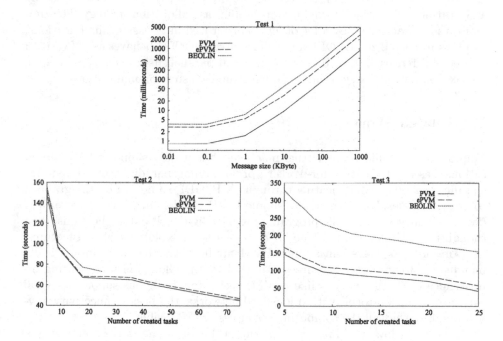

Fig. 4. Comparison tests on PVM, ePVM and BEOLIN.

Test 1 of Figure 4 measures the round trip delay affecting the pvm_send for different message sizes. The communications are sequenced, and so the difference in time between PVM and ePVM just measures the overhead introduced by the interactions among *pvmd*s and *epvmd*s. The test shows that PVM behaves always better than ePVM. However, ePVM exhibits a better performance than BEOLIN, and this is essentially due to the inter-cluster communications among *epvmd*s, which are implemented as an effective UDP variant based on a *selective repeat with cumulative acknowledges* algorithm, while BEOLIN adopts a simple and ineffective *stop and wait* communication algorithm.

In Test 2 of Figure 4 a program for exploring details of the Mandelbrot set [3] is executed. The program has been run with iteration limit set to 20,000, and task allocation has been carried out according to a round robin strategy involving all the available hosts of the two clusters. The graphs show that there are no significant differences among PVM, ePVM and BEOLIN when the task number is low, and this because the program is essentially CPU bound. However, when the task number increases, BEOLIN worsens its performance, while PVM and ePVM exhibit a similar behavior, with differences within 5÷10%. This is due to the ineffective implementation of the BEOLIN daemons managing the clusters' front-end nodes, which penalizes inter-cluster communications.

In the third test, a dynamic simulation package designed for simulating polymers [4] and whose mathematical formulation is given by Langevin equations is executed. The graphs shown in Test 3 of Figure 4 refer to runs characterized by an iteration limit set to 500 and a round robin task allocation strategy. The simulation is characterized by a lot of inter-cluster and inter-task communications, and this has penalized BEOLIN. On the contrary, ePVM behaves as PVM, with limited differences. This is due to the internal parallelism of *epvmd*s, which are able to mask the latency occurring during inter-cluster communications.

5   Related Work

Condor [5] can build PVM virtual machines, but it only supports PVM applications based on the "master-worker" programming paradigm and it requires specific daemons communicating through TCP/UDP to be run on the grabbed computing nodes. Thus, the nodes should be IP visible. On the contrary, ePVM does not impose any specific programming paradigm to PVM applications, even though they should be lightly modified to be able to exploit the ePVM features.

Condor-G [6] users can submit PVM applications that are automatically matched and run on the resources managed by the Globus metacomputing infrastructure [7]. However, unlike ePVM, even though Globus supports several communication protocols within localized networks, the basic connectivity anyway requires that all the computing resources are IP visible.

Legion [8] allows PVM programs to run in the Legion network-computing environment by emulating the PVM library on top of the Legion run-time system. However, unlike ePVM, this approach is fairly complex from an implementation standpoint, does not support the complete PVM API, and requires that all the nodes belonging to the computing environment are IP visible.

PUNCH [9] allows users to run PVM applications via a web interface accessible from standard browsers. Users specify the number and types of machines required for a given run via menus and text-boxes in HTML forms. PUNCH then finds and allocates the necessary resources using the user-supplied information.

Harness [10] can build extensible and reconfigurable virtual machines. It introduces a modular architecture made up by software components, called "plugins". In particular, a plugin emulating PVM has been developed. However, as regards the resource management, Harness behaves as PVM.

The Sun Microsystems' Grid Engine [11] focuses on harnessing the distributed computing power within an organization. It can schedule PVM applications on the grabbed resources, which however must be IP reachable.

It is worth noting that users of the systems reported above must often specify a large amount of information to run their applications, and thus they must also familiarize themselves with a broad range of details as well as the syntax required to specify this information [5–9, 11]. Moreover, users must also adapt their applications in order to utilize the advanced features, such as checkpointing, security, etc., offered by most of the presented systems [5–8, 11].

6   Conclusions

ePVM enables PVM applications to run on platforms made up by high power/cost ratio computing resources widely available within localized networks, such as cluster networks, but that cannot be grabbed by the current software systems for grid computing because not provided with valid IP addresses.

The limited modifications to the PVM library and the development of an optimized multiprocess daemon able to connect distinct PVM virtual machines, each managing a different cluster, have enabled ePVM to achieve a good performance in all the executed tests, and this demonstrates that existing PVM applications can run on cluster grids without having to be rearranged or penalized by the use of complex software systems for grid computing.

References

1. Geist, A., Beguelin, A., et al.: PVM: Parallel Virtual Machine. A Users' Guide and Tutorial for Networked Parallel Computing. MIT Press (1994)
2. Springer, P.L.: PVM Support for Clusters. In: Procs of the 3rd IEEE Int Conf. on Cluster Computing. Newport Beach, CA, USA (2001)
3. http://gmandel.sourceforge.net
4. http://kali.pse.umass.edu/ shulan/md.html
5. Tannenbaum, T., Wright, D., et al.: Condor - A Distributed Job Scheduler. In: Sterling, T. (ed): Beowulf Cluster Computing with Linux. MIT Press (2002)
6. Frey, J., Tannenbaum, T., Foster, I., et al.: Condor-G: A Computation Management Agent for Multi-Institutional Grids. In: Procs of the 10th IEEE Symp. on High Performance Distributed Computing. San Francisco, CA, USA (2001) 55-63
7. Foster, I., Kesselman, C.: Globus: A Metacomputing Infrastructure Toolkit. Int Journal of Supercomputer Applications and High Performance Computing 11(2) (1997) 115–128
8. Grimshaw, A., Wulf, W., et al.: Legion: The Next Logical Step Toward a Nation-wide Virtual Computer. Tech. Rep. CS-94-21, University of Virginia (1994)
9. Royo, D., Kapadia, N.H., Fortes, J.A.B.: Running PVM Applications in the PUNCH Wide Area Network-Computing Environment. In: Procs of the Int Conf. on Vector and Parallel Processing. Porto, Portugal (2000)
10. Kurzyniec, D., Sunderam, V., Migliardi, M.: PVM emulation in the Harness meta-computing framework – design and performance evaluation. In: Procs of the 2nd IEEE Int Symp. on Cluster Computing and the Grid. Berlin, Germany (2002)
11. http://www.sun.com/software/gridware

An Initial Analysis of the Impact of Overlap and Independent Progress for MPI

Ron Brightwell, Keith D. Underwood, and Rolf Riesen

Sandia National Laboratories*
PO Box 5800, MS-1110
Albuquerque, NM 87185-1110
{rbbrigh,kdunder,rolf}@sandia.gov

Abstract. The ability to offload functionality to a programmable network interface is appealing, both for increasing message passing performance and for reducing the overhead on the host processor(s). Two important features of an MPI implementation are independent progress and the ability to overlap computation with communication. In this paper, we compare the performance of several application benchmarks using an MPI implementation that takes advantage of a programmable NIC to implement MPI semantics with an implementation that does not. Unlike previous such comparisons, we compare identical network hardware using virtually the same software stack. This comparison isolates these two important features of an MPI implementation.

1   Introduction

Two desirable qualities of an MPI implementation are the ability to efficiently make progress on outstanding communication operations and the ability to overlap computation and communication. Unfortunately, it is difficult to isolate these features in order to determine how much of an effect they have on overall performance or to use them to characterize applications that may be able to take advantage of the opportunity for performance improvement that they offer. This is especially difficult when comparing application performance with different networks using identical compute hardware. For example, there have been several extensive comparisons of high-performance commodity interconnects, such as Myrinet [1], Quadrics [2], and InfiniBand [3]. These comparison typically use micro-benchmarks to measure raw performance characteristics and application benchmarks or applications to measure whole system performance. Unfortunately, this type of analysis can still leave many important questions unanswered.

Implementations of MPI typically strive to utilize the capabilities of the underlying network to the fullest and exploit as many features as possible. In this paper, we do the opposite. We purposely limit the capabilities of an MPI implementation in order to try to better understand how those specific capabilities impact application performance. Rather than simply using identical compute hardware, we use identical networking hardware and a significant portion of the same network software stack. This

---

* Sandia is a multiprogram laboratory operated by Sandia Corporation, a Lockheed Martin Company, for the United States Department of Energy's National Nuclear Security Administration under contract DE-AC04-94AL85000.

D. Kranzlmüller et al. (Eds.): EuroPVM/MPI 2004, LNCS 3241, pp. 370–377, 2004.

approach allows us to compare two different MPI implementations that have nearly identical micro-benchmark performance only on the basis of the features that differ.

The rest of this paper is organized as follows. The following section provides additional background on the features that we have isolated in this experiment. Section 3 provides an overview of research that motivated this effort while Section 4 discusses how this work complements other previous and ongoing projects in this area. Section 5 describes the specific software and hardware environment used in this study, and Section 6 presents an analysis of the data. The conclusions of this paper are presented in Section 7, which is followed by an outline of future work.

2 Background

The MPI Standard [4] defines a Progress Rule for asynchronous communication operations. The strict interpretation of this rule states that once a non-blocking communication operation has been posted, a matching operation will allow the operation to make progress regardless of whether the application makes further library calls. For example, if rank 0 posts a non-blocking receive and performs an infinite loop (or a significantly long computation) and rank 1 performs a matching blocking send operation, this operation will complete successfully on rank 1 regardless of whether or not rank 0 makes another MPI call. In short, this rule mandates non-local progress semantics for all non-blocking communication operations once they have been enabled.

Unfortunately, there is enough ambiguity in the standard to allow for a weak interpretation of this rule. This interpretation allows a compliant implementation to require the application to make library calls in order to make progress on outstanding communication operations. Independent of compliance, it can be easily observed that an implementation that adheres to the strict interpretation offers an opportunity for a performance advantage over one that supports the weak interpretation. One of the goals of this work is to better understand the impact of independent progress, where an intelligent and/or programmable network interface is responsible for making progress on outstanding communications independent of making MPI library calls.

It is possible to support overlap without supporting independent MPI progress. Networks capable of performing RDMA read and write operations can fully overlap communication with computation. However, the target address of these operations must be known. If the transfer of the target address depends on the user making an MPI library call (after the initial operation has begun) then progress is not independent. If the transfer of the target address is handled directly by the network interface, or by a user-level thread, then independent progress can be made. Conversely, it is possible to have independent progress without overlap. An example of this is the implementation of MPI for ASCI Red [5], where the interrupt-driven nature of the network interface insures that progress is made, but the host processor is dedicated to moving data to the network (at least in the default mode of operation where the communication co-processor is not used).

3 Motivation and Approach

Previous work[6] has indicated that a number of the NAS parallel benchmarks use excessively long posted receive and unexpected message queues in the MPI library. We hypothesized that this could degrade performance on platforms that offload the traversal of these queues onto much slower network interface hardware. The Quadrics environment provides the ideal opportunity to evaluate this hypothesis as it offers a native MPI implementation that offloads the traversal of these queues onto a 100 MHz processor (in the case of the ELAN-3 hardware) processor. Quadrics also offers a Cray SHMEM [7] compatibility interface, which can be used as a lightweight layer for building an MPI library without offloading any of the traditional MPI queue management onto the NIC [8]. Micro-benchmarks[9] indicate that, indeed, message latency is dramatically smaller on an MPI built over SHMEM when queue lengths grow. This work seeks to determine if application performance correlates to that finding.

A second motivation for comparing MPI libraries built on SHMEM and TPorts is that it allows the comparison of a library that provides capabilities for computation and communication overlap as well as independent progress in MPI to a library that does not. The MPI implementation using Quadrics Tports provides independent progress and overlap, since a thread running on the NIC is able to respond to incoming MPI requests without host processor intervention. In contrast, the SHMEM interface still allows messages to be deposited directly into user memory without intervention by the host, but a host CPU must still be used to handle MPI queue management and protocol messages. Using the host processor removes the opportunity for significant overlap and does not allow for independent progress of outstanding communication requests. Thus, a single platform with a similar network software stack can be used to evaluate both approaches.

Overlap and independent progress are believed to be important performance enhancements, but it is seldom possible to evaluate the relative merits of each on identical hardware. By reducing the differences in the system to the MPI implementation, this work is able to evaluate these issue. This has important implications for newer parallel computer networks, such as InfiniBand [3], that use RDMA for MPI implementations. Most implementations of MPI for InfiniBand [10–12] do not efficiently support overlap or independent progress [13].

Finally, benchmarks are notorious for their insensitivity to critical system parameters that ultimately impact system performance; however, "real" applications can require person months of effort to port to new platforms. In contrast, benchmarks are typically easy to port to new environments. Thus, the comparison of the NAS parallel benchmarks on two different types of MPI implementations on a single hardware platform will help to reveal whether they test certain system parameters of interest (in this case, the impact of queue usage and the impact of independent progress in MPI).

4   Related Work

The work in this paper spans the areas of performance analysis, NIC-based protocol offload, and characterizing the behavior of parallel applications. However, we know of no work that characterizes the benefits and drawbacks of NIC offload using identical hardware and nearly identical software stacks.

There is an abundant amount of work that compares different network technology using identical compute hardware in order to evaluate networks as a whole. The most recent and comprehensive study for MPI and commodity HPC networks is [13]. The work we present here attempts to isolate specific properties of a network - host processing versus NIC offload - in order to evaluate the two strategies.

It is typical for papers that describe MPI implementations to explore different strategies within the implementation. An example of this is [14], where polling versus blocking message completion notification is explored. This kind of MPI implementation strategy work explores different methods of using the capabilities offered by a network, but limiting network capabilities in order to characterize specific network functionality has not been explored.

A desciption of the MPI implementation for Cray SHMEM as well as communication micro-benchmark performance is described in [8].

5   Platform

The machine used for our experiments is a 32-node cluster at Los Alamos National Laboratory. Each node in the cluster contains two 1 GHz Intel Itanium-2 processors, 2 GB of main memory, and two Quadrics QsNet (ELAN-3) network interface cards. The nodes were running a patched version of the Linux 2.4.21 kernel. We used version 1.24-27 of the QsNet MPI implementation and version 1.4.12-1 of the QsNet libraries that contained the Cray SHMEM compatibility library. The SHMEM MPI library is a port of MPICH [15] version 1.2.5. All applications were compiled using Version 7.1 Build 20031106 of the Intel compiler suite.

All of our experiments were run using only one process per node using only one network interface, and all results were gathered on a dedicated machine. In order to provide a fair comparison, we disabled use of the optimized collective operations for the Tports MPI runs so that both implementations used the collective operations layered over MPI peer communication operations provided by default with MPICH. Ultimately, this change had negligible impact on the performance of the benchmarks using the Tports MPI.

6   Results

Here we compare the performance of Tports MPI and SHMEM MPI for the class B NAS parallel benchmarks. EP (the embarassingly parallel benchmark) was excluded since it does not do any significant communications. Each benchmark was run 4 times for each number of processors per job. The average of the four runs is reported in Figure 1.

Figure 1(a) graphs the performance of BT for both MPI implementations. Tports MPI has a 2-4% advantage over SHMEM MPI, depending on the number of nodes involved. While this may seem like it could be attributed to noise in the measurements, the maximum difference in the minimum and maximum run times was 0.5% for each implementation. Figure 1(b) shows the performance of CG, which is one of the few benchmarks where SHMEM MPI appears to have advantages. For small numbers of nodes, SHMEM MPI has up to a 3% performance advantage over Tports MPI with measurements that vary by less than 0.5%. This performance advantage is surprising

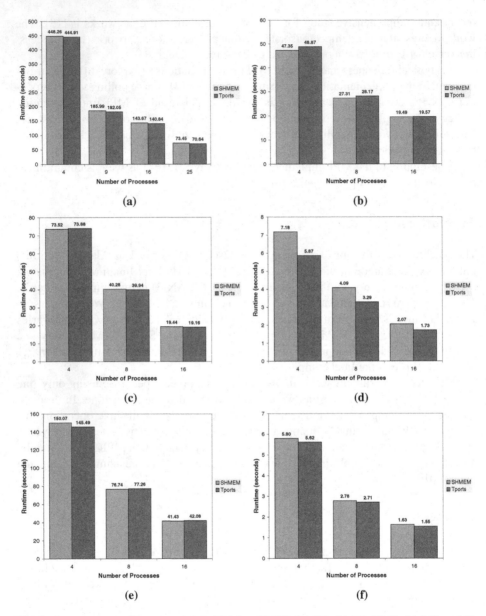

Fig. 1. A comparison of Tports and SHMEM for: **(a)** BT, **(b)** CG, **(c)** FT, **(d)** IS, **(e)** LU, and **(f)** MG

since since CG was found to be a "well-behaved" application in terms of MPI queue usage[6]. The cause is likely to be the use of messages that are predominantly smaller than 2 KB as discussed in [13]. However, this advantage evaporates at 16 nodes.

The FT and IS benchmarks were expected to pose a significant performance issue for an MPI implementation that offloads queue handling to a slower processor on the

NIC (e.g. MPI over Tports). Both of these benchmarks have long unexpected queues, long posted receive queues, and long average traversals of each[6]. Traversing these queues on an embedded processor should have proved costly; however, Figures 1(c) and (d) clearly indicate otherwise. Performance for FT is effectively equivalent between the two implementations and Tports MPI has a significant advantage (20%) for IS. Figures 1(e) and (f) continue to show roughly equivalent performance for the two MPI implementations (on the LU and MG benchmarks), but Figure 2 once again shows a significant margin of performance improvement for the Tports implementation. SP performs 5-10% better simply by using offload that provides overlap and independent progress.

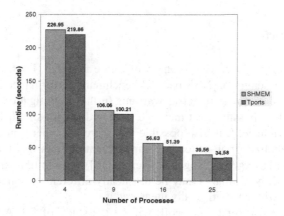

Fig. 2. A comparison of Tports and SHMEM for the SP benchmark

7 Conclusions

This study set out to evaluate the impacts of an MPI implementation that provides independent progress and overlap capabilities by using NIC based MPI offload. The expectation was that both advantages and disadvantages would be uncovered – advantages for "well-behaved" applications that had short MPI queues and disadvantages for "poorly-behaved" applications that used long MPI queues. In the process, we sought to answer three questions: does offload negatively affect applications with long MPI queues? do overlap and independent progress yield an advantage to applications? and, can the NAS benchmarks answer these questions. The answers were surprising.

The Tports MPI (the offloading MPI) wins almost uniformly (and sometimes by a significant margin) despite the fact that the SHMEM MPI has almost identical performance. In the rare cases where SHMEM MPI is faster, it is by a very small margin. In contrast, Tports MPI has a surprisingly large margin of victory over SHMEM MPI (20%) on the IS benchmark, which is "poorly-behaved". Also surprising was the fact that SHMEM MPI had a slight edge in performance for CG (a "well-behaved" application). In summary, "poorly-behaved" applications did not suffer from the slow NIC processor that offloads MPI queue handling.

Overall, the results in this paper point to a significant advantage for the combination of independent progress and overlap in an MPI implementation. The SHMEM MPI only has an advantage that could be argued to be statistically signficant in 3 of the 23 data points compared. The largest of these margins is 3% for one data point for CG. In contrast, the Tports MPI has 5-10% advantages for all of the SP measurements, 2-4% advantages for the BT measurements, and 20% advantages for IS. Since the difference in microbenchmark performance is always smaller than this, offloading and independent progress clearly have performance benefits.

Finally, it is clear that at least 3 of 7 the NAS benchmarks see performance impacts from overlap and independent progress. This is good news for those seeking to evaluate new platforms.

8   Future Work

These results have produced two avenues of future research. First, the seeming lack of impact of queue length on NAS parallel benchmark performance for networks that perform offloading of queue traversal warrants further investigation. There are three possible causes. It is possible that message latency does not significantly affect NAS benchmark performance. It is also possible that there are other aspects of system behavior that are penalizing non-offloading implementations in a way that is not captured by microbenchmarks (e.g. the "Rogue OS" effect[16]). As a third possibility, queue behavior for the NAS benchmarks may be drastically different on Quadrics than on the Myrinet platform where we originally measured it.

The second avenue for future work will be the study of full ASCI applications. Studying full ASCI applications takes significanly more work than studying benchmarks because they can easily include 100,000 lines of code, use 2 or more programming languages, use several difficult to port third party librarys, and require non-automated adaptation of platform specific build scripts; thus, each application can take an application specialist just to build it on a new platform. Nonetheless, we believe that these preliminary results justify further study into ASCI application behavior.

Acknowledgments

The authors would like to gratefully acknowledge the CCS-3 group at Los Alamos National Laboratory, especially Fabrizio Petrini, for providing access to the cluster used for our experiments.

References

1. Boden, N., Cohen, D., Felderman, R.E., Kulawik, A.E., Seitz, C.L., Seizovic, J.N., Su, W.: Myrinet-a gigabit-per-second local-area network. IEEE Micro **15** (1995) 29–36
2. Petrini, F., Feng, W.C., Hoisie, A., Coll, S., Frachtenberg, E.: The Quadrics network: High-performance clustering technology. IEEE Micro **22** (2002) 46–57
3. Infiniband Trade Association: http://www.infinibandta.org. (1999)

4. Message Passing Interface Forum: MPI: A Message-Passing Interface standard. The International Journal of Supercomputer Applications and High Performance Computing **8** (1994)
5. Brightwell, R.B., Shuler, P.L.: Design and implementation of MPI on Puma portals. In: Proceedings of the Second MPI Developer's Conference. (1996) 18–25
6. Brightwell, R., Underwood, K.D.: An analysis of NIC resource usage for offloading MPI. In: Proceedings of the 2002 Workshop on Communication Architecture for Clusters, Santa Fe, NM (2004)
7. Cray Research, Inc.: SHMEM Technical Note for C, SG-2516 2.3. (1994)
8. Brightwell, R.: A new MPI implementation for Cray SHMEM. In: Proceedings of the 11th European PVM/MPI Users' Group Meeting. (2004)
9. Underwood, K.D., Brightwell, R.: The impact of MPI queue usage on message latency. In: Proceedings of the 2004 International Conference on Parallel Processing. (2004)
10. Liu, J., Wu, J., Kini, S.P., Wyckoff, P., Panda, D.K.: High performance RDMA-based MPI implementation over InfiniBand. In: International Conference on Supercomputing. (2003)
11. Liu, J., Jiang, W., Wyckoff, P., Panda, D.K., Ashton, D., Buntinas, D., Gropp, W., Toonen, B.: Design and implementation of MPICH2 over InfiniBand with RDMA support. In: International Parallel and Distributed Processing Symposium. (2004)
12. Rehm, W., Grabner, R., Mietke, F., Mehlan, T., Siebert, C.: An MPICH2 channel device implementation over VAPI on InfiniBand. In: Workshop on Communication Architecture for Clusters. (2004)
13. Liu, J., Chandrasekaran, B., Wu, J., Jiang, W., Kini, S., Yu, W., Buntinas, D., Wyckoff, P., Panda, D.K.: Performance comparison of MPI implementations over InfiniBand, Myrinet and Quadrics. In: The International Conference for High Performance Computing and Communications (SC2003). (2003)
14. Dimitrov, R., Skjellum, A.: Impact of latency on applications' performance. In: Proceedings of the Fourth MPI Developers' and Users' Conference. (2000)
15. Gropp, W., Lusk, E., Doss, N., Skjellum, A.: A high-performance, portable implementation of the MPI message passing interface standard. Parallel Computing **22** (1996) 789–828
16. Petrini, F., Kerbyson, D.J., Pakin, S.: The case of the missing supercomputer performance: Identifying and eliminating the performance variability on the ASCI Q machine. In: Proceedings of the 2003 Conference on High Performance Networking and Computing. (2003)

A Performance-Oriented Technique
for Hybrid Application Development

Emilio Mancini[1], Massimiliano Rak[2], Roberto Torella[2], and Umberto Villano[1]

[1] RCOST, Università del Sannio
Pal. Bosco Lucarelli, Benevento, Italia
{epmancini,villano}@unisannio.it
[2] Dipartimento di Ingegneria dell'Informazione, Seconda Università di Napoli
Via Roma 29, Aversa (CE) Italia
{massimiliano.rak,r.torella}@unina2.it

Abstract. In SMP clusters it is not always convenient to switch from pure message-passing code to hybrid software designs that exploit shared memory. This paper tackles the problem of restructuring an existing MPI code through the insertion of OpenMP directives. The choice of the best code is carried out with a performance-oriented approach, predicting the effect of application hybridization in the MetaPL/HeSSE simulation environment, without writing and running any hybrid software. The technique is validated by applying the devised changes to the code, and comparing the predicted results to actual running time measurements.

1 Introduction

Symmetrical multiprocessor (SMP) hardware currently is pervasive in high performance computing environments. The availability of relatively low-cost SMP nodes, originally targeted at the demanding high-end server market, has given a boost to the diffusion of networks of SMP workstations and clusters of SMPs networked by high-speed switches, or by inexpensive commodity networks [1].

Now that computing nodes made up of multiple processors sharing a common memory are commonly available, it is natural to wonder if it is worth to switch to software designs exploiting shared memory, given that most current software was developed with shared-nothing architectures in mind. The literature reports many experiences of adoption of a two-tier programming style, based on shared memory for communication between processes running in the same node, and on message-passing for *outer-world* communications. Even if special programming models have been purposely developed for such systems [2], the joint use of MPI and OpenMP is emerging as a *de facto* standard. In the hybrid MPI-OpenMP model, a single message-passing task communicating using MPI primitives is allocated to each SMP processing element, and the multiple processors with shared memory in a node are exploited by parallelizing loops using OpenMP directives and runtime support.

Neither the development from scratch of hybrid MPI-OpenMP applications, nor the restructuring of existing MPI code by the addition of OpenMP directives

D. Kranzlmüller et al. (Eds.): EuroPVM/MPI 2004, LNCS 3241, pp. 378–387, 2004.

are particularly complex. The problem is to understand, before any significant development or code restructuring step is taken, if the resulting performance is worth the effort. There is a wide body of literature pointing out that sometimes a standard single-layer MPI application is the best performing solution [3, 4]. In fact, exploiting coarse parallelism at task level and fine- or medium-grain parallelism at loop level is an attractive solution, but its performance depends heavily on architectural issues (type of CPUs, memory bandwidth, caches, ...) and, above all, on the application structure, code and data dimension. Therefore it may be sometimes preferable to use a canonic MPI decomposition, allocating on each node a number of tasks equal to the number of CPUs. By the way, this means simply not to restructure existing MPI applications.

Due to its higher architectural complexity and to the multiplicity of software layers, a hybrid OpenMP/MPI environment makes particularly hard to cope with the wide range of computational alternatives to be explored to obtain high performance [5, 6]. In a previous paper [7], we have shown that simulation-based tools can be successful to model and to predict the performance of *existing* (i.e., fully developed) hybrid OpenMP/MPI applications. The focus in this paper will instead be on the restructuring of traditional MPI code. The objective is to apply performance prediction techniques to MPI code to understand *from the start* if the use of hybrid programming techniques will lead to a performance improvement. The decision to switch to a hybrid software design will possibly be taken by examining the performance results obtained in a performance-oriented development environment. This environment, based on the use of the MetaPL description language [8], will make it possible to insert interactively OpenMP directives in the code and to evaluate immediately the final effect on overall system performance by running HeSSE [9] simulations of the modified code.

After a brief overview of the HeSSE/MetaPL modeling technique, this paper will go on to describe their application to an existing MPI program, showing how to predict the effects of the insertion of OpenMP directives (*hybridization*) in the target software. The application model (not the real code) will be modified and simulated, predicting the effects of the hybridization, and helping the developer to choose the best way to parallelize. Then the real application code will be modified, following the indications given by the analysis, and the predicted application performance will be compared to the actual results obtained in the real cluster environment. After a discussion on the accuracy of the model used, the paper closes with an examination of related work and the conclusions.

2   The HeSSE/MetaPL Modeling Technique

In order to predict the effects of hybridization, i.e., of the insertion of OpenMP directives in an existing MPI code, the very first step is to set up a model of the given application and of the target computing environment. The adopted simulation framework, HeSSE (Heterogeneous System Simulation Environment), is a component-based simulation environment, which builds simulation models by composition of more specialized simulation models, modeling subsystems of a heterogeneous distributed system [9–11]. Simulation components are able to reproduce both the functional (services offered to the system) and behavioral (time spent in performing actions) actions of the different portions of the system. Components can be passive (waiting for service requests from other simulation components), or active (at simulation start, they send service requests to the environment). Active components are mainly used to describe system evolution; they are usually fed by pre-collected or synthetic trace files. Trace files are essentially a low-level description of the system inner behavioral characteristics.

A very interesting possibility offered by software development in simulation environments [8] is the iterative refinement and performance evaluation of program prototypes. Prototypes are incomplete program designs, skeletons of code where the computations interleaved between concurrent process interactions are not fully specified. In a prototype, these "local" computations are represented for simulation purposes by delays equal to the (expected) CPU time that will be spent in the actual code. The use of prototypes has shown that this synthetic way of describing the behavior of a parallel program is very powerful: it is language- and platform-independent, shows only essential features of the software, and can successfully be used for performance analysis at the early development stages.

These considerations led to the development of the MetaPL language [8, 12], a notation system designed to be the evolution of the concept of prototypes. MetaPL is an XML-based Tag language, and provides predefined elements for the description at different levels of detail of generic parallel programs. Using XML extension characteristics, the capabilities of the language can be expanded whenever necessary. The use of a single, flexible notation system may help the development of CASE tools and data interchanging between them. Furthermore, its suitability for simulation promotes performance analysis techniques in the early stages of the software development cycle. MetaPL promotes the use of a transformational approach: a description can be translated into different ones, less general but more detailed, or in other types of system descriptions, such as diagrams or documents. We call the transformation engines *MetaPL filters*; depending on the target of the transformation, filters can operate silently, or can query the user for the information required to perform the translation process.

The link between the MetaPL description language and the HeSSE simulation environment is the production of trace files through a suitable filter [8, 12]. In fact, the joint use of MetaPL and HeSSE works as follows:

- the behavior of the system (software application, middleware, ...) is described in MetaPL language;
- the system is described as a simulator configuration (this is carried out using the HeSSEgraphs tool [12]);
- the behavioral level description is filtered, translating it into a set of trace files, representing the executions to be analyzed;
- simulation is carried out and the obtained performance results are used to drive the further development or the optimization (depending on whether the software is only in prototypal form or has been fully developed).

3   Application Analysis and Hybridization

As mentioned in the introduction, the objective of this paper is the analysis of an existing MPI program, finalized at understanding if the insertion of OpenMP directives will improve the application performance or not. The use of an integrated modelling/simulation environment makes it possible to predict the effects of hybridization, without any change in the real code and with limited use of the target system, which is involved in the development process only for the initial measurement of timing data.

Throughout the paper, we will adopt as case study an existing MPI code [13] solving the well-known N-body problem, which simulates the motion of n particles under their mutual attraction. Of course, the objective of this paper is not to present an ultimate solution for the N-body problem, but just to use an existing, not trivial, MPI code as the basis for possible OpenMP parallelization steps. It should be explicitly noted that these steps are not necessarily trivial, and that sometimes they require clever code restructuring.

The proposed approach is made up of the following steps, described orderly in the following subsections:

1. application and system modeling;
2. model enrichment with timing data;
3. model hybridization and performance analysis.

At the end of these three steps, on the basis of performance data obtained only by simulation, it can be taken the decision if to adopt a hybrid code, or to use an MPI-only code with multiple tasks per SMP node.

3.1   Application and System Modeling

The system used as testbed for the chosen application is the Cygnus cluster at the Parsec Laboratory, 2nd University of Naples. This is a Linux cluster of biprocessor Xeon SMP nodes, with a Fast Ethernet Switched network. Figure 1 shows a screenshot of the description of four nodes of the system within the visual tool HeSSEgraphs. The description can be automatically translated into a HeSSE configuration file. The configuration needs only to be tuned with parameters taken from the real environment. In [10] we showed how to obtain this information, whether the system is available for measurements, or not.

Simulation of hybrid applications in HeSSE needs the adoption of specific components, which reproduce the services that OpenMP offers to the application. It is out of the scope of this paper to describe these simulation components. It is only worth to point out that, when OpenMP applications are simulated, HeSSE uses an additional trace file containing all the OpenMP directives. Process trace files contain references to the OpenMP specific trace file. During the simulation, when a simulated process finds an OpenMP directive, asks for the proper service to an OpenMP Simulator Component, which manages the local threads according to the semantics of OpenMP [7].

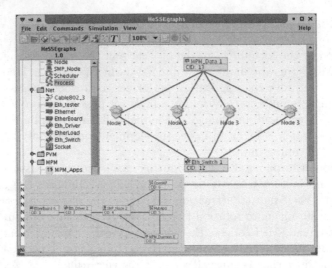

Fig. 1. Cygnus model in the HeSSEgraph environment

The MetaPL application model can be built using the Eclipse environment with suitable XML plugins. This environment can be of great help to ease the analysis process. Figure 2 shows a screenshot of a high level MetaPL description of the target MPI application. After the addition of timing information to the original MPI code description, it will be possible by means of MetaPL filters (composition of XSLTs) to generate simulation traces that describe the application execution on the target cluster. In the chosen environment an Ant project was built, which applies the required transformations in the right order, producing the trace files that can be directly simulated, as the simulator is considered an external tool.

3.2   Model Enrichment with Timing Data

In the following, the MetaPL model will be used to generate the traces used by the simulator for the generation of the application performance predictions. Traces contains the timed sequence of call for services issued by each application process (MPI primitives, OpenMP directives, O.S. calls, ...) and of CPU bursts. They do not contain communication timing, which will be generated at simulation time taking into account the actual communication workload. Hence the trace generation process requires service parameters (e.g., the dimensions of data for an MPI send primitive), which can be easily obtained from the model, and the duration of CPU bursts, which have to be evaluated by direct measurement. Careful measurement and tuning make it possible to take (partially) into account the effects of the target memory system.

The CPU burst measurements have been carried out by instrumenting the relevant sections of code (roughly speaking, the long computation sequences between two MPI communications) by a simple library developed on the top of

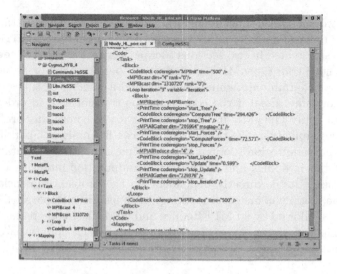

Fig. 2. Nbody MetaPL description in the Eclipse environment

```
<Loop iteration="3" variable="iteration">
<Block>
   <MPIBarrier></MPIBarrier>
   <CodeBlock coderegion="ComputeTree_1" time="229.354"/>
   <CodeBlock coderegion="ComputeTree_2" time="106.354"/>
   <MPIAllGather dim="411928" msgtag="1" />
   <CodeBlock coderegion="ComputeForces" time="146.422">
   <MPIAllReduce dim="4"   />
   <CodeBlock coderegion="NewV" time="0.545">
   <CodeBlock coderegion="NewX" time="0.580/>
   <MPIAllGather dim="229376" />
</Block>
</Loop>
```

Fig. 3. MetaPL description of the Nbody main loop body

PAPI [14]. This just made it possible to evaluate the (CPU) time spent in each application code region. It should be explicitly noted that the model contains no timing information about the duration of MPI communications (protocol software overhead, transmission over a loaded network media, ...). These depend on the hardware system configuration and will be instead automatically generated by simulating the given target configuration driven by the traces that contain only "local" timing information.

Figure 3 shows a sample view of the resulting MetaPL model, namely the main application loop. The application is made up of three basic computation steps, to be iterated a fixed number of times: ComputeTree, ComputeForce, Update. Before and after each of these steps, there is an MPI communication among the worker tasks. The main loop, repeated three times, starts with a barrier synchronization, followed by the first code region (`ComputeTree`), which in its turn is subdivided in two sections (`ComputeTree_1`, `ComputeTree_2`). Both of them are not fit for OpenMP parallelization, because they or do not contain loops at all, or contain only while loops iterated a number of times not known at compilation time. Even if the code could be transformed, in order to make it possible to apply OpenMP parallelization, we will ignore this problem here, assuming for simplicity's sake that `Compute_tree` is simply not parallelizable.

The second code region (`ComputeForces`), performs the computation of the positions of the n bodies and is the most computation-intensive section. Its actual response time is directly dependent on the problem dimension. Unlike `ComputeTree`, `ComputeForces` is made up of loops directly parallelizable with OpenMP. It is followed by an MPI Reduce and by the last code region (`Update`), which is composed of two steps (`NewV` and `NewX`), which update particle speed and positions. Also these code sections can be easily parallelized with OpenMP.

3.3   Model Hybridization and Performance Analysis

The objective of the third and last modeling step is to understand whether it is useful to insert OpenMP directives in the MPI code and to resort to a hybrid OpenMP-MPI code, or it is better to launch as many MPI tasks as available CPUs in each computing node (two tasks per node, in our case). The most important point is that this comparative analysis is performed only on simulated data. In other words, the development decisions will be taken by running only scaled-down code fragments (to obtain the model timing data). Neither the actual MPI code with multiple tasks per node, nor the hybrid OpenMP-MPI code will be executed. To be precise, the latter does not even exist, since it can be developed *after*, if simulation data have proven the validity of the hybrid approach. Further details on this technique and on its limits are discussed in [7].

From the code description, we have seen that it is possible to parallelize with OpenMP `Forces`, `NewX`, and `NewY`. It should be noted that the last three sections have response times dependent on the number of particles assigned to each task. Hence, the use of multiple tasks per node and the consequent reduction of the number of particles per task will surely reduce the overall response time. Stated another way, the analysis performed in the simulator should point out if it is better to have a lot of particles per node and to parallelize loops, or to reduce the their number and to tolerate a higher degree of intra-node communications.

In order to judge if OpenMP parallelization is useful or not, we will modify the application model introducing one by one possible OpenMP loop parallelization directives (at MetaPL description level, not writing actual code), and evaluating by simulation the application response time when *only one* of the loops shown is parallelized. Table 1 shows the times spent in the code regions by the hybrid and MPI (8 tasks) versions. The first column contains the name of the region in which the OpenMP directive was introduced, the second one the response time for MPI with eight tasks, the last one the hybrid (4 tasks) response time. All these times were obtained by simulation.

The analysis of the figures in the table shows that the introduction of the OpenMp directives in Tree_2 reduces the response time by less then 20%. Considered that we can expect a measurement error of about 10% (more on this later), the parallelization of Tree_2 does not appear convenient by itself (it can be convenient if applied along with further OpenMP parallelization, of course). On the other hand, Forces performs slightly better in the MPI-only version, whereas both NewV and NewX benefit from OpenMP parallelization.

The net result (adding all contributions to response time) is that the hybrid model seems the best solution. However, the performance improvement obtained is likely to increase if the overall response time is considered. In fact, the analysis of the response time of the single sections of code does not take into account the effect of communication, which is surely higher for a 8-task decomposition than for a 4-task one. The simulation of the whole application (with the effect of communication) leads to the results shown in the first and third column of Table 2 for several problem dimensions. These point out that communications heavily affects the MPI-only version of the code, and that the hybrid version is a good choice.

Table 1. Response time of Hybrid and MPI-only code versions (simulated, ms)

Code Region	MPI-only (8 tasks)	Hybrid (4 tasks)
Tree_2	50.9	40
Forces	72.5	90
NewV	344	150
NewX	255	150

The analysis steps should actually stop here, at the decision to adopt a hybrid code structure. However, in order to evaluate the accuracy of the simulated results obtained, we have actually written the hybrid code and executed it, along with the MPI-only version on the real target hardware. Table 2 also shows the performance results obtained by measurements on actual code executions (columns 2 and 4). The relative error of the simulated figures compared to the real ones, plotted in Fig.4, is always below 10%. This is a particularly interesting result, considered that the OpenMP simulations have been performed on fictitious code (e.g., not yet developed).

4   Related Work

There is a wide body of literature discussing pros and cons of the hybrid OpenMP-MPI programming model [3, 4, 7, 15]. Most existing work compares hybrid and traditional code by measurements taken on fully-developed code on the actual target execution environments. In our knowledge, the approach followed here, i.e., the use of a development environment to find if hybridization may be

Table 2. Overall response time (measurements on actual execution and simulated, ms)

Dim.	MPI-only, 8 tasks (simulated)	MPI-only, 8 tasks (measured)	Relative Error	Hybrid (simulated)	Hybrid (measured)	Relative Error
512	1700	1800	5,5%	994	908	9.4%
4096	2500	2400	4,1%	1300	1270	2.3%
32768	7600	8500	10.5%	3573	3692	3.2%

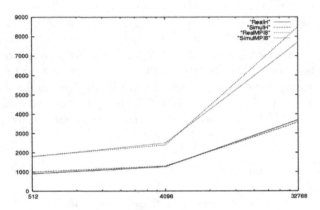

Fig. 4. Real and Simulated response time for MPI-only and hybrid code versions

useful and where, is completely new. A companion paper [7] is focused on the details of the hybrid OpenMP-MPI simulation process, details that have been omitted here. Performance-driven development [16] and prediction techniques [7, 9] are not new, but have been mainly applied for "traditional" code, hardly ever in the context of high performance computing.

5   Conclusions

In this paper we have presented a performance-oriented technique that helps the developer to decide if the use of hybrid OpenMP-MPI code is performance-effective for execution on an SMP cluster. The available literature shows that this is not always the case, and that application/system characteristics have to be considered to choose between hybrid and MPI-only solutions.

Our technique is composed of two main steps: modeling of hardware and MPI-only software, and prediction of the effect of hybridization. The modeling approach is based on the use of the MetaPL description language and environment, while the simulations are carried out by means of the HeSSE simulator. We have shown that the effects of the introduction of OpenMP parallelism in different code sections can be evaluated with errors less then 10%, without writing and executing the actual code. In light of the above, we think that this contribution stands as a viable and effective technique for analysis and possible restructuring of existing MPI code.

References

1. Woodward, P.R.: Perspectives on supercomputing: Three decades of change. Computer **29** (1996) 99–111
2. Bader, D.A., Jaja, J.: SIMPLE: A methodology for programming high performance algorithms on clusters of symmetric multiprocessors. J. of Par. and Distr. Comput. **58** (1999) 92–108
3. Cappello, F., Etiemble, D.: MPI versus MPI+OpenMP on the IBM SP for the NAS benchmarks. In: Proc. of the 2000 ACM/IEEE conf. on Supercomputing. Number 12, Dallas, Tx, USA, IEEE Comp. Soc. (2000) 51–62
4. Chow, E., Hysom, D.: Assessing performance of hybrid MPI/OpenMP programs on SMP clusters. Submitted to J. Par. Distr. Comput. (2001)
5. Boku, T., Yoshikawa, S., Sato, M., Hoover, C.G., Hoover, W.G.: Implementation and performance evaluation of SPAM particle code with OpenMP-MPI hybrid programming. In: Proc. EWOMP 2001, Barcelona (2001)
6. Nakajima, K., Okuda, H.: Parallel iterative solvers for unstructured grids using an OpenMP/MPI hybrid programming model for the GeoFEM platform on SMP cluster architectures. In: Proc. WOMPEI 2002, LNCS. Volume 2327. (2002) 437–448
7. Aversa, R., Di Martino, B., Rak, M., Venticinque, S., Villano, U.: Performance simulation of a hybrid OpenMP/MPI application with HeSSE. In: Proc. of ParCo2003 Conference, Dresda, Germany (2003)
8. Mazzocca, N., Rak, M., Villano, U.: The MetaPL approach to the performance analysis of distributed software systems. In: 3rd International Workshop on Software and Performance (WOSP02), IEEE Press (2002) 142–149
9. Mazzocca, N., Rak, M., Torella, R., Mancini, E., Villano, U.: The HeSSE simulation environment. In: ESMc'2003, Naples, Italy (2003) 270–274
10. Mancini, E., Rak, M., Torella, R., Villano, U.: Off-line performance prediction of message passing application on cluster system. LNCS **2840** (2003) 45–54
11. Mancini, E., Mazzocca, N., Rak, M., Villano, U.: Integrated tools for performance-oriented distributed software development. In: Proc. SERP'03 Conference. Volume 1., Las Vegas (NE), USA (2003) 88–94
12. Aversa, R., Iannello, G., Mazzocca, N.: An MPI driven parallelization strategy for different computing platforms: A case study. LNCS **1332** (1996) 401–408
13. London, K., Moore, S., Mucci, P., Seymour, K., Luczak, R.: The PAPI cross-platform interface to hardware performance counters. In: Dept. of Def. Users' Group Conf. Proc. (2001)
14. Jost, G., Jin, H., Labarta, J., Gimenez, J., Caubet, J.: Performance analysis of multilevel parallel applications on shared memory architectures. In: Proc. IPDPS'03, Nice, France (2003) 80–89
15. Smith, C.U., Williams, L.: Performance Solutions: A Practical Guide to Creating Responsive, Scalable Software. Addison Wesley (2001)
16. Labarta, J., Girona, S., Pillet, V., Cortes, T., Gregoris, L.: DiP: a parallel program development environment. In: Proc. Euro-Par 96. Volume 2. (1996) 665–674

A Refinement Strategy
for a User-Oriented Performance Analysis

Jan Lemeire, Andy Crijns, John Crijns, and Erik Dirkx

Parallel Systems lab, Vrije Universiteit Brussel
Pleinlaan 2, 1000 Brussels, Belgium
{jan.lemeire,erik.dirkx}@vub.ac.be
http://parallel.vub.ac.be

Abstract. We introduce a refinement strategy to bring the parallel performance analysis closer to the user. The analysis starts with a simple high-level performance model. It is based on first-order approximations, in terms of the logical constituents of the parallel program and characteristics of the system. This model is then progressively refined with more detailed low-level performance aspects, to explain divergences from a 'normal', linear regime. We use a causal model to structure the relations between all variables involved. The approach intends to serve as a link between detailed performance data and the developer. It is demonstrated with a parallel matrix multiplication algorithm.

1 Introduction

This paper investigates how the non-expert developer should be given clear insight in the performance of its parallel program. For efficient parallel processing, the developer must master the various performance aspects, ranging from high-level software issues to low-level hardware characteristics. The performance analysis is nowadays supported by various tools that automatically analyze parallel programs. The current challenge however is to give the software developer understandable results with a minimum of learning overhead [APART: http://www.fz-juelich.de/apart], as the tools seem hard to sell to the user community [8].

Most profiling tools like SCALEA [13], AIMS [14], Pablo [10], KOJAK [6] or VAMPIR [7] (semi-) automatically instrument the parallel program and use hardware-profiling to measure very detailed performance data. In the post-mortem analysis, they automatically filter out relevant parts, like bottlenecks, situations of inefficient behavior or performance losses, from the huge amount of low-level information and try to map them onto the developer's program abstraction, like code regions. Our approach works in the opposite direction; it is based on a simple performance model, using the terminology of a non-expert developer. The high-level model is used as a first-order approximation for explaining the performance. Experimental data will then support the model, or indicate the need for a more in-depth analysis when divergences appear. In this manner, the model is extended with low-level characteristics.

D. Kranzlmüller et al. (Eds.): EuroPVM/MPI 2004, LNCS 3241, pp. 388–396, 2004.

We present all variables or fluents in a causal model to indicate their dependencies. The refinements gradually add extra low-level fluents to the model.

The next section explains the parallel algorithm for matrix multiplication, section 3 defines the causal performance model and section 4 our tool EPPA. Section 5 shows the experimental results and the refinement strategy.

2 Parallel Matrix Multiplication

We illustrate our approach with the analysis of parallel multiplication of matrices, $C = A \times B$. The sequential runtime to multiply two dense $n \times n$ matrices is of $O(n^3)$, what makes it worth for being computed in parallel for high values of n. The parallel algorithm uses with a checkerboard partitioning, the matrices are divided in r strips of contiguous rows and c strips of contiguous columns, where $r \times c = p$ and r close to $\sqrt{p}$ [5]. Then, blocks of size $n/r \times n/c$ of matrices A and B are attributed to each processor. The p processes are labeled from $p_{0,0}$ to $p_{r-1, c-1}$. This is the starting point of our parallel algorithm. Each process $p_{i,j}$ will compute submatrix $C_{i,j}$ of the result matrix. Therefore, it requires all submatrices $A_{i,k}$ and $B_{l,j}$ for $0 \le k < r$ and $0 \le l < c$ (Fig. 1). Each process first multiplies its local submatrices and then broadcasts its submatrix $A_{i,j}$ to processes of the same row and $B_{i,j}$ to the processes of the same column [5].

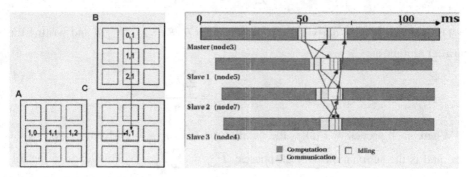

Fig. 1. Parallel Matrix Multiplication: partitioning (left, $p=9$) and execution profile (right, $n=150$, $p=4$).

The experiments are performed on a cluster of 9 dedicated 333MHz Pentium II processors with 256MB RAM, connected by a 100Mb/s non-blocking switch. Figure 1 shows the time line diagram, displaying the three types of phases: computation, communication and idling.

3 The High-Level Performance Model

The performance analysis should consider the impact of each phase, whether it is useful work or overhead, of the parallel program on the speedup. This is reflected by the ratio of the time T^i_{phase} of a phase on processor i with the sequential runtime T_{seq} divided by the number of processors p, what we call the *overhead ratio*:

$$Ovh^i_{phase} = \frac{T^i_{phase}}{T_{seq}/p}. \tag{1}$$

and totalized over all processors, it gives the global impact:

$$Ovh_{phase} = \frac{\sum_i^p T^i_{phase}}{T_{seq}} = \frac{T_{phase}}{T_{seq}}. \tag{2}$$

These definitions differ slightly from the *normalized performance indices* used by AIMS, defined as $Index_{phase} = T_{phase}/T_{par}$ [11], which are always less than one. The overhead ratios become more than one if an overhead surpasses the run time of the useful work. Our choice for this definition is motivated by the direct relation with the efficiency E:

$$E = \frac{1}{\sum_j^{phases} Ovh_j}. \tag{3}$$

This can easily been derived from its definition $E=S/p=T_{seq}/T_{par}/p$ and writing the parallel runtime as:

$$T_{par} = T^1_{par} = T^2_{par} = ... = T^p_{par} = \frac{\sum_i T^i_{par}}{p}. \tag{4}$$

Where each processor i takes the parallel run time T_{par} to perform its part of the job, and is the summation over all phases: $T^i_{par} = \sum_j T^{i,j}_{phase}$. This leads to the following equation, which is equivalent to Eq. 3:

$$Speedup = \frac{T_{seq}}{T_{par}} = \frac{p}{\dfrac{\sum_i \sum_j T^{i,j}_{phase}}{T_{seq}}}. \tag{5}$$

Fig. 2 shows the overhead ratios of all phases for the parallel matrix multiplication. Note that in the ideal case, each processor processes an equal part of the useful work without any overhead. This would result in overhead ratios of 100% and the speedup will equal p (Eq. 4).

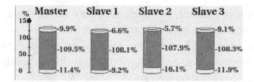

Fig. 2. Overhead ratios of parallel matrix multiplication with $n=150$, $p=4$.

We identify three top-level phase types: the computation, the communication and the idle times. This classification can easily be refined by the developer to identify logical phases in its program and to subdivide different overheads, as done by several authors [1, 13]. The communication is defined as the overhead time not overlapping with computation: the computational overhead due to the exchange of data between processes, in the sense of loss of processor cycles [2].

To structure all variables and to show their dependencies, we use a causal model. See Pearl [9] for an overview of current theory about causality and statistics. The initial performance model is shown in Fig. 3, where direct cause-effect relations are indicated by directed links.

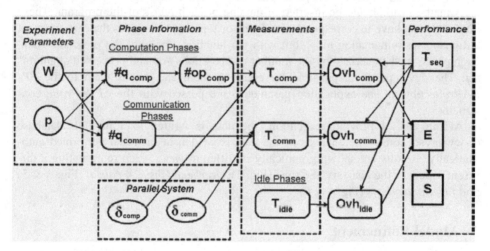

Fig. 3. Causal Performance Model of First-Order Approximation.

The goal of the model is to bring it to simple, mostly linear, relations by introducing useful intermediate variables. Instead of trying to directly estimate the functional relation between for example the speedup S or the computation time T_{comp} and the number of processors p, we add relevant intermediate characteristics of each phase. Each computational phase can be characterized by the number of *operations* and processed *quantums*. The number of operations *#op* relate directly to the processing time as the number of identical operations that are performed during that phase (Eg. the number of compare and swap operations for a sort algorithm). The number of quantums *#q* relate directly to the problem size W (Eg. the number of elements sorted or communicated). In this manner, complex relations can be unraveled and symbolically interpreted more easily, as will be shown in section 5.

Experimental runs are identified by the application parameters, like the problem size W, number of processors p and additional algorithm-specific parameters. The parallel system is initially characterized by δ_{comp} and δ_{comm}.

4   EPPA Tool Overview

EPPA (Experimental Parallel Performance Analysis) is implemented in C++ and is independent of the parallel communication layer [3, http://parallel.vub.ac.be/eppa]. A parallel program should be instrumented manually with one *EPPAProbe* object per process, which will collect all relevant program data. We envisage that the extra work needed for manual code instrumentation is compensated with the advantage of having additional high-level information and all program information understood and controlled by the user. The programmer should identify the logical parts of its program, specify phase variables and the parameters of the experiment.

A method call should be inserted at the end of each *phase* of the program. These *phases* do not have to correspond with loops or function calls, as is the case for most automatic instrumentation tools, but with the functional parts of the program. Then, each phase is characterized by the number of operations *#op* and processed quantums *#q*. The developer is responsible for counting them and passes them to the *EPPAProbe* object. The application parameters are passed with the *EPPAProbe* constructor.

At the end of program execution, the data is written to a mySQL database (www.mysql.com). The analysis of the data is written in Java and performed automatically. Results are shown graphically in different *views*, each representing a different aspect of the analysis: the *time* (Fig. 1), *overhead* (Fig. 2), *causal* (Fig. 3 & 5) and *functional* views (Fig. 4 & 6).

5   Model Refinement

The relations between the variables are analyzed by finding the best predicting equations, using standard regression analysis [4, 5]. We use the LOOCV (Leave-One-Out Cross Validation) method to choose among polynomial equations of different degree. This method overcomes overfitting by testing how each individual observation can be predicted by the other observations.

5.1   Computation

As a first order approximation, we take the computational runtime as proportional to the number of identical operations:

$$T_{computation} = \delta_{comp} . \# operations . \tag{6}$$

Clearly, this equation is not realistic, as for example, in the presence of superscalar architectures with multiple arithmetic units and hardware pipelines. It is even not applicable for a simple PC, when the memory usage exceeds the RAM capacity, as is the case in our experiments. However, we argue that it still is a useful relation. It can serve as a first approximation, but it should be extended when going out of the 'normal' linear regime. Figure 4 shows the experimental results for the matrix multiplication. Up to $\pm 54.10^9$ operations (n=3800) Eq. 6 holds firmly and reveals a δ_{comp} of 0.25µs/operation (the straight line). After that point, the processors start using the swap memory and the runtime increases super linear. Eq. 6 then fails to explain the performance. These divergences should then be supported by a more refined performance model, as in [12], namely measurement of the memory usage (Fig. 5). The extra runtime, caused by memory swapping T_{swap}, gives a roughly linear trend with the swap memory with δ_{swap} = 0.27ms/#q_{swap}. It means that T_{swap} is 'caused' by the data in swap memory #q_{swap}. Note that this relation is only valid for this specific program.

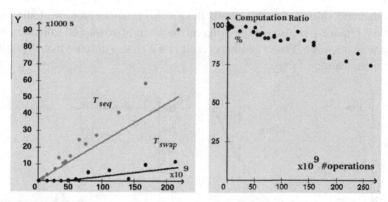

Fig. 4. Sequential run time (left) and computation ratio (right) vs. number of operations (p=4).

The right curve of Fig. 4 is a result of the smaller memory requirements of the parallel algorithm. When the sequential runtime starts to saturate, the sum of the parallel computation times on all processors is less. The computation ratio $\dfrac{\sum_{i}^{p} T_{comp}^{i}}{T_{seq}}$ drops to 75% for $n > 6000$ and a better than ideal speedup of 5.4 is reached.

5.2 The Communication Overhead

Cannon's algorithm requires $max(r,c)-1$ communication steps, where each process sends 2 submatrices of $n/r \times n/c$, containing n^2/p elements or quantums. The communication time is then

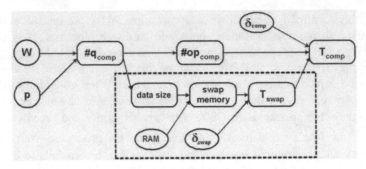

Fig. 5. Refined performance model for computation phases.

$$\sum_j^p T_{comm}^j \sim p.2.\frac{n^2}{p}.\max(r,c) \sim n^2.\max(r,c) \qquad (7)$$

For quadratic values of p, this results in $O(n^2\sqrt{p})$ relation. Non-quadratic values of p however require partitions with different number of rows and columns, hence a higher communication. Prime numbers result in a 1 row, p column matrix partitioning with communication of $O(p)$.

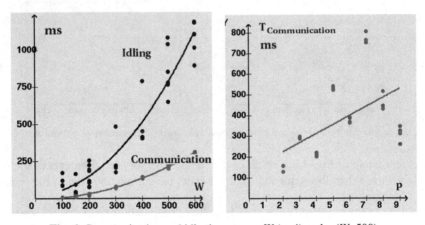

Fig. 6. Communication and idle time versus W ($p=4$) and p ($W=500$).

The experimental results (Fig. 6) for the communication confirm this theoretical expression. The idle time follows the same trend as a function of W. As the load imbalances are very low, the idle time is mainly caused by the message delays, which are also proportional to the number of elements being sent.

The right curve of Fig. 6 right shows communication time as a function of the number of processors p. It can be seen that the points are more difficult to interpret symbolically, because of the non-polynomial relation between data size and p (Eq. 7). This illustrates the advantage of using an intermediate variable $\#q_{comm}$ to get 2 curves that can be interpreted more easily. The curve $\#q_{comm}$ versus p would reveal Eq. 7, while T_{comm} versus $\#q_{comm}$ size could reveal non-linear trends in message delays, which is much more difficult for the global relation T_{comm} versus p of Fig. 6 (right curve).

6 Conclusions

Our approach presents a comprehensible performance analysis to the user, which starts with a high-level performance model. We introduced a causal model to show the relations between all relevant characteristics. In cases the simple model cannot reveal the nature of the performance, hardware characteristics and additional over-heads are added to refine the performance model. In this way, the user is only con-fronted with low-level characteristics if they affect the overall performance.

To facilitate the manual code instrumentation, it could be *guided* by a visual tool, as done by several performance tools [12, 13]. Furthermore, the so created causal diagrams could be exploited to *reason* about the performance [9].

References

1. Bull, J.M.: A Hierarchical Classification of Overheads in Parallel Programs. In: Proceedings of First IFIP TC10 International Workshop on Software Engineering for Parallel and Distributed Systems, Chapman Hall, (March 1996) 208-219.
2. Crovella, M. E. and Leblanc, T.J.: Parallel Performance Prediction using Lost Cycles Analysis. In: Proc. of Supercomputing '94, IEEE Computer Society (1994).
3. Crijns, J. and Crijns, A. Automatische Experimentele Analyse van Systeem en Algoritmeparameters op Parallelle Performanties. *Thesis*, Vrije Universiteit Brussel (VUB), Brussels, 2003.
4. Keeping, E.S., Introduction to Statistical Inference, Dover Publications Inc, New York, 1995.
5. Kumar, V., Grama, A., Gupta, A. and Karypsis, G.: Introduction to Parallel Computing. Design and Analysis of Algorithms. Benjamin Cummings, California (1994).
6. Mohr, B., and Wolf, F. KOJAK - A Tool Set for Automatic Performance Analysis of Parallel Programs. *Euro-Par Conf.* 2003: 1301-1304.
7. Nagel, W.E., Arnold, A., Weber, M., Hoppe, H.-C. and Solchenbach, K. VAMPIR: Visualization and analysis of MPI resources. *Supercomputer*, 12(1):69-80, January 1996.
8. Pancake, C.M.: Applying Human Factors to the Design of Performance Tools. In: Proc. of the 5th Euro-Par Conf., Springer (1999).
9. Pearl, J. Causality. Models, Reasoning and Inference. Cambridge University Press, Cambridge, 2000.
10. Reed, D.A., Aydt, R.A., Noe, R. J., Roth, P.C., Shields, K.A., Shwartz, B.W., and Tavera, L.F. Scalable Performance Analysis: The Pablo Performance Analysis Environment. In *Proc. Scalable Parallel Libraries Conf.*, IEEE Computer Society, 1993.

11. Sarukkai, S. R., Yan, J., Gotwals and J. K.. Normalized performance indices for message passing parallel programs. *In Proc. of the 8th international conference on Supercomputing, Manchester*, England, 1994.
12. Snavely, A. et all., A framework for performance modeling and prediction. In Proc. of the 2002 ACM/IEEE conference on Supercomputing, Baltimore, Maryland (2002) 1-17.
13. Truong, H-L and Fahringer, T.: Performance Analysis for MPI Applications with SCALEA. *In Proc. of the 9th European PVM/MPI Conf.*, Linz, Austria (September 2002).
14. Yan, J. C., Sarukkai, S. R., and Mehra, P.: Performance Measurement, Visualization and Modeling of Parallel and Distributed Programs using the AIMS Toolkit. *Software Practice & Experience*, April 1995.

What Size Cluster Equals a Dedicated Chip

Siegfried Höfinger

Novartis Institutes for Biomedical Research, Vienna
Unit: IK@N, In Silico Sciences
Brunnerstr. 59, A-1235, Vienna, Austria
siegfried.hoefinger@pharma.novartis.com
http://www.nibr.novartis.com

Abstract. Poisson-Boltzmann calculations within the framework of the Boundary Element Method [5] are parallelized with MPI. The approach is shown to be numerically accurate and of reasonable computational cost. Parallel performance is investigated on two different types of architectures, a cluster of Alphas ES40, 667 MHz EV67, and an IBM Regatta 690, 1.3 GHz Power 4. The obtained scaling characteristics deviate only slightly from the theoretical scaling due to Amdahl's Law. An alternative version is also implemented that makes use of a specialized chip for the core-calculational part. The raw-performance of this specialized ASIC, MD-GRAPE-2, is compared to the parallel performance gained via MPI. It is estimated that for a medium-sized test-case one would have to operate a parallel machine composed of roughly 330 CPUs to equal the performance of 1 single specialized device.

1 Introduction

Many scientific calculations suffer from limited computational performance. Especially in the fields of biomolecular, pharmaceutical and medicinal research the extreme complexity of the studied systems often renders a detailed theoretical approach either impossible or at least very cumbersome. Nevertheless the demand for precise computational investigations of typical structural features in all areas of biomedical research is constantly growing and therefore all disciplines in computational sciences are nowadays challenged to come up with drastically improved methods and algorithms that eventually can cope with the steadily growing dimensionality and complexity of today's key-topics in science and engineering.

In certain cases a specific type of calculation has been identified to be of such central importance that computer scientists have implemented the corresponding algorithm on a dedicated chip, a so-called *ASIC (Application Specific Integrated Circuit)*, that cannot do any ordinary type of arithmetics, but is largely superior to any general-purpose computer when it comes to that specific type of application it has been designed for. A very impressive example for such an ASIC has been given with the development of the *MDM (Molecular Dynamics Machine)* [1]. Luckily this ASIC is also commercially available for exploratory research in

D. Kranzlmüller et al. (Eds.): EuroPVM/MPI 2004, LNCS 3241, pp. 397–404, 2004.

the form of the standard PCI-cards *MD-GRAPE-2*. However, due to the very specialized design of an ASIC its applicability to alternative topics beyond the originally intended class of problems appears to be a rather tough issue.

The present communication is about an application that makes use of MD-GRAPE-2 cards. The general question is how well it will perform compared to a parallelized version using MPI [3]. While the original focus of MD-GRAPE-2 is on computing non-bonded types of pairwise interactions, e.g. electrostatic forces or van der Waals forces between all the atoms in large molecules, the concept has been extended within the current application and the ASIC is used alternatively to compute solutions to the *Poisson Boltzmann equation (PB)* [4] within the framework of the *Boundary Element Method (BEM)* [5]. In a very simplified view PB may be regarded as a means of estimating the influence coming solely from the environment, e.g. solvent, dielectric medium, glass etc., when a certain molecule of interest is dissolved.

1.1   Description of PB/BEM

A detailed technical description of the individual steps necessary to perform a PB/BEM calculation goes beyond the scope of this communication and will become subject to a future article. Here a brief summary of the most important key-steps involved in the process shall be given.

1. Suppose one has a molecule, that is composed of a number of atoms, say M, which all of them carry a certain charge Q^{at}. We are interested in the energetic effect we face when this molecule becomes embedded in a dielectric medium, e.g. water. The first thing one has to take care of, is the definition of a dielectric boundary, that separates molecular space ($\varepsilon = 1$) from dielectric space ($\varepsilon \approx 80$ in the case of water), the continuum. Typically to the BEM this boundary may be described in terms of triangles encompassing the molecular space. For the present case we have used the molecular surface program by Connolly [6] with which a dissection into a set of N small, flat triangles comprising the dielectric boundary is achieved.

2. All the atomic charges will give rise to an initial induction of polarization charges occuring at all the triangles of the boundary $\{Q^{at}\}_M \rightarrow \{q^{pol}\}_N^{1=ini}$.

3. Each of the polarization charges at each of the triangles will cross-react with all the other one charges on all the other triangles. Formally this process may be described as

$$\underline{\underline{D}}_{N,N} \cdot \underline{q}_N^{pol} = \underline{q}_N^{ini} \tag{1}$$

where $\underline{\underline{D}}_{N,N}$ is a matrix of dimension N x N, $\underline{q}_N^{ini}$ is a vector of dimension N closely related to the expression shown above that takes into account just the action of the atoms, and finally $\underline{q}_N^{pol}$ is again a vector of dimension N that represents all the unknowns.

4. The solution to equation (1) may be obtained by direct matrix-inversion [5] [7]. However, already for medium-sized molecules the number of triangles forming the dielectric boundary becomes a limiting factor and conventional techniques for matrix-inversion cannot be used any longer. Fortunately

equation (1) may be recast and an iterative approach may be pursued – $q_{\underline{N}}^{pol,i} \rightarrow q_{\underline{N}}^{pol,i+1}$. The big advantage of this iterative approach is that one never has to store a matrix of dimension $N \times N$, but instead can get by with just vectors of dimension N [1]. In addition, the number of necessary iterations i to achieve a certain degree of accuracy may be kept small and a well-known convergence acceleration scheme may be applied to increase the success-rate of the iterative algorithm [8]. Finally it must be pointed out that this iterative approach could only be run on an ASIC like MD-GRAPE-2 because all the considered interactions are essentially of Coulombic nature, hence of one of the few types of specialized calculations the chip was made for.

1.2   Studied Model Case

Endoglucanase (code 8A3H) was taken from the PDB database [9] and only the protein-structure (no hetero-components, no solvent) was taken into account. The protein consists of 300 amino acid residues, or 4570 atoms (parameter M in section 1.1) and exhibits a net charge of -32. Non-bonded parameters of the AMBER-96 force field [10] were assigned to each of the atoms in the protein and the molecular surface was determined via Connolly's MSROLL program [6] resulting in a triangulation of the molecular surface of 26230 triangles (parameter N in section 1.1) as shown in Fig. 1. A quick check of proper functionality is to sum up the set of induced polarization charges and see if this sum can counter-balance the net charge of the protein and indeed in all the calculations presented next this sum always turned out to be +32.002 (without any applied scaling). Variation in the final results of hydration free energies $\Delta G^{H_2 O}$ was $\pm$ 0.0003 kcal/mol on different platforms (Alpha, IBM) and $\pm$ 0.399 kcal/mol when comparing MD-GRAPE-2 results to conventional platform results (Alpha, IBM). The iterative approach was used throughout and the calculations always terminated within 64 cycles[2] thereby using DIIS [8], a method to improve the convergence rate in iterative sequences. This way the absolute accuracy for each of the 26230 polarization charges was $\pm 1.0 \ 10^{-5}$ a.u. .

2   Parallelization via MPI

Since the iterative approach in PB/BEM as outlined in section 1.1 adopts a repetitive scheme, it was straightforward to attempt a parallelization strategy in which the outermost loop in each of the iterations became distributed over a number of parallel operating nodes. Thus in the beginning a "master"-process sends out all the current actual polarization charges $q_{\underline{N}}^{pol,i}$ to each of the nodes, which all solve a subset of lines of eq. (1) and afterwards send back their subset

---

[1] where N up to about 10^6 can be solved in one single call to the MD-GRAPE-2 board.

[2] which is a result of the convergence threshold criterion chosen somewhat arbitrary.

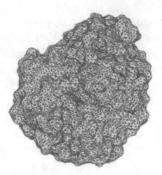

Fig. 1. Molecular Surface of the protein Endoglucanase (PDB code 8A3H) generated with the program MSROLL [6]. The protein consists of 4570 atoms, has a net charge of -32 and the surface is made up of 26230 triangles.

of improved polarization charges $q_N^{pol,i+1}$ to the "master"-process, which assembles all the pieces of incoming partial results. Next the "master"-process checks the degree of convergence, applies DIIS and either terminates or sends out the improved charges again to repeat the parallel section until the degree of accuracy is sufficient. MPI has been used for parallelization and the code was tested on a cluster of Alphas ES40, 667 MHz EV67, as well as on the SMP machine IBM Regatta p690, 1.3 GHz Power 4. According to Amdahl's Law the theoretical speedup of some parallel application running on N processing nodes is expressed as $SpeedUp = \frac{T(1)}{T(N)}$ with T(1) and T(N) being the execution times on 1 or N CPUs respectively. Since $T(1)$ may be split into a serial and a parallel fraction – $T(1) = f_s \cdot T(1) + f_p \cdot T(1)$ – any general scaling expressed in terms of speedup will be determined by the magnitude of the serial fraction f_s, in particular $SpeedUp = \frac{1}{f_s + \frac{f_p}{N}}$. For the present application f_s has been determined from independent timings of appropriate code-sections to be equal to 0.007143. A series of measurements was performed with increasing numbers of parallel operating nodes on the two different platforms mentioned above and a corresponding plot of the obtained scaling together with a comparison to ideal behavior (Amdahl) is shown in Fig. 2. Absolute calculation speed as well as obtained speedup values are also summarized in Table 1. The chosen problem size is reflected in the number of boundary elements taken into account – 26230 in particular (which is parameter N in section 1.1).

3   MD-GRAPE-2 Implementation

The iterative PB/BEM approach was further ported to MD-GRAPE-2 (please see reference [2] for a general overview). Tetsu Narumi's [1] MR1-library offered a convenient solution for an appropriate API. Each of the MD-GRAPE-2 boards is equipped with 4 ASICs. A maximum number of 4 MD-GRAPE-2 boards could be tested. In addition to the MD-GRAPE-2 boards so-called booster-cards were

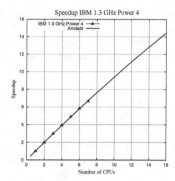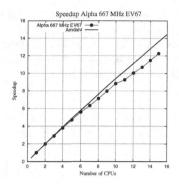

Fig. 2. Measured speedup on the IBM Regatta p690, 1.3 GHz Power 4, as well as on the cluster of Alphas ES40, 667 MHz EV67, for the case of Poisson Boltzmann calculations done in parallel. Ideal behavior is also shown as Amadahl's Law.

employed that mainly enhance the board's raw-performance by supplying it with additional power. A series of PB/BEM calculations on the system mentioned in section 1.2 was performed with increasing numbers of MD-GRAPE-2 cards. Additional improvements resulting from the co-operation of booster-cards was tested likewise. All the obtained performance data is listed in Table 1.

4   Discussion

A set of different versions of PB/BEM calculations has been implemented and particular attention has been paid to optimization of the run-time performance of the computations. Several conclusions may be drawn. At first, when directly comparing the actual performance on the two architectures, the Alpha ES40 system seems to be superior to the IBM Regatta p690 by roughly a factor of 1.6 (compare first-line entries of columns 6 and 5 in Table 1). Either version of parallel code on both architectures exhibits rather good scaling-characteristics (see Fig. 2). However, despite a small advantage of the Alpha-cluster in terms of absolute performance, the scaling of the parallel version on the IBM p690 is close to optimal and hence at a certain point when employing a larger number of CPUs the IBM p690 should actually become superior to the Alpha-cluster. This effect is solely a function of better scaling on the IBM p690.

The actual parallel performance with increasing numbers of CPUs was measured and the data became subject to a least-squares fit. A function similar to the type of Amdahl's Law but with the two fitted parameters of serial fraction and parallel fraction was deduced. The two resulting speedup curves read $SpeedUp_{Fit}^{Alpha} = \frac{1}{0.0168 + \frac{0.9847}{N}}$ and $SpeedUp_{Fit}^{IBM} = \frac{1}{0.0091 + \frac{0.9814}{N}}$ and a graphical representation of both scaling-approximations is given in Fig. 3. One can clearly see that beyond a certain critical number of CPUs a plateau is reached upon which any additional enlargement of the parallel architecture could bring only marginal improvement of the absolute performance any more.

Table 1. Speedup (left side of the table) and absolute performance (right side of the table) of Poisson Boltzmann calculations carried out in parallel or on the ASIC MD-GRAPE-2. The two parallel platforms were an SMP, IBM Regatta p690, 1.3 GHz Power 4, as well as a cluster of Alphas ES40, 667 MHz EV67. Theoretical behavior due to Amdahl's Law is also included.

# CPUs or # MD-GRAPE-2s	Amdahl Theor. Speedups	Alpha 667 MHz Exper. Speedups	IBM 1.3 GHz Exper. Speedups	Alpha 667 MHz Abs. Perf. [sec/it]	IBM 1.3 GHz Abs. Perf. [sec/it]	MD-GRAPE-2 no Booster Cards Abs. Perf. [sec/it]	MD-GRAPE-2 incl. Booster Cards Abs. Perf. [sec/it]
1	1.000	—	—	58.6	95.4	2.38	1.16
2	1.986	1.976	1.988	29.7	48.0	1.48	0.87
3	2.958	2.888	2.963	20.3	32.2	1.23	0.83
4	3.916	3.801	3.923	15.4	24.3	1.10	0.80
5	4.861	4.696	4.887	12.5	19.5	—	—
6	5.793	5.583	5.830	10.5	16.4	—	—
7	6.712	6.327	6.664	9.3	14.3	—	—
8	7.619	7.120	—	8.2	—	—	—
9	8.514	7.966	—	7.4	—	—	—
10	9.396	8.828	—	6.6	—	—	—
11	10.267	9.264	—	6.3	—	—	—
12	11.126	10.032	—	5.8	—	—	—
13	11.974	10.690	—	5.5	—	—	—
14	12.810	11.474	—	5.1	—	—	—
15	13.636	12.261	—	4.8	—	—	—

Next it was of particular interest to compare the MPI-versions to the MD-GRAPE-2 variant. The absolute calculational speed of a PB/BEM computation performed on 1 single MD-GRAPE-2 board including a booster-card was determined (see initial entry in column 8 of Table 1). A putative speedup-factor was derived, that would be required to obtain comparable absolute performance between the parallel platforms and the MD-GRAPE-2-variant. From this together with the data shown in Fig. 3 a theoretical number of CPUs could be estimated that would be required to achieve equal performance between MPI-versions and the ASIC. This number turned out to be surprisingly large and reached a value of around 330 for both architectures (also indicated in Fig. 3). Thus it appears that for certain well-selected core-processes in numerical science an appropriate hardware-solution can actually be of considerable beneficial character. This is even more so true if one thinks of a large assembly of specialized ASICs, since from looking again into column 8 of Table 1 it seems that the absolute limit in peak-performance on the set of 4 MD-GRAPE-2 cards has not even yet been reached with the test-case investigated. In this context it is particularly interesting to read that the ever-increasing raw-performance of general purpose chips seems to face a hard-limit in manufacturing once in future [11].

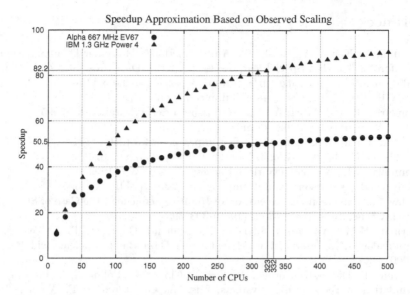

Fig. 3. Estimated Scaling on two different architectures and derived parallel computing requirements (# CPUs) to equal the performance of 1 single MD-GRAPE-2 card with attached booster card for the case of Poisson Boltzmann calculations.

5   Conclusion

PB/BEM calculations could be efficiently parallelized with MPI and the obtained scaling was shown to reach from quite fair to up to rather strong on both parallel platforms tested, a cluster of Alphas ES40, 667 MHz EV67, as well as an IBM Regatta p690, 1.3 GHz Power 4. A second variant was implemented based on the ASIC MD-GRAPE-2 and a comparison showed that one would have to operate a parallel machine composed of roughly 330 CPUs of the types mentioned above to equal the performance of 1 single MD-GRAPE-2 board with attached booster-card. The type of application considered here – PB/BEM – is particularly suited for MD-GRAPE-2 but has not originally been considered to be a target for the ASIC. Thus it was shown that an application other than plain Molecular Dynamics can effectively profit from the specialized hardware MD-GRAPE-2.

Acknowledgment

The author would like to thank Tetsu Narumi from the RIKEN institute in Japan for help with MD-GRAPE-2 and the MR1 library, Benjamin Almeida, Wolfgang Michael Terényi from IK@N Vienna and Pascal Afflard from IK@N Basel for support with the HPC-infrastructure and András Aszódi, head of IK@N ISS for help with the manuscript.

References

1. Narumi, T., Kawai, A., Koishi, T.: An 8.61 TFlop/s Molecular Dynamics Simulation for NaCl with a Special-Purpose Computer: MDM. Proc. SC. 2001, Denver, (2001) http://atlas.riken.go.jp/~narumi/paper/mdm/index.html
2. MD-GRAPE. http://www.research.ibm.com/grape/
3. Dongarra, J. et.al.: MPI: A Message-Passing Interface Standard; (1995) http://www-unix.mcs.anl.gov/mpi/
4. Luo, R., David, L. Gilson, M.K.: J. Comp. Chem. **23** (2002) 1244–1253
5. Zauhar, R.J., Morgan, R.S.: J. Mol. Biol. **186** (1985) 815–820
6. Connolly, M.L.: Molecular Surface Package; Version 3.8 (1999)
7. Höfinger, S., Simonson, T.: J. Comp. Chem. **22**, 3, (2001) 290–305
8. Pulay, P.: Convergence Acceleration of Iterative Sequences, the Case of SCF- Iteration. Chem. Phys. Lett. **73**, 2, (1980) 393–398
9. Berman, H.M., Westbrook, J., Feng, Z., Gilliland, G., Bhat, T.N., Weissig, H., Shindyalov, I.N., Bourne, P.E.: The Protein Data Bank. Nucleic Acid Res. **28** (2000) 235–242
10. Kollman, P., Dixon, R., Cornell, W., Fox, T., Chipot, C., Pohorille, A.: in Computer Simulation of Biomolecular Systems, Eds. van Gunsteren, W.F., Weiner, P.K., Wilkinson, A.J. **3**
11. Först, C.C., Ashman, C.R., Schwarz, K., Blöchl, P.E.: The interface between silicon and a high-k oxide. Nature **427**, (2004) 53–56

Architecture and Performance
of the BlueGene/L Message Layer

George Almási[1], Charles Archer[2], John Gunnels[1], Philip Heidelberger[1],
Xavier Martorell[1], and José E. Moreira[1]

[1] IBM Thomas J. Watson Research Center
Yorktown Heights, NY 10598-0218
{gheorghe,gunnels,philip,xavim,jmoreira}@us.ibm.com
[2] IBM Systems Group
Rochester, MN 55901
archerc@us.ibm.com

Abstract. The BlueGene/L supercomputer is planned to consist of 65,536 dual-processor compute nodes interconnected by high speed torus and tree networks. Compute nodes can only address local memory, making message passing the natural programming model for the machine. In this paper we present the architecture and performance of the BlueGene/L message layer, the software library that makes an efficient MPI implementation possible. We describe the components and protocols of the message layer, and present microbenchmark based performance results for several aspects of the library.

1 Introduction

The BlueGene/L supercomputer is a new massively parallel system being developed by IBM in partnership with Lawrence Livermore National Laboratory (LLNL). BlueGene/L uses system-on-a-chip integration [5] and a highly scalable architecture [2] to assemble 65,536 dual-processor compute nodes. When operating at its target frequency of 700 MHz, BlueGene/L will deliver 180 or 360 Teraflops of peak computing power, depending on its mode of operation. BlueGene/L is targeted to become operational in early 2005. Each BlueGene/L compute node can address only its local memory, making message passing the natural programming model for the machine.

This paper describes the design and implementation of the BlueGene/L *message layer*, the library we used to implement MPI [8] on this machine. In last year's PVM/MPI paper [3] we presented a high level design of the system based on **simulations** we had run. We made predictions about the kind of performance tuning that would be necessary to achieve good performance in the MPI subsystem.

This year's paper presents actual performance results along with a low-level architecture design of the BlueGene/L message layer. It is organized as follows. Section 2 is a short presentation of BlueGene/L hardware and software. Section 3 introduces and describes the message layer. Section 4 describes the point-to-point messaging protocols implemented in the message layer. Section 5 motivates the need for collective operations in the message layer. Section 6 presents performance results based on several microbenchmarks that we used to guide the development process. Section 7 concludes by presenting the challenges we face in the near future.

D. Kranzlmüller et al. (Eds.): EuroPVM/MPI 2004, LNCS 3241, pp. 405–414, 2004.

2   A Short Discussion of BlueGene/L Hardware and Software

The BlueGene/L hardware [2] and system software [4] have been extensively described previously. In this section we remind the reader of the hardware features most relevant to the discussion to follow.

BlueGene/L Processors: the 65,536 compute nodes of BlueGene/L are based on a custom system-on-a-chip design that integrates embedded low power processors, high performance network interfaces, and embedded memory. The low power characteristics of this architecture permit a very dense packaging. One air-cooled BlueGene/L rack contains 1024 compute nodes (2048 processors) with a peak performance of 5.7 Teraflops.

Cache Coherency: The standard PowerPC 440 cores are not designed to support multiprocessor architectures. The L1 caches are not coherent and the processor does not implement atomic memory operations. Software must take great care to insure that coherency is correctly handled in software. Coherency handled at the granularity of the CPUs' L1 cache lines: 32 bytes. This means that objects not delimited by 32 byte boundaries cannot be shared by the CPUs. To overcome these limitations BlueGene/L provides a variety of custom synchronization devices in the chip such as the lockbox (a limited number of memory locations for fast atomic test-and-sets and barriers) and 16 KB of shared SRAM.

Floating Point: Each processor is augmented with a dual floating-point unit consisting of two 64-bit floating-point units operating in parallel. The dual floating-point unit contains two 32×64-bit register files, and is capable of dispatching two fused multiply-adds in every cycle, i.e. 2.8 GFlops/s per node at the 700 MHz target frequency. When both cores are used, the peak performance is doubled to 5.6 GFlops/s.

The Torus Network is the main network used for user communication. Each compute node is connected to its 6 neighbors through bi-directional links. The 64 racks in the full BlueGene/L system form a $64 \times 32 \times 32$ three-dimensional torus. The network hardware guarantees reliable, deadlock free delivery of variable length packets. Routing is done on an individual basis, using one of two routing strategies: a *deterministic* routing algorithm, all packets between two nodes follow the same path along the x, y, z dimensions (in this order); and a minimal *adaptive* routing algorithm, which permits better link utilization but allows consecutive packets to arrive at the destination out of order.

Network Efficiency: The torus packet length is between 32 and 256 bytes in multiples of 32. The first 16 bytes of every packet contain destination, routing and software header information. Therefore, at most 240 bytes of each packet can be used as payload. For every 256 bytes injected into the torus, 14 additional bytes traverse the wire with CRCs etc. Thus the efficiency of the torus network is at most $\eta = \frac{240}{270} = 89\%$.

Network Bandwidth: Each link delivers two bits of raw data per CPU cycle (0.25 Bytes/cycle), or $\eta \times 0.25 = 0.22$ bytes/cycle of payload data. Adding up the raw bandwidth of the 6 incoming + 6 outgoing links on each node, we obtain $12 \times 0.25 = 3$ bytes/cycle per node. The corresponding bidirectional payload bandwidth is 2.64 bytes/cycle/node or 1848 MB/s at 700 MHz clock speed.

Network Ordering Semantics: Adaptively routed network packets may arrive out of order, forcing the message layer to reorder them before delivery. On the other hand, deterministically routed packets lead to network congestion.

BlueGene/L Communication Software: BlueGene/L is designed from ground up as a machine to run MPI on. For this purpose we ported the MPICH2 package from ANL [1, 6]. The MPI port is based on the BlueGene/L *message layer*.

3   Message Layer Architecture

Figure 1 shows the structural and functional composition of the message layer. It is divided into three main categories - basic functional support, point-to-point communication primitives (or protocols) and collective communication primitives. The base layer acts as a support infrastructure for the implementation of all the communication protocols.

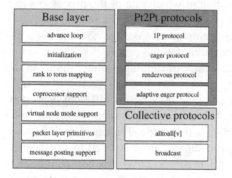

Fig. 1. Structural composition of the message layer

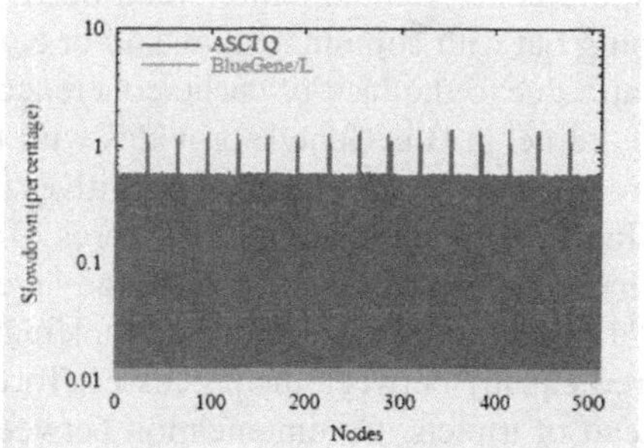

Fig. 2. Comparison of noise levels on the ASCI Q and BlueGene/L supercomputers

Initialization: Startup of the message layer implies initialization of the state machines, rank mapping subsystem and operating mode based on input from the user. The message layer assumes network hardware to be in a clean state at startup, and takes control of it. While the message layer is operating no other software packages should read or write network packets.

Advance Loop: The message layer's design is centered around an *advance loop* that services the torus hardware (i.e. sends and receives packets). The advance loop is designed to be explicitly called by the user, because although the network hardware supports interrupt driven operation, handling hardware interrupts is time consuming. In addition, interrupts tend to introduce "jitter" into the system and degrade the performance of MPI [7]. On BlueGene/L there are **no** timer interrupts or system interrupts of any kind, leading to two orders of magnitude less jitter than on comparable systems. Figure 2 compares the noise level on a running BlueGene/L system with that measured on the ASCI Q supercomputer.

The price for polling based operation is that the network makes no progress without direct help from the user. We plan to add interrupt-driven operation to the message layer in order to support overlapping of computation and communication in certain cases.

Mapping: The correct mapping of MPI applications to the torus network is a critical factor influencing performance and scaling. The message layer, like MPI, has a notion of process ranks, ranging between 0 and $N - 1$ where N is the number of processes participating. Message layer ranks are the same as the COMM_WORLD ranks in MPI. The message layer allows arbitrary mapping of torus coordinates to ranks. This mapping can be specified via an input file listing the torus coordinates of each process in increasing rank order. If no mapping is specified, torus coordinates are mapped in lexicographic order.

3.1 Coprocessor and Virtual Node Mode Support

Coprocessor mode describes a mode of operation in which one of the processors in a compute node runs the main thread of the user's program, with the second processor helping out with communication and/or computation tasks. Coprocessor mode is problematic due to the lack of cache coherence between the two processors. The compute node kernel in BlueGene/L provides with the ability of running a single thread in the coprocessor and cache flushing primitives.

In virtual node mode the two cores of a compute node act as different processes. Each has its own rank in the message layer. The message layer supports virtual node mode by providing correct torus to rank mapping in this mode. The hardware FIFOs are shared equally between the processes. Torus coordinates are expressed by quadruplets instead of triplets. Communication between the two processors in a compute node is done via a region of memory, called the scratchpad, that both processors have access to. *Virtual FIFOs* make portions of the scratchpad look like a send FIFO to one of the processors and a receive FIFO to the other.

3.2 Packet Layer Primitives

The message layer services network hardware using three functions. It needs to check the status of the hardware FIFOs, and it injects and extracts packets from the FIFOs. The message layer's performance depends a great deal on the number of CPU cycles spent handling each individual network packet. The absolute limits of packet read/write times are about 100 CPU cycles for writes and 204 cycles for reads.

We keep packet processing time low by avoiding memory copies in the software stack. Outgoing packets are sent directly from the send buffer and incoming packets are read from the network directly to their final destination address when possible.

We use *partial packets* to help us avoid unnecessary memory copies during packet reads. Instead of reading a whole packet into a temporary buffer, the message layer uses reads out only the header and calls a *handler*. The rest of the packet is left in the FIFO until the handler decides, based on the header information, where to put it.

BlueGene/L network hardware accesses are restricted to quad word aligned memory buffers. The alignment restriction is caused by the double floating point load and store instructions we use to access the network devices. Unfortunately not all user buffers are aligned to 16 byte boundaries. When forced to read into non-aligned memory buffers, we have discovered that we can use a portion of the 204 cycles spent reading a packet from the network to perform an in-memory realignment of the already available data. Thus we effectively overlap the network read and the realignment copy, resulting in up to 50% savings in processing cycles for certain packet alignments.

4 Point-to-Point Protocols in the Message Layer

In this section we discuss the point-to-point messaging capabilities of the message layer. Messages can be sent with a number of protocols depending on the latency and bandwidth requirements.

4.1 The Eager Protocol

The eager protocol is simple both in terms of programmer's interface and implementation. It guarantees ordered delivery of messages by enforcing packet delivery order. All eager protocol packets are sent using deterministic routing. Packing and unpacking eager protocol messages is simple and efficient, with a running counter keeping track of the message offset both at the sender and the receiver.

Every eager message transmits a fixed size memory buffer that may contain message metadata. The contents of this buffer is opaque to the message layer; in the Blue-Gene/L MPI implementation we use it to transmit MPI matching information, such as the sender's MPI rank, the message tag and the context identifier.

4.2 The One-Packet Protocol

The one packet protocol is an even simpler version of the eager protocol for cases when the send buffer fits into a single packet. The one packet protocol saves overhead costs by virtue of a very simple packetizer. The programmer's API is also simpler than eager message's, because there is no need for the `recvdone` callback. The `recvnew` callback carries with it a temporary message buffer, and it is the user's responsibility to copy its contents before the callback returns. It has the lowest overhead and latency costs of all point-to-point messaging protocols.

4.3 The Rendezvous Protocol

The rendezvous protocol is necessary because eager messages are sent using deterministic routing, resulting in reduced network efficiency. The only packet sent via deterministic route in the rendezvous protocol is the initial "scout" packet that essentially asks permission from the receiver to send data. The receiver returns an acknowledgment, followed by the data transfer from the sender.

In our current implementation of the rendezvous protocol message reception is done by the coprocessor, subject to availability, cache coherence and alignment constraints. This results in reduced load on the receiver's main processor.

The rendezvous protocol has a very high per-message latency due to the initial roundtrip and to the involvement of the coprocessor; however, it also has the lowest per-packet overhead cost of all protocols, resulting in high bandwidth.

4.4 The Adaptive Eager Protocol

The adaptive eager protocol is a version of the eager protocol that uses **no** deterministically routed packets; instead it sends a confirmation packet to the sender every time the first packet of a new message is received. The sender can only start sending the next message after it has received a confirmation packet. This ensures that the receiver sees the first packet of each message in order.

We believe that the adaptive eager protocol will become more important as the Blue-Gene/L machine scales beyond 20,000 processors.

5   MPI Collective Operation Support in the Message Layer

It is typical of an MPI implementation to implement collective communication in terms of point-to-point messages. This is certainly the case for MPICH2, the framework used by BlueGene/L MPI. But on the BlueGene/L platform the default collective implementations of MPICH2 suffer from low performance, for at least three reasons.

First, the MPICH2 collectives are written with a crossbar-type network in mind, and not for special network topologies like the BlueGene/L torus network. Thus the default implementation more often than not suffers from poor mapping (see Section 3).

Second, point-to-point messaging in BlueGene/L MPI has a relatively high overhead in terms of CPU cycles. Implementing e.g. MPI broadcast in terms of a series of point-to-point messages will result in poor behavior for short message sizes, where overhead dominates the execution time of the collective.

Finally, some of the network hardware's performance features are hidden when using only standard point-to-point messaging. A good example of this is the use of the *deposit bit*, a feature of the network hardware that lets packets be "deposited" on every node they touch on the way to the destination.

Our work on collective communication in the message layer has just begun. We have message layer based implementations of MPI_Bcast and MPI_Alltoall[v]. The broadcast implementation benefits from all three factors we have enumerated. The alltoall implementation is somewhat immature - although it benefits from lower overhead it has a lower target bandwidth because it uses a type of packet with fewer payload bytes. Our short term future plans include implementations of MPI_Barrier and MPI_Allgatherv as well as MPI_Allreduce. Our primary focus is on these primitives because they are in demand by the people doing applications tuning on Blue-Gene/L today. In particular, broadcast, allgather and barrier are heavily used by the ubiquitous HPL benchmark that determines the TOP500 placement of BlueGene/L.

6   Performance Analysis

In this section we discuss the performance characteristics of the message layer. We are interested only in basic performance numbers, such as nearest-neighbor roundtrip latency, sustained single and multiple link bandwidth and the $\frac{N}{2}$ value, the message size where bandwidth reaches half of its asymptotic value.

Roundtrip latency is typically measured in processor cycles or microseconds of elapsed time. Because BlueGene/L is running at multiple CPU frequencies, in this paper we present results in terms of processor cycles, and we mention the corresponding microsecond measurements in the text. Similarly, we show bandwidth results in terms of bytes transferred per processor cycle; on this machine the ratio of processor and network speeds is fixed, making the Bytes/cycle unit of measure more reliable than MBytes/second.

For measuring performance we used various microbenchmarks, written both on top of the message layer as well as using MPI as a driver for the message layer. These are the very benchmarks we used to tune the message layer before releasing the code to outside users for running large applications. We find that these benchmarks are extremely useful in pinpointing performance deficiencies of the message layer.

6.1 Point-to-Point Message Latency

Figure 3 shows the half-roundtrip latency of 1-byte messages sent with all four point-to-point protocols. All latencies were measured with message layer adaptations of Dave Turner's mpipong program [9], as well as the un-adapted original for measurements from within MPI. Unsurprisingly, the one-packet protocol has the lowest overhead, weighing in at 1600 cycles. The highest overhead by far belongs to the rendezvous protocol, with the two eager variants in the middle of the range. When measured from within MPI, the latency numbers increase due to the additional software overhead. All measurements are shown both in cycles and in microseconds (assuming a 700MHz clock speed).

6.2 Asymptotic Bandwidth

We measured the asymptotic bandwidth of three of the four point-to-point protocols: we set up one node to simultaneously send to and receive from multiple neighbors. We plotted the achieved bandwidth as a curve family, varying the number of senders along the horizontal axis and the number of receivers between the curves.

The absolute bandwidth achieved by the multiple simultaneous send/receive scenario depends on the software overhead necessary to send/receive a packet. Figures 4,5 and 6 show the asymptotic bandwidth achieved by the eager, adaptive eager and rendezvous protocols. Performance in these graphs is measured in bytes transmitted per CPU cycle; at the 700 MHz frequency of the machine, 1 Byte/cycle corresponds to 700 MBytes/s bandwidth.

Of the three protocols, the rendezvous protocol achieves the best asymptotic bandwidth (1.45 Bytes/cycle, or 1 GByte/s, with 4 neighbors sending and 3 receiving). This is because of the lower per-packet reception cost of the protocol, as well as due to the fact that reception is offloaded to the coprocessor.

6.3 Bandwidth vs. Message Size

The BlueGene/L message layer approaches asymptotic bandwidth for relatively low message sizes. The two tables in Figure 7 compare the message sizes when *half* of

Protocol name	msglayer latency		MPI latency	
	cycles	μs	cycles	μs
one-packet	1600	2.29	2350	3.35
eager	2700	3.86	4000	5.71
adaptive eager	3300	4.71	11000	15.71
rendezvous	12000	17.14	17500	25.00

Fig. 3. Roundtrip latency comparison of all protocols

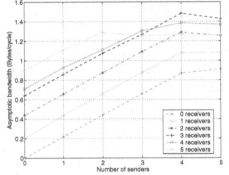

Fig. 4. Asymptotic bandwidth: eager protocol

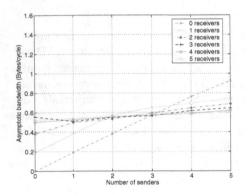

Fig. 5. Asymptotic bandwidth: adaptive eager protocol

Fig. 6. Asymptotic bandwidth: rendezvous protocol

protocol	S/R pairs	$\frac{N}{2}$ Bytes	BW B/cycle	BW MB/s
	1	512B	0.44	308
	2	2048B	0.88	616
rendezvous	3	10KB	1.20	840
	4	32KB	1.38	966
	5	32KB	1.23	861

protocol	S/R pairs	$\frac{N}{2}$ Bytes	BW B/cycle	BW MB/s
	1	512B	0.44	308
	2	1024B	0.72	504
eager	3	1024B	0.78	546
	4	2048B	0.80	560
	5	2048B	0.84	588

Fig. 7. Comparison of $\frac{N}{2}$ values of eager and rendezvous protocols

asymptotic bandwidth is achieved for the eager and rendezvous (coprocessor mode) protocols. The table shows that while the asymptotic bandwidth achieved by the rendezvous protocol is higher, the eager protocol can deliver good performance at much smaller message sizes.

6.4 Message Layer Broadcast

In this section we compare the performance of three implementations of `MPI_Bcast`. The first candidate is the default implementation of `MPI_Bcast` in MPICH2. Second, we have a mesh-aware implementation of broadcast using point-to-point MPI messages. Finally, we have a mesh-aware implementation of broadcast directly inside the message layer, using the torus network hardware's deposit bit feature.

The standard MPICH2 implementation of `MPI_Bcast` builds a binary tree of nodes (regardless of their position in the mesh/torus) to do the broadcast. Because the links connecting the individual nodes may multiply overlap, the expected bandwidth is not too high.

The mesh-aware broadcast implementation has a target bandwidth of two links worth on a 2D mesh. The message is cut into two roughly equal pieces and redistributed along two non-overlapping sets of links (as shown in Figure 8. On a 2D *torus* the optimal broadcast bandwidth would be 4 links worth, although the message layer would not be able to sustain that without help from the coprocessor.

Figure 9 compares the performance of the three broadcast implementations mentioned earlier, measured on a 4×4 mesh. The standard broadcast tops out at 0.1 Bytes/cycle, or about half of a single link's bandwidth. The mesh-aware MPI based implementation reaches about 0.3 Bytes/cycle, but the $\frac{N}{2}$ message size is inordinately large at about 200 KBytes. The message layer based implementation reaches exactly 0.44 Bytes/cycle, or 2 links worth of bandwidth, and reaches half that bandwidth for 5 KByte messages - clearly a large win over the other two implementations.

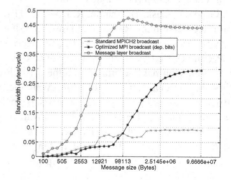

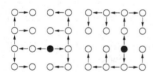

Fig. 8. Mesh-aware broadcast algorithm

Fig. 9. Comparison of `MPI_Bcast` implementations

7 Conclusions and Plans for the Future

The architecture discussion and the performance results presented in Section 6 are meant to showcase the tools we used for performance tuning in order to achieve reasonable MPI latency and bandwidth numbers. In our experience the micro-benchmark performance numbers presented in this paper are good predictors of the performance of large applications. Once we reached the phase where point-to-point messaging characteristics were deemed acceptable, it quickly became obvious that further fine-tuning of point-to-point protocols requires steadily increasing effort and yields smaller and smaller performance increments. At this stage optimization of collective performance, specifically optimization of MPI_Barrier, MPI_Bcast, MPI_alltoall etc. seems to yield the most immediate results. We are also working on deploying optimized collectives on the tree network; this work will be addressed in a future paper.

References

1. The MPICH and MPICH2 homepage.
   http://www-unix.mcs.anl.gov/mpi/mpich.
2. N. R. Adiga et al. An overview of the BlueGene/L supercomputer. In *SC2002 – High Performance Networking and Computing*, Baltimore, MD, November 2002.
3. G. Almási, C. Archer, J. G. C. nos, M. Gupta, X. Martorell, J. E. Moreira, W. Gropp, S. Rus, and B. Toonen. MPI on BlueGene/L: Designing an Efficient General Purpose Messaging Solution for a Large Cellular System. Lecture Notes in Computer Science. Springer-Verlag, September 2003.
4. G. Almási, R. Bellofatto, J. Brunheroto, C. Caşcaval, J. G. C. nos, L. Ceze, P. Crumley, C. Erway, J. Gagliano, D. Lieber, X. Martorell, J. E. Moreira, A. Sanomiya, and K. Strauss. An overview of the BlueGene/L system software organization. In *Proceedings of Euro-Par 2003 Conference*, Lecture Notes in Computer Science, Klagenfurt, Austria, August 2003. Springer-Verlag.
5. G. Almasi et al. Cellular supercomputing with system-on-a-chip. In *IEEE International Solid-state Circuits Conference ISSCC*, 2001.
6. W. Gropp, E. Lusk, D. Ashton, R. Ross, R. Thakur, and B. Toonen. MPICH Abstract Device Interface Version 3.4 Reference Manual: Draft of May 20, 2003.
   http://www-unix.mcs.anl.gov/mpi/mpich/adi3/adi3man.pdf.
7. F. Petrini, D. Kerbyson, and S. Pakin. The Case of the Missing Supercomputer Performance: Achieving Optimal Performance on the 8,192 Processors of ASCI Q. In *ACM/IEEE SC2003*, Phoenix, Arizona, Nov. 10–16, 2003. Available from http://www.c3.lanl.gov/˜fabrizio/papers/sc03_noise.pdf.
8. M. Snir, S. Otto, S. Huss-Lederman, D. Walker, and J. Dongarra. *MPI - The Complete Reference, second edition*. The MIT Press, 2000.
9. D. Turner, A. Oline, X. Chen, and T. Benjegerdes. Integrating new capabilities into NetPIPE. Lecture Notes in Computer Science. Springer-Verlag, September 2003.

Special Session of EuroPVM/MPI 2004

Current Trends in Numerical Simulation for Parallel Engineering Environments

ParSim 2004

Carsten Trinitis[1] and Martin Schulz[2]

[1] Lehrstuhl für Rechnertechnik und Rechnerorganisation (LRR)
Institut für Informatik
Technische Universität München, Germany
Carsten.Trinitis@in.tum.de

[2] Center for Applied Scientific Computing
Lawrence Livermore National Laboratory
Livermore, CA
schulz5@llnl.gov

Simulating practical problems in engineering disciplines has become a key field for the use of parallel programming environments. Despite remarkable progress in both CPU power and network technology as well as extensive developments in numerical simulation and software integration, this field still provides – and will continue to provide – challenging problems for both computer scientists and engineers in the application disciplines. In addition, the appearance of new paradigms like Computational Grids or E-Services has introduced new opportunities and challenges for parallel computation. This rich and complex environment demands an intensive cooperation between scientists from engineering disciplines and from computer science.

Since its introduction at EuroPVM/MPI 2002, ParSim is dedicated to providing a forum for interdisciplinary cooperations in this important field. It brings together researches with different backgrounds to discuss current trends in parallel simulation. It is our hope that this offers the opportunity to establish new contacts, to open up new perspectives, and to foster cooperations across disciplines. The EuroPVM/MPI conference series, as one of Europe's prime events in parallel computation, serves as an ideal surrounding for ParSim. This combination enables the participants to present and discuss their work within the scope of both the session and the host conference.

This year, ten papers were submitted to ParSim and we selected five of them. They cover both computer science aspects, including cache optimization, as well as experience with special applications from various fields of engineering and physics. We are confident that this resulted in an attractive program and we hope that this session will be an informal setting for lively discussions.

Several people contributed to this event. Thanks go to Jack Dongarra, the EuroPVM/MPI general chair, and to Peter Kacsuk and Dieter Kranzlmüller, the

D. Kranzlmüller et al. (Eds.): EuroPVM/MPI 2004, LNCS 3241, pp. 415–416, 2004.
© Springer-Verlag Berlin Heidelberg 2004

PC chairs, for their encouragement and support to continue the ParSim series at EuroPVM/MPI 2004. We would also like to thank the numerous reviewers, who provided us with their reviews in such a short amount of time and thereby helped us to maintain the tight schedule. Last, but certainly not least, we would like to thank all those who took the time to submit papers and hence made this event possible in the first place.

We hope this session will fulfill its purpose to provide new insights from both the engineering and the computer science side and encourages interdisciplinary exchange of ideas and cooperations. We hope that this will continue ParSim's tradition at EuroPVM/MPI.

Parallelization of a Monte Carlo Simulation for a Space Cosmic Particles Detector[*]

Francisco Almeida[1], Carlos Delgado[3],
Ramón J. García López[2,3], and Francisco de Sande[1]

[1] Depto. de Estadística, Investigación Operativa y Computación
Univ. de La Laguna, 38271–La Laguna, Spain
{falmeida,fsande}@ull.es
[2] Departamento de Astrofísica, Univ. de La Laguna
38206–La Laguna, Spain
rgl@iac.es
[3] Instituto de Astrofísica de Canarias (IAC)
38205–La Laguna, Spain
delgadom@iac.es

Abstract. Work in progress in the design of a parallel simulation of a
multipurpose cosmic rays detector is presented. The detector is part of an
experiment whose main goal is to study the spectrum and composition
of charged cosmic rays with unprecedent sensibility.
A sequential version of the simulator based on Monte Carlo method is
currently available, but it is ineffective for the study of large samples
of moderately high charged particles. Parallelism seems to be the most
suitable approach to increase its performance.
Driven by different reasons we have decided to take the sequential code
as a starting point instead of producing a parallel version from scratch.
In this work we present the preliminary computational results that we
have obtained for our first prototype of the parallel MPI `master-slave`
simulator on a PC cluster platform.

1 Introduction

Current analysis methods on astroparticle physics require detailed models of
the detector response for realistic experimental conditions. Due to the difficulty
of the problem this is usually achieved by Monte Carlo simulation, pipelined
to the reconstruction software designed for the real detector. This problem has
fundamental implications in theoretical physics and astrophysics [1]. The signal
search nature of this kind of experiments makes necessary to simulate a number
of particles crossing the detector close or larger than the real number of particles
detected. Indeed this ensures that the statistical uncertainty on any signal is not
dominated by the statistical fluctuations of the simulated data.

---

[*] This work has been partially supported by the EC (FEDER) and the Spanish MCyT
(Plan Nacional de I+D+I, TIC2002-04498-C05-05 and TIC2002-04400-C03-03).

D. Kranzlmüller et al. (Eds.): EuroPVM/MPI 2004, LNCS 3241, pp. 417–424, 2004.

A large amount of simulated cosmic rays is needed in advance to understand the detector response. However, the simulation of the particles are very high time consuming in the current sequential code. Furthermore, the nature of the random number generation in Monte Carlo simulations difficult their execution by independent runs using different machines. Parallelism appears as the natural choice to speed up the simulation while keeping control of the random number generation.

The remaining of the paper is organized as follows. Section 2 describes the structure, objectives and difficulties found in the design of the sequential simulator. The main challenges involved in the parallelization are presented in section 3. A centralized `master-slave` parallel approach is presented as the first parallelization scheme. In section 4 we present preliminary computational results. Although an important speed up is achieved still some work can be done to reduce the running time of the sequential simulation. We finalize the paper in section 5 with some concluding remarks and future lines of work.

2   The Problem

In the experiment under consideration we deal with the first large superconducting cosmic rays detector designed to study an unexplored region of the spectrum. It has very important implications in fundamental physics and astrophysics [1] and is supported by NASA [2] and DOE [3] among others.

We deal with the simulation of a multipurpose particle detector designed to spend a long period of time taking data on space. Among others, the experiment goal is to study the spectrum and composition of charged cosmic rays with unprecedent sensibility. This is achieved by using state of the art detector technology and large amounts of collected data ($\sim 10^{10}$ detected particles are expected). The present simulation code designed by the experiment collaboration is based on the GEANT v3.21 simulation package [4]. In its current state the code execution is fully sequential, so mass production of Monte Carlo data is performed by executing it on a large set of machines. Each simulation is fed with an initial random seed separated enough on the random number sequence of any other seed as to ensure the absence of correlations between the simulations in different machines.

The simulation code is composed of a main *Triggering* process where the direction and characteristics of the input particles are simulated (figure 1). This main process is composed of two basic procedures:

- *Propagation (or Tracking):* Simulates the propagation of the particle. This propagation involves the simulation of the interaction with the detector and the recursive propagation and interactions of the resulting elements.
- *Reconstruction:* Simulates the analogical and digital response of the detector to the mentioned interactions. This information is used to reconstruct the properties of the particle generated in the *Triggering* process.

Due to the physics of the interaction of charged particles with matter, the amount of random numbers needed to simulate a single particle, scales roughly as

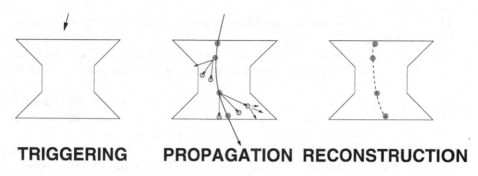

TRIGGERING   PROPAGATION RECONSTRUCTION

Fig. 1. The simulation process

the charge (the number of protons in the nuclei, Z) of the cosmic ray squared, as well as the simulation time and the amount of reconstructed information. Taking into account that the simulation time for a charge $Z = 1$ particle ranges from one to several seconds using a single CPU, the current mass production scheme becomes ineffective for simulation of large samples of moderately high charged particles. This ineffectiveness has its roots in the prohibitive long simulation time for a small set of incoming cosmic rays using a single machine and in the difficulty for avoiding correlations between different machines due to the amount of random numbers needed.

Parallelization seems to be the adequate solution to the exposed problem for moderate to high Z simulation as long as the core routines, specially the random number generation and physics simulation ones, are kept unchanged, as they have been tested for years independently in many particle physics experiments at CERN [5], [6], [7], [8]. The solution has to be such that the work is spread over a number of processors, to deal with the execution time problem, and the consistency of the random number is kept. Moreover it has to manage with the large quantity of processed data per particle, which amounts up to several tenths of Kb for $Z = 1$ cosmic rays. Due to the difficulties involved in the development of such code, at the present time, there are no previous parallel versions of the simulation available.

Finally a last requirement has to be met: the parallel solution has to be available as soon as possible since the the Monte Carlo data production should be large enough before the beginning of the physical data acquisition. A large amount of high Z simulated cosmic rays is needed in advance to understand the detector response.

3   Parallelization

From the high performance computing point of view, the parallelization of the code is a challenge because it involves many different issues that make it difficult:

- It is a *real life* code and therefore the size is not easy to manage (about 200000 lines).

- The code is written using two different programming languages: it mixes C++ and Fortran77.
- The physical simulation requires very specialized knowledge in the field, not easy to achieve for a computer scientist: an interdisciplinary effort is mandatory.
- It is a collaborative work where many programmers have been implied, and therefore the structure of the application is not clear even for some one not involved in the development, even if she is an expert programmer.
- The program uses several external libraries (more than 10 in its simplest mode), including threaded code. The incorporation of the parallel context has to be carefully analyzed to avoid non desired interactions.
- The code is highly sensitive to numerical arithmetic precision, so that it strongly relies on the compilers used.
- The very large amount of data that have to be managed by the parallel code is also a challenge. One terabyte of data can be produced by the simulator for an average input.

These difficulties force us to avoid the parallel design starting from scratch. One of our constraints is to introduce the minimum amount of changes in the original code. Once the most time consuming code sections have been identified, our goal is to parallelize the application intervening only on them.

In the last years OpenMP [9] and MPI [10] have been universally accepted as the standard tools to develop parallel applications. Some reasons brought us to discard OpenMP and the shared memory model as the initial option for our parallelization:

- To reach acceptable performance results we foresee that we will need a huge number of processors. Nowadays this is a limitation for shared memory architectures.
- To precisely identify the accessibility of the variables in the code is a difficulty when dealing with such large amount of data. An extra complication comes from the number of Fortran `common` blocks in the program.
- The nature of the code makes it sensitive to load imbalance. We hesitate if the OpenMP dynamic schedule clause is mature enough to deal with it efficiently [11].

In the parallelization of the code, we are following a two level incremental approach: in the first level we will broach the different sections of the code independently, while in the second level we will focus our attention in different target platforms. Currently we have parallelized the *Triggering* section and a version for PC clusters is now under development. *Propagation* and *Reconstruction* sections of the code are also candidates to be parallelized. The parallel platforms that we are considering for future development range from PC clusters and proprietary parallel platforms to grid resources, combining both of them.

Figure 2 illustrates the structure of the sequential code and the parallelization of the *Triggering* section. In a centralized `master-slave` approach, the `master` tracks the subproblem generation and takes charge of the load balance. The

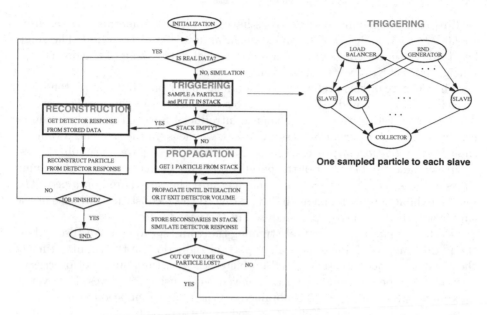

Fig. 2. Structure of the parallel `master-slave`

`collector` process manages the huge amount of output data generated by the `slaves`. Special consideration has to be deserved to the random number generation for the Monte Carlo simulation. We have chosen to use a centralized random number generator to avoid anomalies in the convergence of the simulation. The large amount of random numbers needed suggest to follow a strategy that overlaps computation and communication. This should allow that the `slaves` get the random numbers as soon as they are required.

In the scheme depicted in figure 2, the `slave` processes track the interactions produced by one particle when it goes through the detector producing new particles. This process intensively demands random numbers to the `generator` process and the result of these interactions are delivered to the `collector` process.

4   Computational Results

Currently we are working in the computational experiments. In this section we report the performance achieved for this preliminary parallelization of the simulation on a PC cluster platform using the `mpich` [12] implementation of MPI. The computational experiment has been carried out in a parallel machine with 16 nodes. Each node is a 2 GB memory shared memory bi-processor composed of Intel Xeon 2.80GHz processors interconnected with a 100 Mbit/sec. fast Ethernet switch and running Linux.

We have simulated 10000 He4 nuclei (alpha radiation) impinging onto the detector with an uniform spatial distribution and with an isotropic distribution on directions. The simulated particles have an initial momentum ranging from 400GeV/c to 3700GeV/c with an exponential decaying distribution. The simulated flux is equivalent to several minutes of real data taking of the detector at a height of 350 Km over the sea level.

The purpose of this preliminary experiment is not only to focus on the speed up achieved but to understand the sources of overhead in the parallel version. The analysis studies the load imbalance introduced for each of the procedures involved in the *Triggering* process. As usual, we measure the sequential running time for the *Triggering* process and compare it with the parallel running times. To understand the behavior of the parallel program we also isolated the running times for the *Tracking* procedure and measure the speed ups obtained. The results exhibit some anomalous behavior due to the reconstruction phase of the simulation that will be deeper analyzed in the near future.

Table 1 and figure 3 show these running times and speed ups. We observe how the parallel approach introduces an important reduction on the running time of the *Tracking* procedure. The speed up increases with the number of processors increase. An speed up of 20.69 is achieved when using 31 slaves. However, a maximum speed up of 13.78 is obtained for the global simulation process when using 21 slaves. No increment is observed after this point when the number of processors increase.

Table 1. Running Times (seconds) and Speed ups for the original code and for the tracking procedure

Number of processors	Running Time	Tracking Time	Total Speed up	Tracking Speed up
1	35623	12126		
2	25799	13011	1.38	0.93
6	9284	4154	3.83	2.91
11	4663	2166	7.63	5.59
16	3191	1117	11.16	10.85
21	2584	1022	13.78	11.86
26	2943	880	12.10	13.77
31	2604	585	13.67	20.69

Figure 4 shows the load imbalance introduced in the parallelization of the *Tracking* procedure when using 11 slaves. The load imbalance introduced is no higher than a 20%. We consider it as acceptable for this preliminary version. However, the load imbalance introduced in the *Reconstruction* phase is prohibitive (figure 5). The analysis is developed using 31 slaves. This is the main reason for the low performance obtained with a large number of processors. The *Reconstruction* process degenerates for some events and consume all the CPU time of the assigned processor. This is a well known phenomenon that also appears in the sequential simulation. However, the number of *anomalous* events in the parallel simulation are much higher than those appearing in the sequential

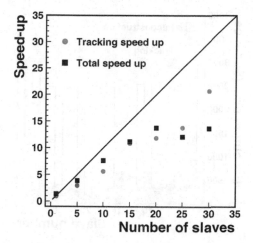

Fig. 3. Speed up curves for the global simulation process and for the *Tracking* procedure

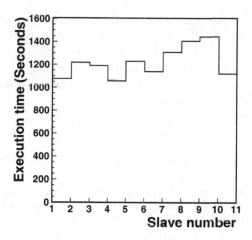

Fig. 4. Load balance in the *Tracking* procedure

one. This *anomalous* behavior of the parallel simulation is the main reason for the decrease of the speed up when a large number of processors is introduced.

5   Conclusions

We conclude that the scientific aim of applying high performance computing to computationally-intensive codes in particle physics is being successfully achieved. The relevance of our preliminary results come both from the scientific relevance of this code and also from stating that parallel computing techniques are the key to broach large size real problems in the mentioned scientific field. Although the

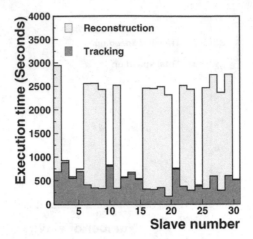

Fig. 5. The load balance for the *Tracking* and *Reconstruction*

speed up curves are not too impressive, an important reduction of the sequential running time is obtained with the number of processors considered.

The sources of overhead in the current parallel code will be deeper analyzed in the near future and new levels of parallelization will be introduced. With this objectives in mind, we look out for an improved parallel version scaling to a much larger number of processors.

References

1. M. S. Longair, High Energy Astrophysics, Cambridge University Press, Cambridge, England, 1992.
2. National Aeronautics and Space Administration, http://www.nasa.gov/
3. U.S. Department of Energy, http://www.energy.gov/engine/content.do
4. R. Brun et. al, GEANT Users Guide, CERN (1994).
5. ALEPH Collab., Nucl. Inst. and Meth. A (294) (1990) 121.
6. DELPHI Collab., Nucl. Inst. and Meth. A (303) (1991) 233.
7. L3 Collab., Nucl. Inst. and Meth. A (289) (1990) 35.
8. OPAL Collab., Nucl. Inst. and Meth. A (305) (1991) 275.
9. OpenMP Architecture Review Board, OpenMP Fortran Application Program Interface, OpenMP Forum,
   http://www.openmp.org/specs/mp-documents/fspec20.pdf (Nov 2000).
10. MPI Forum, The MPI standard, http://www.mpi-forum.org/.
11. J. M. Bull, Measuring synchronisation and scheduling overheads in openmp, in: Proc. First European Workshop on OpenMP (EWOMP 2002), Lund, Sweden, 1999, pp. 99–105.
12. W. Gropp, E. Lusk, N. Doss, A. Skjellum, A high-performance, portable implementation of the MPI message passing interface standard, Parallel Computing 22 (6) (1996) 789–828.

On the Parallelization of a Cache-Optimal Iterative Solver for PDEs Based on Hierarchical Data Structures and Space-Filling Curves

Frank Günther, Andreas Krahnke, Markus Langlotz, Miriam Mehl, Markus Pögl, and Christoph Zenger

Technische Universität München, Institut für Informatik
{guenthef,krahnke,langlotz,mehl,poegl,zenger}@in.tum.de
http://www5.in.tum.de

Abstract. Competitive numerical simulation codes solving partial differential equations have to tap the full potential of both modern numerical methods – like multi-grid and adaptive grid refinement – and available computing resources. In general, these two are rival tasks. Typically, hierarchical data structures resulting from multigrid and adaptive grid refinement impede efficient usage of modern memory architectures on the one hand and complicate the efficient parallelization on the other hand due to scattered data for coarse-level-points and unbalanced data trees. In our previous work, we managed to bring together high performance aspects in numerics as well as in hardware usage in a very satisfying way. The key to this success was to integrate space-filling curves consequently not only in the programs flow control but also in the construction of data structures which are processed linearly even for hierarchical multi-level data. In this paper, we present first results on the second challenge, namely the efficient parallelization of algorithms working on hierarchical data. It shows that with the same algorithms as desribed above, the two main demands on good parellel programs can be fulfilled in a natural way, too: The balanced data partitioning can be done quite easily and cheaply by cutting the queue of data linearized along the space-filling curve into equal pieces. Furtheron, this partitioning is quasi-optimal regarding the amount of communication. Thus, we will end up with a code that overcomes the quandary between hierarchical data and efficient memory usage and parallelization in a very natural way by a very deep integration of space-filling-curves in the underlying algorithm.

1 The Sequential Algorithm

In this section we give a very short description of the underlying sequential program, which is proven to combine numerical and hardware efficiency on a very high level. It was developed to show that modern numerical methods based on hierarchical data representation like multi-grid and adaptive grid refinement are no contradiction to a very efficient usage of modern hardware architectures like memory hierarchy[6].

D. Kranzlmüller et al. (Eds.): EuroPVM/MPI 2004, LNCS 3241, pp. 425–429, 2004.
© Springer-Verlag Berlin Heidelberg 2004

The basic idea is to deeply integrate the concept of space-filling curves [10] in the flow control of a recursive cell-oriented iterative program for solving PDEs. It was known before that – due to locality properties of the curves – reordering grid cells according to the numbering induced by a space-filling curve improves chache-efficiency (see e.g. [1]). We go one step further and – in addition to the reordering of cells – construct appropriate data structures wich are processed strictly linear. The numerical tasks are to implement a Finite-Element-Method with adaptive grid refinement and an additive multi-grid method, which leads in a natural way to hierarchical generating systems for the FEM-spaces to be constructed. Bringing together these generating systems with space-filling curves, we could show that all data, e.g. coefficients in the generating systems, can be handled by a small number of stacks which are processed strictly linear. The number of stacks only depends on the dimensionality (2D or 3D) of the problem but not on the depth of the underlying spacetree. The space-filling curve defines a set of locally deterministic rules for the stack-access. Thus, all prerequisites for high efficiency of memory access in the memory hierarchy of modern processors as well as for a good exploitation of prefetching techniques are fulfilled: Our data access is highly local both in time and in space[7].

As presented in [6], [7] and [8] we obtain hit-rates on the L2-cache of modern processors beyond 99,0%. We can show that the measurable amount of cache-misses is in the same order of magnitude as the minimum for every imaginable iterative method. So the time spent in memory access is no longer the limiting factor to the runtime of the program. In addition, we also overcame the second aspect of memory-boundedness: Typically, the grid-resolution is limited by the size of main memory in a given computer. In our algorithm, the space-filling curve gives an absolutly precise prediction, at which point during an iteration a special part of the linear data is needed. So it is possible to load data stored on hard disk "just in time" to the main memory without losing efficiency. With a fixed number of relatively small buffers we can (in principle) process grids, where the maximal resolution is limited only by the size of hard disks. "In principle" means, that now the time needed for calculations is the limiting factor, which of course becomes drastic for resolutions far beyond the size of main memory.

In consequence, the next thing to do is to parallelize the code to break through the barriers resulting from the calculation time of large problems.

2   Parallelization of the Algorithm

It is well known that the concept of space-filling curves [10] fits the demands of data parallel approaches based on domain decomposition very well: Data are linearized along the space-filling curve and can therefore be partitioned in a balanced way very easily and cheaply (see figure 1 for a simple two-dimensional example). In particular, the balanced partitioning is much cheaper than via graph algorithms. Further on, the resulting domain decomposition can be shown to be quasi-optimal regarding the size of interfaces of subdomains and, thus, communication costs [1–5, 9, 11, 12]. Another advantage is the negligible amount

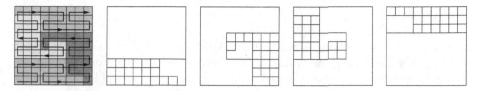

Fig. 1. Partitioning of a regular two-dimensional grid using the Peano-curve (left-hand side: partitioning, right-hand side: grids assigned to the resulting four processes)

of memory that is necessary to store the partitioning. Generally, a process has to know the location of its boundary. In the three dimensional case, we would have to store a two dimensional boundary. Using the space-filling curve linearisation we can represent the partitioning by the points, where the space-filling curve is cut into parts (*seperators*). Hence, we only need $p-1$ seperators to store all information on the domain decomposition. These seperators can be determined very cheaply by counting the cell numbers during one traversal of the grid along the Peano-curve (for a definition og the Peano-curve see [10]).

Difficulties for the parallelization are caused by the fact, that the grid data are logically associated to the vertices of the grid cells (this is nessecary to allow a cell-oriented operator evaluation) but stored pointwise on several stacks. Usually, a data parallel approach divides the domain and the corresponding data structures in the same way. Our approach uses more sophisticated techniques to realize the data partitioning in order to preserve the data structures of the sequential algorithm and its linear processing, and, thus, the high cache efficiency. The main difficulty is to build up the stack structure, which is needed to start the traversal of the grid along the Peano's curve at any point in the domain. This can be done by introducing some extra information for the coarse boundary nodes, which is stored on the stacks. As the number of coarse boundary nodes is small in comparision to the total number of nodes, the additional need in memory is small, too. As a result, the parallel implementation still passes through the whole domain in each process, but stays as coarse as possible within these parts of the domain, which do not belong to the process. Together with the fact that no compuations are performed outside the subdomain assigned to the process, the additional work ist very small.

In the context of multigrid, the additional storage of coarse level cells in each process even turns into an advantage as (cell-parts) of the residual can be restricted locally on one process and will be exchanged between processes like other 'boundary' data. Thus, for the additive multigrid algorithm, we can do completely without any master process collecting and updating coarse grid data.

Finally, we need an efficient method for exchanging the data of the boundary nodes. We implemented an efficient algorithm to determine (once per (re-)partitioning), which boundary node has to be sent to which processor. This leads to a minimal communication effort as each boundary node is only exchanged between the two associated processes and the number of boundary nodes is quasi-minimal.

3   Results

The above mentioned changes were implemented and tested against the original sequential version. The parallel implementation executed within one process led to a cache efficiency of above 99,0%. Running the same example on multiple processors showed no deterioration. All processes still have hitrates of more than 99,0%. Hence, one can see the linear access to the data, which is responsible for the cache optimality still works fine in the parallel case. Furthermore, the parallelization overhead doesn't influence the overall L2 cache efficiency.

As the needed effort for data exchange is proportional to the number of boundary nodes one can expect, that the time used for communication is small in comparision to the time needed for the calculation. This is the basic requirement to reach a linear speedup. Up to now, we couldn't ascertain the speedup, as the efficient version of the algorithm, which is responsible for the data exchange of the boundary values has not been debugged yet.

4   Conclusion

We could show, that the parallel implementation has the following properties:

- It extends the cache-optimal sequential core to the possibility of parallel execution without decaying the cache-optimality for an individual instance of the program on a single processor.
- Due to our concept based on space-filling curves, both the overhead for partitioning the domain and for data communication is quasi-optimal.
- Coarse level data require no extra handling by a master process.

With the final (not yet debugged) version we expect a linear speedup-property because of minimized communication-work. We are sure that we are able to present the complete transfer of the sequential code's advantages to a cluster in combination with the optimal parallelization strategy at the time of the conference. In addition, we will present more detailed results on runtime, scalability and parallelization overhead.

References

1. M. J. AFTOSMIS, M. J. BERGER, AND G. ADOMAVIVIUS, *A Parallel Multilevel Method for adaptively Refined Cartesian Grids with Embedded Boundaries*, AIAA Paper, 2000.
2. W. CLARKE, *Key-based parallel adaptive refinement for FEM*, bachelor thesis, Australian National Univ., Dept. of Engineering, 1996.
3. M. GRIEBEL, S. KNAPEK, G. ZUMBUSCH, AND A. CAGLAR, *Numerische Simulation in der Moleküldynamik. Numerik, Algorithmen, Parallelisierung, Anwendungen*. Springer, Berlin, Heidelberg, 2004.
4. M. GRIEBEL AND G. W. ZUMBUSCH, *Parallel multigrid in an adaptive PDE solver based on hashing and space-filling curves*, Parallel Computing, 25:827-843, 1999.

5. M. GRIEBEL AND G. ZUMBUSCH, *Hash based adaptive parallel multilevel methods with space-filling curves*, in Horst Rollnik and Dietrich Wolf, editors, NIC Symposium 2001, volume 9 of NIC Series, ISBN 3-00-009055-X, pages 479-492, Germany, 2002. Forschungszentrum Jülich.

6. F. GÜNTHER, M. MEHL, M. PÖGL, C. ZENGER *A Cache-Aware Algorithm for PDEs on hierarchical data structures based on space-filling curves*, SIAM Journal on Scientific Computing, submitted.

7. F. GÜNTHER *Eine cache-optimale Implementierung der Finite-Elemente-Methode*, Dissertation, TU München, 2004.

8. M. PÖGL, *Entwicklung eines ccheoptimalen 3D Finite-Element-Verfahrens für große Probleme*, Dissertation, TU München, 2004.

9. S. ROBERTS, S. KLYANASUNDARAM, M. CARDEW-HALL, AND W. CLARKE, *A key based parallel adaptive refinement technique for finite element methods*, in Proc. Computational Techniques and Applications: CTAC '97, B. J. Noye, M. D. Teubner, and A. W. Gill, eds. World Scientific, Singapore, 1998, p. 577-584.

10. H. SAGAN, *Space-Filling Curves*, Springer-Verlag, New York, 1994.

11. G. ZUMBUSCH, *Adaptive Parallel Multilevel Methods for Partial Differential Equations*, Habilitationsschrift, Universität Bonn, 2001.

12. G. W. ZUMBUSCH, *On the quality of space-filling curve induced partitions*, Z. Angew. Math. Mech., 81:25-28, 2001. Suppl. 1, also as report SFB 256, University Bonn, no. 674, 2000.

Parallelization of an Adaptive Vlasov Solver

Olivier Hoenen[1], Michel Mehrenberger[2], and Éric Violard[1]

[1] Université Louis Pasteur, Laboratoire LSIIT, Groupe ICPS,
Boulevard Sébastien Brant, F-67400 Illkirch, France
{hoenen,violard}@icps.u-strasbg.fr
[2] Université Louis Pasteur, Laboratoire IRMA,
7 rue René Descartes, F-67084 Strasbourg, France
mehrenbe@math.u-strasbg.fr

Abstract. This paper presents an efficient parallel implementation of
a Vlasov solver. Our implementation is based on an adaptive numerical
scheme of resolution. The underlying numerical method uses a dyadic
mesh which is particularly well suited to manage data locality. We have
developed an adapted data distribution pattern based on a division of
the computational domain into regions and integrated a load balancing
mechanism which periodically redefines regions to follow the evolution
of the adaptive mesh. Experimental results show the good efficiency of
our code and confirm the adequacy of our implementation choices. This
work is a part of the CALVI project[1].

1 Introduction

The Vlasov equation (see e.g. [9] for its mathematical expression) describes the
evolution of a system of particles under the effects of self-consistent electro-
magnetic fields. Most Vlasov solvers in use today are based on the Particle In
Cell method which consists in solving the Vlasov equation with a gridless particle
method coupled with a grid based field solver (see e.g. [2]). For some problems
in plasma physics or beam physics, particle methods are too noisy and it is of
advantage to solve the Vlasov equation on a grid of phase space, i.e., the position
and velocity space $(x, v) \in \mathbb{R}^d \times \mathbb{R}^d, d = 1, .., 3$. This has proven very efficient on
uniform meshes in the two-dimensional phase space (for $d = 1$). However when
the dimensionality increases the number of points on a uniform grid becomes
too important for being performed on a single computer. So some parallelized
versions had been developed (see e.g. [10], for $4D$ phase space Vlasov simulations)
and it is essential to regain optimality by keeping only the 'necessary' grid points.
Such adaptive methods have recently been developed, like in [8],[5],[3] where
the authors use moving distribution function grids, interpolatory wavelets of
Deslaurier and Dubuc or hierarchical biquadratic finite elements. We refer also
to [4] for a summary of many Vlasov solvers.

---

[1] CALVI is a french INRIA project devoted to the numerical simulation of problems
in Plasma Physics and beams propagation.

D. Kranzlmüller et al. (Eds.): EuroPVM/MPI 2004, LNCS 3241, pp. 430–435, 2004.
© Springer-Verlag Berlin Heidelberg 2004

In this project, we had in mind to implement an efficient parallelized version
of an adaptive Vlasov solver. So we have developed a code based on [3], where
the underlying partitions of dyadic tensor-product cells offered a simple way to
distribute data. After describing the numerical method in the adaptive context,
we present a parallelization of this method and its mechanism of load balancing,
and exhibit numerical results.

2   An Adaptive Resolution Scheme
   for the Vlasov Equation

Here is a brief description of the numerical method of resolution. We refer the
reader to [1] for a more detailed presentation. For sake of conciseness, we give the
scheme for a 2-dimensional phase space, but it generalizes to higher dimensions.

The numerical solution at time $t^n = n \Delta t$ is represented by the approximate
$f^n(a)$ of the solution at every nodes a of a *dyadic adaptive mesh* $\mathcal{M}^n$. A dyadic
adaptive mesh forms a possibly non-uniform partition of the phase space: considering the unit square $[0, 1]^2$ as the computational domain, each cell of the
mesh identifies an elementary surface $[k\, 2^{-j}, (k + 1)\, 2^{-j}] \times [l\, 2^{-j}, (l + 1)\, 2^{-j}]$,
where $k, l \in \mathbb{N}$, and $j \in \mathbb{N}$ is the level of the cell. We have $j_0 \le j \le J$, where
j_0 and J stand for the coarsest and finest level of discretization. Each of the
cell has 9 uniformly distributed nodes. Then, going to the next time step t^{n+1}
consists in three steps:

1. **Prediction** of $\mathcal{M}^{n+1}$: for each cell $\alpha \in \mathcal{M}^n$, denoting j its level, compute
   its center c_α and the forward advected point $\mathcal{A}(c_\alpha)$ by following the characteristics of the Vlasov equation (see [1] for more details about characteristics
   and advection operator $\mathcal{A}$). Then add to $\mathcal{M}^{n+1}$ the unique cell $\bar{\alpha}$ of level j
   which fits at that place in $\mathcal{M}^{n+1}$ and all the necessary cells so that $\mathcal{M}^{n+1}$ is
   a dyadic adaptive mesh. Last, if $j < J$, refine $\bar{\alpha}$ of one level, that is, replace
   it by the 4 cells of level $j+1$ which cover the same surface.
2. **Evaluation:** for each node a of $\mathcal{M}^{n+1}$, compute the backward advected
   point $\mathcal{A}^{-1}(a)$ and set $f^{n+1}(a)$ to $f^n(\mathcal{A}^{-1}(a))$: the evaluation $f^n(c)$ of the
   solution at any point $c \in [0, 1]^2$ is obtained by searching the unique cell α of
   the adaptive mesh $\mathcal{M}^n$ where the point is located, using the values at the
   nodes of that cell and computing the local biquadratic interpolation on that
   cell, say $I(c, \alpha, f^n(c))$.
3. **Compression** of $\mathcal{M}^{n+1}$: from $j = J - 1$ to j_0, replace 4 cells of level $j+1$
   by a cell α of level j (do the converse of refining α) when the norm of the
   differences $f^{n+1}(a) - I(a, \alpha, f^n(a))$, for all node a of α, is small enough.

Some other methods represent the solution with less nodes and give a procedure to retrieve the other nodes by computation. This is done for example in
the numerical methods which use a wavelet decomposition framework [5] and
where the compression phase deletes nodes instead of cells. On the contrary, our
method keeps all the nodes, this may loose adaptivity but improve data locality. Moreover, when the global movement of particles is known, our method can
easily be extended to the concept of moving grid presented in [8].

3   Parallel Implementation

Extraction of Parallelism. The numerical method induces a data-parallel algorithm by considering the adaptive mesh as a parallel data structure whose elements are the cells of the mesh with their associated nodes and values. The parallelization then relies on distributing these elements among processors.

Data Distribution. The computational domain is subdivided into *regions*. A *region* is a surface of the computational domain which is defined by an union of mesh cells. Regions are allocated to processors so that each processor owns and computes the mesh cells and nodes which are included in its own *region*. As the mesh adapts to the evolution in time of the physics, the number of cells within a region change and it is then necessary to include a load balancing mechanism which then consists in redefining regions for each processor.

Communications. We implement a specific communication scheme in order to overlap communications with computations during the prediction and evaluation phases: as the number and the source of messages is not known *a priori*, a special *end-of-send* message is used to stop the initialization of receives. Moreover, in order to minimize communications, we apply compression within the region limit only. So the compression phase do not require any communication in our implementation. This is an approximation of the numerical method since we eliminate less cells than in the original method, but it does not hazard convergence.

Data Structure. Each processor owns a local representation of the mesh. The mesh is represented by two hash tables: the cell hash table stores a set of cells which forms a partition of the whole computational domain and associates each cell with its owner identity. The node hash table stores the value at each node within the region. This representation allows cells and nodes to be accessed in constant time while minimizing the memory usage.

Load Balancing. As said previously, our load balancing mechanism consists in redefining regions. The new regions should have the following characteristics: the number of cells in each region should be approximatively the same and each region should have a "good shape" to improve the compression. Moreover, every region should be connex in order to reduce the volume of communications. We use the Hilbert's curve [6] to achieve this last requirement.

We model the global load and its localization onto the computational domain by a quad-tree [7] whose nodes are weighted by the number of leaves in the subtree. Each leaf of this quad-tree identifies one cell of the mesh and the level of a leaf in the tree is the level of the corresponding cell in the mesh.

We then build the new regions by partitioning the quad-tree. Each region is the union of the cells corresponding to the leaves of each part of the quad-tree.

To obtain a good partition, we browse the quad-tree starting from its root to its leaves, and try to make a cut as soon as possible. A part, say $\mathcal{P}$, of the partition is such that $(1-\lambda)*I \leq \|\mathcal{P}\| \leq (1+\lambda)*I$, where $\|\mathcal{P}\|$ is the number of

Table 1. Elapsed time (s) and **speed-up** on a HP cluster

# procs	$J = 7$		$J = 8$		$J = 9$	
1	1089	**1**	1896	**1**	3202	**1**
2	543	**2**	937	**2**	1559	**2**
4	285	**3.82**	494	**3.83**	823	**3.89**
8	167	**6.52**	287	**6.6**	468	**6.84**
16	99	**11**	169	**11.21**	277	**11.55**

Table 2. Elapsed time (s) and **speed-up** on a SGI O3800

# procs	$J = 6$		$J = 7$		$J = 8$	
1	1055	**1**	1827	**1**	3074	**1**
2	527	**2**	908	**2**	1514	**2.03**
4	275	**3.83**	475	**3.84**	797	**3.86**
8	155	**6.8**	274	**6.66**	459	**6.70**
16	89	**11.85**	161	**11.34**	268	**11.47**
32	57	**18.5**	105	**17.4**	177	**17.37**

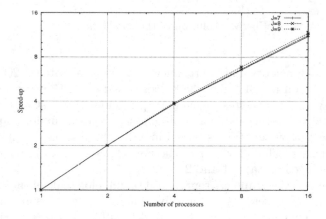

Fig. 1. Speed-up of the parallel code on a HP cluster

leaves of the part, I equals to the total number of cells divided by the number of processors, and $\lambda \in [0, 1]$ is an error factor that permits a certain degree of liberty for finding good parts.

We use this method at initialization, and a less expensive version to update regions at runtime without penalizing performance.

4 Numerical Results

Our parallel code has been written in C++/MPI and tested (1) on a HP cluster, composed of 30 identical Itanium bi-processors nodes running at $1.3Ghz$, with

Fig. 2. Evolution of the particle beam in the phase space

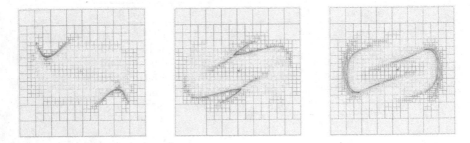

Fig. 3. Evolution of the dyadic mesh

$8GBytes$ of main memory and interconnected through a switched $200MBytes/s$ network and (2) on a SGI Origin 3800, composed of $R14k$ processors running at $500Mhz$, with $512MBytes$ of memory per node. Our test case is a 30 sec simulation (i.e. 160 iterations) of a semi-Gaussian beam in uniform applied electric field. The Vlasov equation is solved in a 2D phase space. We measured the wall-clock time for different values of the mesh finest level (J) and $j_0 = 3$. The results are reported on table 1 and 2.

Figure 1 shows the graphical representation on a log scale of the speed-up on the HP cluster. We observe that, for a fixed number of processors, the speed-up is approximatively constant as the level of details (J) increases which is a quite good property of our code. Table 1 and 2 show that the speed-up is approximately the same for two different parallel architectures.

5   Conclusion and Future Work

In this paper, we presented an efficient parallel implementation of a Vlasov solver using a numerical method based on a dyadic adaptive mesh. Numerical results show the good efficiency of our code for the $2D$ case. We still have some optimizations to implement – grouping multiple sends for example, and we are currently working on extending the load balancing mechanism for greater dimensions. Then, we plan to target the computational grid as execution environment, which will imply new scheduling and data locality constraints to deal with.

Fig. 4. Evolution of the regions for 8 processors

References

1. N. Besse, *Convergence of a semi-Lagrangian scheme for the one-dimensional Vlasov-Poisson system*, SIAM J. Numer. Anal., Vol 42, (2004), pp. 350-382.
2. C. K. Birdshall, A.B. Langdon, *Plasmaphysics via computer simulation*, McGraw-Hill, 1985.
3. M. Campos-Pinto, M. Mehrenberger, *Adaptive numerical resolution of the Vlasov equation* submitted in Numerical methods for hyperbolic and kinetic problems.
4. F. Filbet, *Numerical Methods for the Vlasov equation* ENUMATH'01 Proceedings.
5. M. Gutnic, Ioana Paun, E. Sonnendrücker, *Vlasov simulations on an adaptive phase-space grid* to appear in Comput. Phys. Comm.
6. J. K. Lawder, P. J. H. King, *Using Space-Filling Curves for Multi-dimensional Indexing*, Lecture Notes in Computer Science 1832 (2000).
7. A. Patra, J.T. Oden, *Problem decomposition for adaptive hp finite element methods*, Computing Systems in Eng., 6 (1995).
8. E. Sonnendrücker, F.Filbet, A. Friedman, E. Oudet, J.L. Vay *Vlasov simulation of beams on a moving phase-space grid* to appear in Comput. Phys. Comm.
9. E. Sonnendrücker, J. Roche, P. Bertrand and A. Ghizzo *The Semi-Lagrangian Method for the Numerical Resolution of Vlasov Equations.* J. Comput. Phys. 149(1998), pp. 201-220.
10. E. Violard, F. Filbet *Parallelization of a Vlasov Solver by Communication Overlapping*, Proceedings PDPTA 2002 (2002).

A Framework for Optimising Parameter Studies on a Cluster Computer by the Example of Micro-system Design

Dietmar Fey, Marcus Komann, and Christian Kauhaus

Friedrich-Schiller-University Jena, Institute of Computer Science
Ernst-Abbe-Platz 2, D-07443 Jena, Germany
{fey,komann,kauhaus}@informatik.uni-jena.de

Abstract. We present a framework to carry out optimising parameter studies on a cluster environment. The intention of such computation-intensive studies is to find an optimal parameter set concerning a specific objective function. We applied this framework on an example of an optimised design of a sophisticated composed optoelectronic detector. The characteristics of such a detector depend upon 25 different parameters which are input for a FEM program to simulate the detector's behaviour. Depending on the input data the simulation time varies from 10 minutes up to two days for a single simulation. With our framework it was possible to automate the parallel execution of 9000 simulation runs in three days on a 9 node cluster. On a single 2 GHz PC computer all runs would have taken more than one month. As a result we found a structure which improved the detector's former characteristics by a factor of 25.

1 Introduction

The goal of our work is to develop a generic framework for the optimised design of micro systems with parameter studies on a cluster computer. Optimising parameter studies means that we look for a specific parameter set which optimises the behaviour of a simulated system. It is useful to carry out such studies on a cluster because one has to simulate a high number of different parameter sets what needs a lot of computation resources [1], [2]. The importance to support the automatic execution of scientific computing experiments on parallel computing environments like cluster computers or Grid architectures was already recognised a decade ago in the tool Nimrod [3]. Nevertheless until now there is still low tool support in this context. Only a few solutions exist, e.g. in the ILAB project [4] a graphical user interface is used to control the input files for parameter studies. Management systems like ZOO [5] and ZENTURIO [6] are based on directive-based languages to control the execution of parameter studies.

One of the new approaches in our work is the emphasise on *optimising* scientific computing experiments by combining parameter studies with optimisation methods like for instance evolutionary algorithmic approaches on a cluster system. Furthermore our intention is to offer a solution to users without learning a new language but by specifying the necessary requirements as far as possible through a Web portal.

The framework we present in this paper is based on four components. The first one is a master program which controls the parameter studies, i.e. it initiates single pa-

D. Kranzlmüller et al. (Eds.): EuroPVM/MPI 2004, LNCS 3241, pp. 436–441, 2004.

rameter runs, evaluates their results and creates new parameter sets with improved characteristics. The second component is a job management system which is used by our master program to submit the simulation jobs to our cluster in order to achieve maximal processor utilisation. The third component is a system utility which starts the master program in periodic intervals. The fourth component is the simulation program to carry out a single parameter run.

So far, we have investigated our approach to find an optimised structure for the sophisticated setup of optical detectors which are used in experiments of particle physics. The behaviour of the detectors can be simulated by a finite-elements-method (FEM) simulation [7] program called DIPOG which was implemented by the Weierstraß-Institute in Berlin [8]. The detector consists of a multi-layer structure which is determined by 25 different parameters, being the input for DIPOG. A single run of one parameter set can take from a few minutes up to several days depending on the concrete values of the input parameters. The intention is to find that parameter set which shows the best shape of the generated photo detector current curve. This shape is evaluated by a corresponding objective function.

The remainder of this paper is structured as follows. In section 2 we present our developed framework for the optimised parameter studies on cluster systems. Topic of section 3 is the implementation on the cluster and the received results for the example of the micro system we investigated. Finally, we will conclude with a summary and an outlook of future intended work.

2   Framework for Optimised Parameter Studies

Figure 1 shows the interaction of the main components for a framework for optimised parameter studies. The central part of our framework is the master program which was developed by us. A system utility invokes the master program periodically after a certain amount of time. In comparison to continuous running of the master program this approach increases robustness in case of temporary system errors or system crashes. Furthermore it avoids unnecessary processor usage produced by a most of the time out-of-work master process running in the background.

The first time the master program is called, it creates the initial parameter files for the first simulation runs. Furthermore its task is to generate the corresponding job scripts for the submission of the simulation programs to the job management system. It is possible to insert any simulation program producing accessible numerical results. The only thing such a simulation program has to observe is the syntactic order of the input parameter files produced by the master. If the format of the input parameter file produced by the master is different form the format expected by the simulation program a small converter routine has to be written. The job management system supplies for the optimal distribution of jobs on the available computation resources. For each parameter run parameter files and their corresponding job scripts are written into separate directories as well as result files and possible error files. After each following invocations the master program looks for results of terminated simulation runs. By means of a custom objective function and an evolutionary optimisation algorithm the results are evaluated, new parameter sets are generated and the corresponding jobs are submitted.

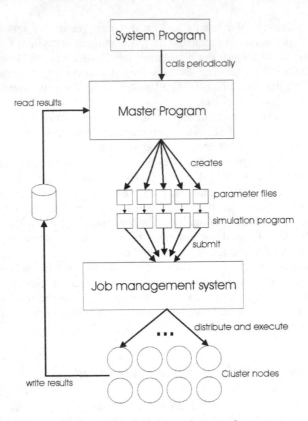

Fig. 1. Flowchart of our framework.

In order to avoid idle cycles in the processors of the cluster nodes we implemented the following schedule for the master program's inspection scheme. At the beginning the master program submits $2 \times N$ jobs, if there are N nodes in the cluster. In addition the invocation period of the master program matches the lowest possible simulation run time which has to be determined in advance. Therefore within one period only a maximum number of N jobs can terminate. But in this worst case there are still N jobs available in the queue. Due to these facts the next event must be the invocation of the master program which refills the queue. In order to avoid overrun the number of new created jobs matches exactly the number of old ones that have finished.

A serious problem which can occur in carrying out optimised parameter studies is that some parameter values cause huge simulation times but improve only poorly the results concerning the objective function. Then it is recommendable to confine the variation of these parameters to a tolerable range in the master program. Since an automatic determination of such a range is difficult we prefer so far manually carried out preliminary simulation runs. This holds also for the determination of the lowest possible simulation run time of a single run.

3 Implementation on a Cluster System and Simulation Results

We implemented our optimisation method on a Beowulf style cluster [9]. This cluster consists of eight equal nodes and one master node. The eight nodes have Intel Pentium 4 processors with 2 GHz frequency and 512 Mbytes of RAM each, while the master has a 2.4 GHz hyperthreading Pentium 4 processor with 1 Gbyte RAM. The whole system is interconnected via a Gbit-Ethernet with matching switch. Since the simulation runs are independent of each other the varying computing powers of master and slaves are non-relevant for the effectiveness of our approach.

The components of our framework system have been the following ones: as system utility for the automatic invocation of our master program we used CRON. The job management program on our system was PBS (Patch Portable System) [10], and the simulation program was the above mentioned DIPOG.

By means of DIPOG it is possible to calculate the generated photo current in special optical photo detectors along the lateral extent of the detector's substrate. The set-up of such a detector is characterised by a multi-layered embedded periodic lattice structure which is mounted on top of the detector. The lattice structure consists of a sequence of semiconductor, metal, and plastic layers. This sequence is irradiated with an electro-magnetic wave emitted by an optical radiation source, e.g. a laser. The goal is to find a lattice structure which shows an optimum shape of the generated lateral photo current. This optimum shape, which is described below, is mathematically defined with a corresponding objective function. DIPOG needs for its computation as input a parameter file which is determined by the user. This parameter file specifies the lattice structure of the detector and some optical parameters, e.g. the wavelength of the light source and the illumination angle.

The results of our computing experiments address two statements: first the quality of our found parameter set and second the speed up we achieved by carrying out the simulation on a cluster in comparison to executing it on a single computer. We proved the effectivity and efficiency of our solution by applying our approach to the design of an optoelectronic detector. In this computing experiment we received a satisfying result parameter set (see Figure 2).

Both curves show the generated photo current in the detector versus the lateral displacement of the lattice which is mounted on top of the detector. The application of the detector requires a low current at the left edge and a high current at the right edge of the curve. In addition the graph should look sinusoidal due to the constraint that the photo current should have a low increase at the borders and a high increase in the middle. The green graph shows the behaviour of the initial parameter set which was the best one found out in manually carried out simulations. The red curve presents the graph produced by the finally received parameter set which was found after running 9000 simulations within three days. A significant improvement from the green to the red graph can be seen. The curve is more sinusoidal and the value of transmitted light is twice as high as initially. Those both features, sinusoidal shape of the curve and difference of maximum and minimum photo current, are weighted inputs for the objective function which measures the quality of an investigated parameter set. The parameter set corresponding to the red graph improved the quality by a factor of 25 compared to the initial quality value of the parameter set of the green curve.

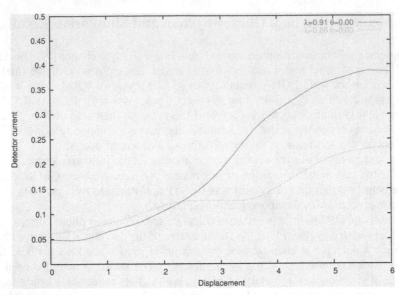

Fig. 2. Comparison of initial and found optimised parameter set. The red curve corresponds to the found optimised parameter set in which the optical wavelength λ was 910 nm and the incident angle θ to the detector substrate was 0°. The green curve corresponds to an initial parameter set with λ = 860 nm and θ = 0°.

Furthermore we compared two evolutionary optimisation methods exploiting a global and a limited population for the generation of new parameter sets. After each simulation cycle parameter sets were taken out of the population and mixed with each other or mutated to create a new one. Global population means that in each step all generated parameter sets in the past were taken into account for the refining of new parameters. However, the probability to be selected is larger for parameter sets with better quality. In case of limited population only those parameter sets which belong to the 20 best in the past were considered for the generation of a new population. Both methods were implemented in the master program. The approach with a limited population method showed only slight advantages concerning the quality of the results. But by analysing the computing times we found out that parameter sets with high quality were found in one third less computing time by using a limited population.

4 Summary and Outlook

We presented first results concerning a generic framework for optimising parameter studies on cluster computers. To prove the feasibility of our solution we used an evolutionary optimisation method on a Beowulf-style cluster to find an optimal parameter set for the design of a photo detector which is characterised by complex optical and electrical interaction. The calculated results improved the characteristics of a given start parameter set by a factor of 25 with respect to an appropriate defined objective function.

The calculation on a cluster was very useful because the computation time would have required more than one month on a single machine compared to about three days on the cluster computer. Finally our solution presents a framework for generic optimising parameter studies on cluster computers.

Further steps include a generalization of the framework, e.g. by means of a Web portal. Such a portal should offer the possibility to define a custom objective function and a converter for translating a custom parameter set format to a standardised one. Furthermore, the portal should allow to set a link to the simulation program and to define a custom optimisation algorithm. These components will be automatically included in our framework to find an optimum by starting a parallel simulation.

Acknowledgement

This research project was partially carried out as a multi-disciplinary work of physicists and computer scientists. We would like to acknowledge the fruitful discussions with Dr. H. Übensee from CiS Institute, Erfurt, in this context. Dr. Übensee works on the design and manufacturing of special multi-layered detector structures. He was also the supplier of the initial parameter set.

References

1. Abdalhaq, B., Cortés, A., Margalef, T., and Luque, E.: Evolutionary Optimization Techniques on Computational Grids. In: Sloot, P.M.A. et. al. (Eds.): ICCS 2002, LNCS 2329, Springer-Verlag Berlin Heidelberg (2002) 513-522.
2. Linderoth, J. and Wright S.J: Computational Grids for Stochastic Programming, Optimization. Technical Report 01-01, UW-Madison, Wisconsin-USA (2001).
3. Abramson, D., Sosic, R., Giddy, R., and Hall, B.: Nimrod: A tool for performing parameterised simulations using distributed workstations high performance parametric modeling with nimrod/G: Killer application for the global grid? In *Proceedings of the 4th IEEE Symposium on High Performance Distributed Computing (HPDC-95)*, 520–528, Virginia, Aug. 1995. IEEE Computer Society Press.
4. Yarrow, M., McCann, Biswas, K.M., R., and der Wijngaart, R.F.V.: Ilab: An advanced user interface approach for complex parameter study process specification on the information power grid. In *Proceedings of Grid 2000: International Workshop on Grid Computing*, Bangalore, India, Dec. 2000. ACM Press and IEEE Computer Society Press.
5. Ioannidis, Y.E., Livny, M., Gupta, S., and Ponnekanti, N.: ZOO: A desktop experiment management environment. In T. M. Vijayaraman, A. P. Buchmann, C. Mohan, and N. L. Sarda,editors, *VLDB'96, Proceedings of 22th International Conference on Very Large Data Bases*, pages 274–285, Mumbai (Bombay), India, 3–6 Sept. 1996. Morgan Kaufmann.
6. Prodan, R., Fahringer, T.: ZENTURIO: An Experiment Management System for Cluster and Grid Computing. *In Proceedings of the 4th International Conference on Cluster Computing (CLUSTER 2002)*, Chicago, USA, September 2002. IEEE Computer Society Press.
7. Melosh, R.J.: Structural Engineering Analysis by Finite Elements: Prentice-Hall, Englewood Cliffs (1990).
8. Bänsch, E., Haußer, F., Voigt, A: Finite Element Method for Epitaxial Growth with thermodynamic boundary conditions. *WIAS Preprint No. 873*, (2003), www.wias-berlin.de
9. Savarese, D.F., Sterling, T.: Beowulf. In: Buyya, R. (ed.): *High Performance Cluster Computing: Architectures and Systems, Vol.1*. Prentice Hall, New York (1999) pp. 625-645l.
10. Veridian Systems: PBS: The Portable Batch System. http://www.openpbs.org.

Numerical Simulations
on PC Graphics Hardware

Jens Krüger, Thomas Schiwietz, Peter Kipfer, and Rüdiger Westermann

Technische Universität München, Munich, Germany
{jens.krueger,schiwiet,kipfer,westermann}@in.tum.de
http://wwwcg.in.tum.de

Abstract. On recent PC graphics cards, fully programmable parallel geometry and pixel units are available providing powerful instruction sets to perform arithmetic and logical operations. In addition to computational functionality, pixel (fragment) units also provide an efficient memory interface to local graphics data.
To take full advantage of this technology, considerable effort has been spent on the development of algorithms amenable to the intrinsic parallelism and efficient communication on such cards. In many examples, programmable graphics processing units (GPUs) have been explored to speed up algorithms previously run on the CPU. In this paper, we will demonstrate the benefits of commodity graphics hardware for the parallel implementation of general techniques of numerical computing.

1 Introduction

Over the last couple of years, the evolution of GPUs has followed a tripled Moores law, currently providing up to 222 million transistors compared to 50 million on an Intel P4 Northwood CPU. Recent GPUs can be thought of as stream processors, which operate on data streams consisting of an ordered sequence of attributed primitives like vertices or fragments. GPUs can also be thought of as SIMD computers, in which a number of processing units simultaneously execute the same instructions on stream primitives. At various stages in the rendering pipeline, GPUs provide parallel and fully programmable processing units that act on the data in a SIMD-like manner. Each of these units itself is a vectorized processor capable of computing up to 4 scalar values in parallel.

In recent years, a popular direction of research is leading towards the implementation of general techniques of numerical computing on such hardware. The results of these efforts have shown that for compute bound applications as well as for memory bandwidth bound applications the GPU has the potential to outperform software solutions [1–3, 6].

To initialize computations, the application program specifies polygons to be rendered by sending a vertex stream to the GPU. This stream is automatically distributed to the vertex units, where currently up to 6 of these units work in parallel on the specified vertex information. The result of this computation is sent to the rasterizer. The rasterizer maps the input vertex stream to a fragment

D. Kranzlmüller et al. (Eds.): EuroPVM/MPI 2004, LNCS 3241, pp. 442–449, 2004.

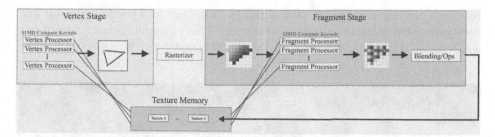

Fig. 1. This figure illustrates the rendering pipeline as it is realized on recent PC graphics cards.

stream by computing the coverage of polygons in screen space. For each covered pixel one fragment is generated thus increasing the number of stream primitives in general. The fragment stream is processed by currently up to 16 fragment units, which work in parallel on up to 4 scalars at a time. Vertex and fragment units can concurrently access shared DDR3 memory. Such fetches can be executed parallel to other operations, as long as these operations do not depend on the retrieved value. In this way, memory access latency can often be hidden. Figure 1 shows an overview of the basic architecture.

Both vertex and fragment processing units provide powerful instruction sets to perform arithmetic and logical operations. C-like high level shading languages [4, 5]in combination with optimizing compilers allow for easy and efficient access to the available functionality. In addition, GPUs also provide an efficient memory interface to local data, i.e. random access to texture maps and frame buffer objects. Not only can application data be encoded into such objects to allow for high performance access, but rendering results can also be written to such objects, thus providing an efficient means for the communication between successive rendering passes.

2   Numerical Simulation on GPUs

To realize numerical simulations on the GPU, we have implemented a general linear algebra framework that is based on the internal representation of vectors and matrices as 2D textures. On top of these representations, intrinsically parallel and streamable vector-vector and matrix-vector operations have been implemented very efficiently. To simulate computations on a 2D grid, only one quadrilateral covering as many fragments in screen space as there are grid cells has to be rendered. The rasterizer generates a stream of exactly this number of fragments, one for every vector/matrix component. In the fragment units, arithmetic operations between pairs of input values are performed, and the result is directly written to the output stream.

To combine the elements of one vector, we employ the well known reduce-operation as implemented on parallel architectures. Therefore, quadrilaterals at ever smaller size are rendered in consecutive rendering passes. In every pass, each fragment reads four adjacent data values from the previous rendering result and combines them using the specified operation. This process is repeated until one single value is left, thus enabling the reduction of one vector in logarithmic time.

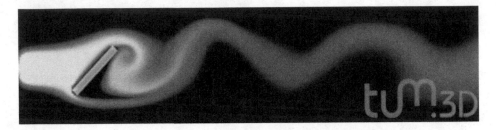

Fig. 2. Simulation of the Karman Vortex Street on a 256x64 Grid is shown. The simulation and rendering runs at 400 fps using our GPU framework on current ATI hardware.

The described linear algebra operations have been encapsulated into a C++ class framework, on top of which we have implemented implicit solvers for systems of algebraic equations. Using this framework, we demonstrate the solution to the incompressible Navier-Stokes equations in 2D at quite impressive frame rates, including boundary conditions, obstacles and simulation of velocities on a staggered grid (see Figure 2).

Besides the fact that the GPU-based simulation framework outperforms CPU-based solvers of about a factor of 10-15, running the simulation on the GPU has another important advantage: Because simulation results already exist on the GPU, they can be rendered immediately. In the following example these results are rendered by injecting dye into the flow and by advecting the dye according to the simulated flow field. Having the data already in graphics memory avoids any data transfer between the CPU and the GPU.

Incompressible Navier-Stokes Equations

Fluid phenomena like water and gaseous media can be simulated by the incompressible Navier-Stokes equations. We solve for the velocity $V = (u, v)^T$ governed by the these equations

$$\frac{\partial u}{\partial t} = \frac{1}{Re}\nabla^2 u - V \cdot \nabla u + f_x - \nabla p$$
$$\frac{\partial v}{\partial t} = \frac{1}{Re}\nabla^2 v - V \cdot \nabla v + f_y - \nabla p$$
$$\nabla \cdot V = 0$$

in two passes. First, by ignoring the pressure term an intermediate velocity is computed. The diffusion operator is discretized by means of central differences, and, as proposed in [11], we solve for the advection part by tracing the velocity field backward in time. To make the resulting intermediate vector field free of divergence, pressure is used as a correction term. Mass conservation of incompressible media leads to a Poisson-Equation for updating this pressure term. This equation is solved using a Conjugate-Gradient method build upon the numerical framework described above.

In addition to the plain solution to the Navier Stokes equations as described above we have integrated the following extensions to improve the simulation.

- We have integrated a mechanism that allows one to arbitrarily specify inflow regions and characteristics. Therefore, the current inflow settings are stored in the color components of a 2D texture map, which is interpreted as external forces acting on the flow in every iteration of the simulation process.
- Instead of a collocated grid the entire simulation is run on a staggered grid as shown in Figure 3. In this way, centered space derivatives use successive points of the same variable, and the dispersion characteristics is improved because the effective grid length is halved. Furthermore, we can now handle arbitrarily positioned obstacles exhibiting special boundary conditions very effectively.
- To preserve vorticity on the regular staggered grid we have integrated vorticity confinement [12] into our simulation code. At each grid point, a fragment shader computes

$$v_c = \tilde{n} \times (\nabla \times \omega)$$

where $\tilde{n} = \nabla|\omega|/|\nabla|\omega||$ is a unit vector pointing towards the centroid of the vortical region, and ω is the vorticity vector. This vector is added to the velocity to convect vorticity towards the centroid.

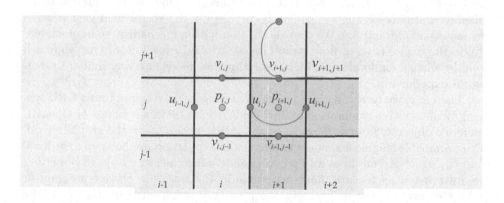

Fig. 3. The image shows a staggered grid with obstacles (grey) and how the velocity values at the obstacle borders are mirrored to match the border conditions.

In addition to the plain numerical simulation on the GPU we will now given an example how the results of this computation, which is stored in the graphics memory can be reused for visualization. This visualization is done without the bus transfer bottleneck.

3   GPU Particle Tracing

In real-world fluid flow experiments, external materials such as dye, hydrogen bubbles, or heat energy are injected into the flow. The advection of these external materials can create stream lines, streak lines, or path lines to highlight the flow patterns. Analogies to these experimental techniques have been adopted by scientific visualization researchers. Numerical methods and three-dimensional computer graphics techniques have been used to produce graphical icons such as arrows, motion particles, stream lines, stream ribbons, and stream tubes that act as three-dimensional depth cues.

Over the last decade, such methods have been investigated intensively in terms of numerical accuracy and grid structures, as well as acceleration, implementation and perception issues. While these techniques are effective in revealing the flow fields local features, they lack in that they can usually not produce *and* display the amount of graphical icons needed to visually convey large amounts of three-dimensional directional information at interactive rates.

This is due to the following reasons: First, both numerical and memory bandwidth requirements imposed by accurate particle integration schemes are too high as to allow for the simultaneous processing of large particle sets. Second, even in case that the graphical icons to visualize the flow characteristics can be computed at sufficient rates, rendering of these icons includes the transfer of data to the graphics system and thus limits the performance significantly.

To overcome the limitations of classical particle based techniques and global imaging techniques we provide a system for real-time integration and rendering of large particle sets. Contrarily to topology or feature based techniques, which aim at extracting relevant flow structures thus reducing the information to be displayed at once [7–10], we attack the problem of 3D vector field visualization by means of interactivity. We provide the user with a mechanism to interactively guide the exploration of flow structures at arbitrary resolution. Our approach enables virtual exploration of high resolution flow fields in a way similar to real-world experiments.

Therefore, we have integrated numerical integration schemes into a GPU system for interactive exploration of flow fields [13]. It takes advantage of OpenGL memory objects (*SuperBuffers*) to store particle positions on the graphics card. Programmable fragment shaders implement interpolation methods up to order 3, and construct stream lines and stream bands. Since memory objects can either be interpreted as texture maps accessible in the fragment shader program or as vertex arrays used as input to the geometry units, particle tracing can be entirely performed on the GPU without any read back to application memory.

Our implementation exploits a feature of recent ATI graphics hardware that allows graphics memory to be treated as a render target, a texture, or vertex data. This feature is presented to the application through an extension to OpenGL called *SuperBuffers*. The interface allows the application to allocate graphics memory directly, and to specify how that memory is to be used. This information, in turn, is used by the driver to allocate memory in a format suitable for the requested uses. When the allocated memory is bound to an *attachment*

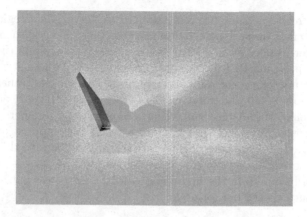

Fig. 4. High performance particle tracing on the GPU.

point (a render target, texture, or vertex array), no copying takes place. The net effect for the application program therefore is a separation of raw GPU memory from OpenGL's semantic meaning of the data. Thus, SuperBuffers provide an efficient mechanism for storing GPU computation results and later using those results for subsequent GPU computations.

In figure 4 we demonstrate the effectiveness of our approach for particle integration and rendering. About 42 millions of particles per second are traced through the flow using an Euler integration scheme. This number is reduced to 21 millions and 12 millions, respectively, using schemes of 2nd and 3rd order accuracy, respectively. It is interesting to note that a CPU version of particle tracing in 3D flow fields roughly takes about a factor of 30 longer than the GPU version. This is mainly due to the following reasons: First, the advection step extremely benefits from the numerical compute power and high memory bandwidth of the parallel fragment units on our target architecture. Second, the transfer of that many particles to the GPU in every frame of the animation imposes a serious limitation on the overall performance.

4   Future Work

In order to process large simulation domains, we can exploit distribution and parallelization functionality present in common multiprocessor architectures For interactive steering of simulations, we envision a distributed approach that makes use of the low-latency node interconnectors and high performance intra-node bus architectures to provide adequate user response times.

The whole system is initialized by computing a regular partition of the simulation domain in parallel that has the least edge-cut property. The obtained information is processed by a program on each node, which assembles the data for all GPUs on this node. A dedicated thread is responsible for uploading the data to the GPUs in order to avoid synchronization delays. Now each GPU processes a part of the sub-domain that is obtained by a regular split. The boundary conditions are read-back and are redistributed to the adjacent regions in a hierarchical way by first updating the boundaries on each node and then between the nodes on a peer-to-peer basis. Note that this perfectly exploits the high bandwidth on the upcoming PCI-Express graphics bus and reduces at the same time the bandwidth needed on the node interconnect. The packets transferred at this level are therefore small and allow efficient inter-node transport across the low latency interconnector. After completing a configurable number of simulation time-steps, a visualization of the simulation result is computed on each client node and the partial results, i.e. images, are transferred back to the master node for display. The packet size of this final communication step is rather big, so we can efficiently maintain the high bandwidth PCI-Express path on the client nodes all through to the master node.

5 Conclusion

In this work, we have described a general framework for the implementation of numerical simulation techniques on graphics hardware. For this purpose, we have developed efficient internal layouts for vectors and matrices. By considering matrices as a set of diagonal or column vectors and by representing vectors as 2D texture maps, matrix-vector and vector-vector operations can be accelerated considerably compared to software based approaches.

In order to demonstrate the effectiveness and the efficiency of our approach, we have described a GPU implementation to numerically solve the incompressible Navier-Stokes equations. The results have shown that recent GPUs can not only be used for rendering purposes, but also for numerical simulation and integration. The combination of shader programs to be executed on the GPU and new concepts like memory objects allow one to carry out numerical simulations efficiently and to directly visualize the simulation results.

References

1. JENS KRÜGER AND RÜDIGER WESTERMANN. Linear algebra operators for gpu implementation of numerical algorithms. In *ACM Computer Graphics (Proc. SIGGRAPH '03)*.
2. BOLZ, J., FARMER, I., GRINSPUN, E., AND SCHROEDER, P. 2003. Sparse matrix solvers on the GPU: Conjugate gradients and multigrid. *Computer Graphics SIGGRAPH 03 Proceedings*.
3. MORELAND, K AND ANGEL, E. 2003. The FFT on a GPU. *SIGGRAPH/ Eurographics Workshop on Graphics Hardware 2003*.
4. MICROSOFT, 2002. DirectX9 SDK. http://www.microsoft.com/DirectX.
5. MARK, W., GLANVILLE, R., AKELEY, K., AND KILGARD, M. 2003. Cg: A system for programming graphics hardware in a C-like language. In *ACM Computer Graphics (Proc. SIGGRAPH '03)*.
6. N. Goodnight, C. Woolley, G. Lewin, D. Luebke, and G. Humphreys. A multigrid solver for boundary-value problems using programmable graphics hardware. In *Proceedings ACM SIGGRAPH/Eurographics Workshop on Graphics Hardware*

7. L. Hesselink and T. Delmarcelle. *Scientific visualization - advances and challenges*, chapter Visualization of vector and tensor data sets, pages 367–390. Academic Press, 1994.
8. R. Peikert and Roth. M. The 'parallel vectors' operator - a vector field visualization primitive. In *Proceedings IEEE Visualization 99*, pages 263–271, 1999.
9. A. Telea and J. Wijk. Simplified representation of vector fields. In *Proceedings IEEE Visualization 99*, pages 35–43, 1999.
10. Frits H. Post, Benjamin Vrolijk, Helwig Hauser, Robert S. Laramee, and Helmut Doleisch. The state of the art in flow visualisation: Feature extraction and tracking. *Computer Graphics Forum*, 22(4):775–792, 2003.
11. J. Stam. Stable fluids. *Computer Graphics SIGGRAPH 99 Proceedings*, pages 121–128, 1999.
12. John Steinhoff and David Underhill. Modification of the euler equations for "vorticity confinement": Application to the computation of interacting vortex rings. *Physics of Fluids Vol 6(8)*, pages 2738–2744, 1994.
13. PETER KIPFER, MARK SEGAL AND RÜDIGER WESTERMANN. UberFlow: A GPU-Based Particle Engine To appear in *Proceedings ACM SIGGRAPH/Eurographics Workshop on Graphics Hardware 2004*.

Author Index

Lecture Notes in Computer Science

For information about Vols. 1–3114

please contact your bookseller or Springer